Test Bank

Precalculus with Limits
A Graphing Approach

SEVENTH EDITION

Ron Larson
The Pennsylvania State University,
The Behrend College

CENGAGE Learning

Australia • Brazil • Mexico • Singapore • United Kingdom • United States

CENGAGE Learning

WCN: 01-100-101

For product information and technology assistance, contact us at **Cengage Learning Customer & Sales Support, 1-800-354-9706**.

For permission to use material from this text or product, submit all requests online at **www.cengage.com/permissions**
Further permissions questions can be emailed to **permissionrequest@cengage.com**.

ISBN: 978-1-305-11757-0

Cengage Learning
20 Channel Center Street
Boston, MA 02210
USA

Cengage Learning is a leading provider of customized learning solutions with office locations around the globe, including Singapore, the United Kingdom, Australia, Mexico, Brazil, and Japan. Locate your local office at: **www.cengage.com/global**.

Cengage Learning products are represented in Canada by Nelson Education, Ltd.

To learn more about Cengage Learning Solutions, visit **www.cengage.com**.

Purchase any of our products at your local college store or at our preferred online store **www.cengagebrain.com**.

Printed in the United States of America
Print Number: 02 Print Year: 2015

Contents

Name: ______________________________ Date: _____________

1. Use function notation to write g in terms of $f(x) = x^3$.

$$g(x) = -\frac{1}{5}(x+4)^3$$

A) $g(x) = -\frac{1}{5}[f(x)]^3 + 4$

B) $g(x) = -\frac{1}{5}[f(x)+4]$

C) $g(x) = -[f(x)]^3 + \frac{64}{5}$

D) $g(x) = -\frac{1}{5}[f(x)]^3 + 64$

E) $g(x) = -\frac{1}{5}[f(x+4)]$

2. Hooke's Law states that the force F required to compress or stretch a spring (within its elastic limits) is proportional to the distance d that the spring is compressed or stretched from its original length. That is, $F = kd$, where k is the measure of the stiffness of the spring and is called the *spring constant*. The table below shows the elongation d in centimeters of a spring when a force of F kilograms is applied.

Force, F	Elongation, d
20	3.5
40	6.3
60	10.0
80	13.3
100	16.5

Find the equation of the line that seems to best fit the data.

A) $F = 12.098d$
B) $F = 3.024d$
C) $F = 6.049d$
D) $F = 4.537d$
E) $F = 7.561d$

3. Find $(fg)(x)$.

$f(x)=\sqrt{-5x}$ $\qquad$ $g(x)=\sqrt{-8x+6}$

A) $(fg)(x)=2x\sqrt{10}-\sqrt{30x}$
B) $(fg)(x)=2x\sqrt{10-30x}$
C) $(fg)(x)=\sqrt{-13x+6}$
D) $(fg)(x)=\sqrt{40x^2+6}$
E) $(fg)(x)=\sqrt{40x^2-30x}$

4. If f is an even function, determine if g is even, odd, or neither.

$g(x)=-f(x-2)$

A) even
B) odd
C) cannot be determined
D) neither

5. Given the following function, $h(x)$, find two functions f and g such that $(f\circ g)(x)=h(x)$.

$h(x)=\sqrt[3]{x^2-11}$

A) $f(x)=\sqrt[3]{x^2},\ g(x)=-11$
B) $f(x)=\sqrt[3]{x^2},\ g(x)=x-11$
C) $f(x)=\sqrt[3]{x},\ g(x)=x-11$
D) $f(x)=\sqrt[3]{x-11},\ g(x)=x^2$
E) $f(x)=\sqrt[3]{x-11},\ g(x)=x+11$

6. Evaluate the following function at the specified value of the independent variable and simplify.

$f(w)=\dfrac{-7w^2+20}{w^2}; \qquad f(0)$

A) 20
B) 0
C) –7
D) 13
E) undefined

7. Determine algebraically whether the following function is one-to-one.

$|x-5|,\ \ x \le 5$

A) $\begin{aligned} |a-5| &= |b-5| \\ 5-a &= 5-b \\ -a &= -b \\ a &= b \end{aligned}$; one-to-one

B) $\begin{aligned} |a-5| &= |b-5| \\ |a|-5 &= |b|-5 \\ |a| &= |b| \\ a &= b \end{aligned}$; one-to-one

C) $\begin{aligned} |a-5| &= |b-5| \\ a+5 &= 5-b \\ a &= -b \end{aligned}$; not one-to-one

D) $\begin{aligned} |a-5| &= |b-5| \\ |-5|-a &= |-5|+b \\ -a &= b \end{aligned}$; not one-to-one

E) $\begin{aligned} |a-5| &= |b-5| \\ |-5|-a &= |-5|-b \\ -a &= -b \\ a &= b \end{aligned}$; one-to-one

8. Determine an equation that may represented by the graph shown below.

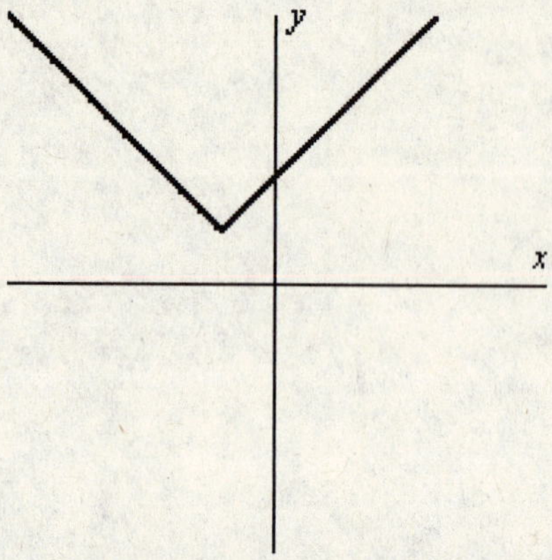

A) $f(x)=|x+1|-1$
B) $f(x)=|x-1|+1$
C) $f(x)=|x+1|+1$
D) $f(x)=|x-1|-1$
E) $f(x)=-|x-1|+1$

9. Determine the domain and range of the inverse function f^{-1} of the following function f.

$f(x)=-|x+6|+2$, where $x>-6$

A) Domain: $[-6,\infty)$; Range: $[2,\infty)$
B) Domain: $(-\infty,2]$; Range: $[-6,\infty)$
C) Domain: $[-6,2]$; Range: $[-6,\infty)$
D) Domain: $(-\infty,-6]$; Range: $[-2,\infty)$
E) Domain: $(-\infty,\infty)$; Range: $(-\infty,\infty)$

10. Find the domain of the function.

$f(y)=\sqrt{9-y^2}$

A) $-3\le y\le 3$
B) $y\le -3$ or $y\ge 3$
C) $y\ge 0$
D) $y\le 3$
E) all real numbers

11. Find the slope-intercept form of the line passing through the points.
    (–1, –6), (0, –2)
    A) $y = 4x + 23$
    B) $y = 4x - 2$
    C) $y = \frac{1}{4}x - \frac{23}{4}$
    D) $y = -\frac{1}{4}x + \frac{1}{2}$
    E) $y = -4x - 10$

12. Write the slope-intercept form of the equation of the line through the given point perpendicular to the given line.
    point: (–4, 7)   line: $-5x - 15y = -6$
    A) $y = \frac{1}{5}x + \frac{39}{5}$
    B) $y = -\frac{1}{3}x + \frac{17}{3}$
    C) $y = 3x + 19$
    D) $y = -5x + 27$
    E) $y = 3x - \frac{5}{3}$

13. Compare the graph of the following function with the graph of $f(x) = |x|$.
    $y = \left|\frac{4}{9}x\right|$
    A) vertical shift of $\frac{4}{9}$ units up
    B) horizontal stretch of $\frac{9}{4}$ units
    C) vertical shrink of $\frac{4}{9}$ units
    D) horizontal shrink of $\frac{4}{9}$ units
    vertical shift of $\frac{9}{4}$ units
    E) horizontal shrink of $\frac{4}{9}$ units

14. Which equation does not represent y as a function of x?
    A) $x = 2y + 5$
    B) $x = 6$
    C) $y = -5x - 7$
    D) $y = |6 + 9x^2|$
    E) $y = \sqrt{-8 + 4x}$

15. Evaluate the function at the specified value of the independent variable and simplify.
    $$q(p) = \frac{-2p}{5p - 2}$$
    $$q\,(x - 9)$$
    A) $\dfrac{-2x + 18}{5x - 47}$
    B) $\dfrac{-2x - 18}{5x - 47}$
    C) $\dfrac{-2p + 18}{5p - 47}$
    D) $\dfrac{18}{43}$
    E) $-\dfrac{18}{47}$

16. Determine the domain of $g(x) = \dfrac{1}{x^2 - 49}$.
    A) $[-7,7]$
    B) $(-7,0]\cup[0,7)$
    C) $(-\infty,-7)\cup(-7,7)\cup(7,\infty)$
    D) $(-\infty,-7]\cup[7,\infty)$
    E) $(-\infty,\infty)$

17. Find the difference quotient and simplify your answer.

$f(w) = -9w^2 + 2w,\qquad \dfrac{f(4+h)-f(4)}{h},\ h \neq 0$

A) $10 + h$
B) $-70 - 9w + \dfrac{16}{w}$
C) $2 - 9w + \dfrac{16}{w}$
D) $2 - 9h$
E) $-70 - 9h$

18. The scatter plots of different data are shown below. Determine whether there is a positive correlation, negative correlation, or no discernible correlation between the variables.

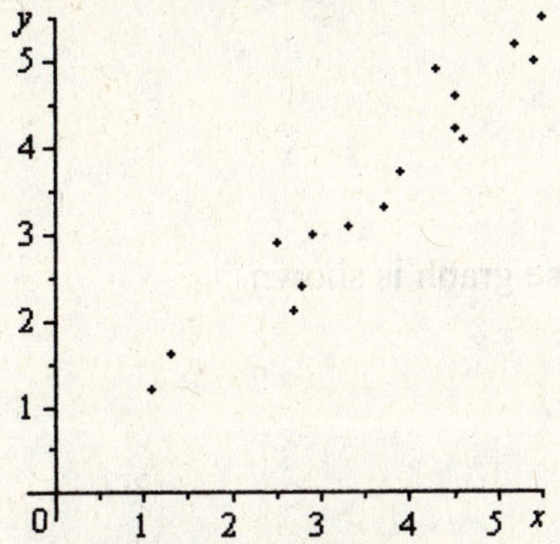

A) positive correlation
B) negative correlation
C) no discernible correlation

19. Evaluate the following function for $f(x) = -2x^2 + 1$ and $g(x) = x + 4$ algebraically.

$\left(\frac{f}{g}\right)(q-4)$

A) $\frac{-2q^2+5}{q+8}$

B) $\frac{-2q^2+8q-31}{q}$

C) $\frac{-2q^2+5}{q}$

D) $\frac{-2q^2+16q-31}{q}$

E) $\frac{-2q^2-3}{q}$

20. Use the graph of

$f(x) = x^2$

to write an equation for the function whose graph is shown.

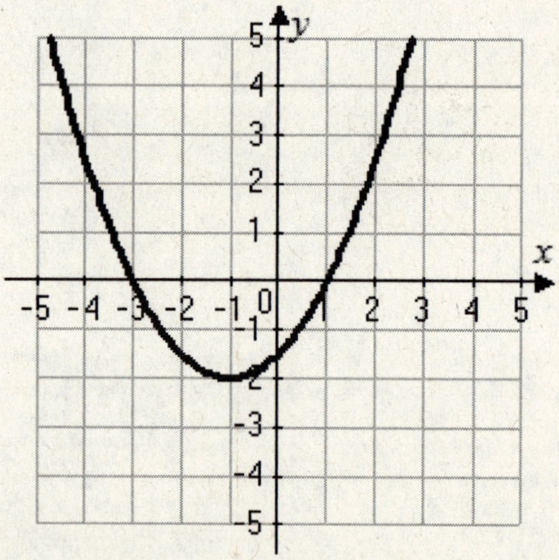

A) $f(x) = (x+1)^2 - 2$

B) $f(x) = (x-1)^2 - 2$

C) $f(x) = (x+1)^2 + 2$

D) $f(x) = \frac{1}{2}(x-1)^2 - 2$

E) $f(x) = \frac{1}{2}(x+1)^2 - 2$

Answer Key

1. E
2. C
3. E
4. C
5. D
6. E
7. A
8. C
9. B
10. A
11. B
12. C
13. B
14. B
15. A
16. C
17. E
18. A
19. D
20. E

Name: ___________________________ Date: _____________

1. Evaluate the indicated function for $f(x) = x^2 - 5$ and $g(x) = x + 9$.
   $(fg)(-1)$
   A) -32
   B) -48
   C) -46
   D) 40
   E) -50

2. Find the value(s) of x for which $f(x) = g(x)$.
   $f(x) = x^2 - 7x + 3$ $\qquad$ $g(x) = -3x + 8$
   A) $3, 10, \frac{8}{3}$
   B) $3, -7, \frac{8}{3}$
   C) $5, -1$
   D) $-5, 1$
   E) $4, \frac{8}{3}$

3. Find $(f - g)(x)$.
   $f(x) = -\dfrac{8x}{4x+7}$ $\qquad$ $g(x) = -\dfrac{4}{x}$
   A) $(f-g)(x) = \dfrac{-8x+4}{3x+7}$
   B) $(f-g)(x) = \dfrac{-8x+23}{4x+7}$
   C) $(f-g)(x) = \dfrac{-8x+9}{4x+7}$
   D) $(f-g)(x) = \dfrac{-8x^2+16x-28}{4x^2+7x}$
   E) $(f-g)(x) = \dfrac{-8x^2+16x+28}{4x^2+7x}$

4. If f is an even function, determine if g is even, odd, or neither.
   $g(x) = f(-x) + 1$
   A) even
   B) odd
   C) cannot be determined
   D) neither

5. Evaluate the function at the specified value of the independent variable and simplify.
   $f(p) = \dfrac{-3p}{4p-3}$
   $f(s+8)$
   A) $\dfrac{-3s-24}{4s+29}$
   B) $\dfrac{-3s+24}{4s+29}$
   C) $\dfrac{-3p-24}{4p+29}$
   D) $\dfrac{24}{35}$
   E) $-\dfrac{24}{29}$

6. Determine the domain of $g(x) = \dfrac{1}{x^2 - 81}$.
   A) $[-9,9]$
   B) $(-9,0]\cup[0,9)$
   C) $(-\infty,-9)\cup(-9,9)\cup(9,\infty)$
   D) $(-\infty,-9]\cup[9,\infty)$
   E) $(-\infty,\infty)$

7. Determine whether lines L_1 and L_2 passing through the pairs of points are parallel, perpendicular, or neither.
   L_1: (7, –4), (–9, –1)
   L_2: (4, –6), (–3, 9)
   A) parallel
   B) perpendicular
   C) neither

8. Algebraically determine whether the function below is even, odd, or neither.

   $f(q) = 2q^{3/2}$

   A) even
   B) odd
   C) cannot be determined
   D) neither

9. Find $f \circ g$.

   $f(x) = x + 2 \qquad g(x) = \dfrac{5}{x^2 - 4}$

   A) $(f \circ g)(x) = \dfrac{5}{x^2}$
   B) $(f \circ g)(x) = \dfrac{5}{x^2 + 4x}$
   C) $(f \circ g)(x) = \dfrac{2x^2 + 3}{x^2 - 4}$
   D) $(f \circ g)(x) = \dfrac{7}{x^2 - 4}$
   E) $(f \circ g)(x) = \dfrac{2x^2 - 3}{x^2 - 4}$

10. Determine an equation that may represented by the graph shown below.

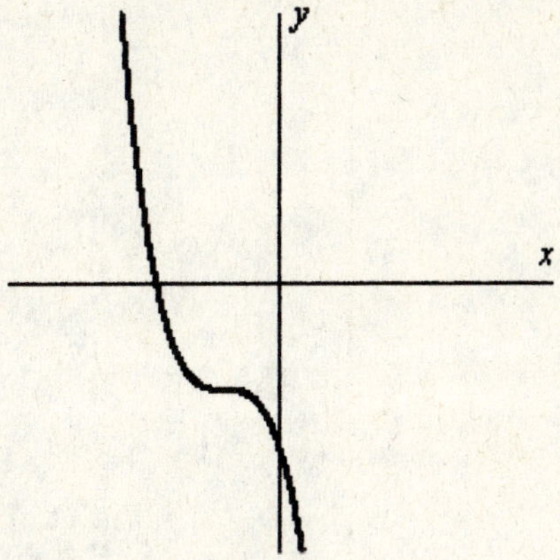

A) $f(x)=(x-1)^3+2$
B) $f(x)=-(x-1)^3+2$
C) $f(x)=-(x-1)^3-2$
D) $f(x)=-(x+1)^3-2$
E) $f(x)=-(x+1)^3+2$

11. Estimate the slope of the line.

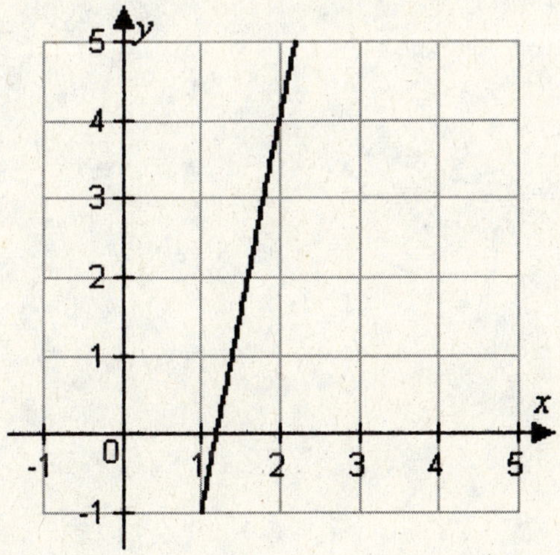

A) -5
B) 0
C) 5
D) $\frac{1}{5}$
E) $\frac{2}{5}$

12. Compare the graph of the following function with the graph of $f(x) = |x|$.

$$y = \left|\frac{1}{9}x\right|$$

A) vertical shift of $\frac{1}{9}$ unit up

B) horizontal stretch of $\frac{9}{1}$ unit

C) vertical shrink of $\frac{1}{9}$ unit

D) horizontal shrink of $\frac{1}{9}$ unit

vertical shift of $\frac{9}{1}$ unit

E) horizontal shrink of $\frac{1}{9}$ unit

13. Use a graphing utility to graph the function and visually determine the intervals over which the function is increasing, decreasing, or constant.

$f(x) = 2x^4 - 4x^2$

A) decreasing on $(0, 0)$

increasing on $(0, \infty)$

B) increasing on $(-\infty, -1)$

decreasing on $(-1, 0)$

increasing on $(0, 1)$

decreasing on $(1, \infty)$

C) decreasing on $(-\infty, -1)$

increasing on $(-1, 1)$

decreasing on $(1, \infty)$

D) increasing on $(-\infty, 0)$

decreasing on $(0, \infty)$

E) decreasing on $(-\infty, -1)$

increasing on $(-1, 0)$

decreasing on $(0, 1)$

increasing on $(1, \infty)$

14. Hooke's Law states that the force F required to compress or stretch a spring (within its elastic limits) is proportional to the distance d that the spring is compressed or stretched from its original length. That is, $F = kd$, where k is the measure of the stiffness of the spring and is called the *spring constant*. The table below shows the elongation d in centimeters of a spring when a force of F kilograms is applied.

Force, *F*	**Elongation, *d***
20	2.8
40	5.0
60	8.0
80	10.6
100	13.2

Find the equation of the line that seems to best fit the data. Use the model to estimate the elongation of the spring when a force of 50 kilograms is applied. Round your answer to one decimal place.
A) 13.2 centimeters
B) 9.9 centimeters
C) 3.3 centimeters
D) 6.6 centimeters
E) 5.0 centimeters

15. Find $f \circ g$.
$f(x) = 3x - 2$ $\qquad$ $g(x) = x - 5$
A) $(f \circ g)(x) = 3x - 17$
B) $(f \circ g)(x) = 3x - 7$
C) $(f \circ g)(x) = 3x^2 - 17x + 10$
D) $(f \circ g)(x) = 2x + 3$
E) $(f \circ g)(x) = 2x - 7$

16. Use the graph of the function to find the domain and range of *f*.

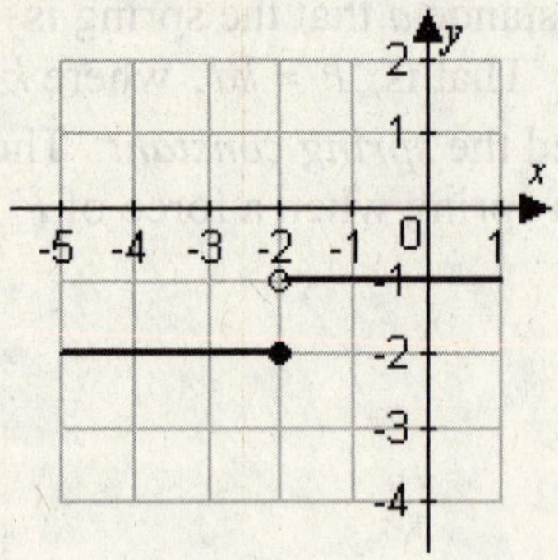

A) domain : $(-\infty,-2)\cup(-2,\infty)$
   range : $(-\infty,-2)\cup(-1,\infty)$

B) domain : $(-\infty,-2)\cup(-2,\infty)$
   range : $\{-2,-1\}$

C) domain : all real numbers
   range : $\{-2,-1\}$

D) domain : $(-\infty,-2)\cup(-2,\infty)$
   range : $(-1,1)$

E) domain : $\{-2,-1\}$
   range : all real numbers

17. Find the inverse function of *f*.

$$f(x)=x^5+5$$

A) $f^{-1}(x)=-\sqrt[5]{x}+5$

B) $f^{-1}(x)=\sqrt[5]{x}+5$

C) $f^{-1}(x)=-\sqrt[5]{x+5}$

D) $f^{-1}(x)=\sqrt[5]{x-5}$

E) $f^{-1}(x)=\sqrt[5]{x}-5$

18. Evaluate the following function at the specified value of the independent variable and simplify.

$$f(u)=\frac{4u^2+12}{u^2}; \qquad f(0)$$

A) 12
B) 0
C) 4
D) 16
E) undefined

19. Find $g\circ f$.

$f(x)=x+2 \qquad g(x)=x^2$

A) $(g\circ f)(x)=x^2+2$
B) $(g\circ f)(x)=x^2-4$
C) $(g\circ f)(x)=x^2+4$
D) $(g\circ f)(x)=x^2+2x+4$
E) $(g\circ f)(x)=x^2+4x+4$

20. Find all real values of x such that $f(x)=0$.

$$f(x)=\frac{-3x-2}{5}$$

A) $-\frac{2}{15}$
B) $\pm\frac{2}{15}$
C) $\pm\frac{2}{3}$
D) $-\frac{2}{3}$
E) $\frac{2}{3}$

Answer Key

1. A
2. C
3. E
4. A
5. A
6. C
7. C
8. D
9. E
10. D
11. C
12. B
13. E
14. D
15. A
16. C
17. D
18. E
19. E
20. D

Name: ______________________ Date: ____________

1. Find the difference quotient and simplify your answer.

$f(s) = -2s^2 - 2s, \quad \dfrac{f(4+h)-f(4)}{h}, h \neq 0$

A) $6 + h$
B) $-18 - 2s - \dfrac{16}{s}$
C) $-2 - 2s - \dfrac{16}{s}$
D) $-2 - 2h$
E) $-18 - 2h$

2. Determine whether the function has an inverse function. If it does, find the inverse function.

$f(x) = x^2 + 5$

A) No inverse function exists.
B) $f^{-1}(x) = \sqrt{x} + 5, \ x \geq 0$
C) $f^{-1}(x) = \sqrt{x} - 5$
D) $f^{-1}(x) = \sqrt{x+5}, \ x \geq -6$
E) $f^{-1}(x) = \sqrt{x-5}$

3. Which equation does not represent y as a function of x?

A) $x = -9y + 2$
B) $x = -1$
C) $y = 7x - 9$
D) $y = |6 - x^2|$
E) $y = \sqrt{-9 + 6x}$

4. Determine the domain and range of the inverse function f^{-1} of the following function f.

$f(x) = -|x+7|-1$, where $x > -7$

A) Domain: $[-7,\infty)$; Range: $[-1,\infty)$
B) Domain: $(-\infty,-1]$; Range: $[-7,\infty)$
C) Domain: $[-7,-1]$; Range: $[-7,\infty)$
D) Domain: $(-\infty,-7]$; Range: $[1,\infty)$
E) Domain: $(-\infty,\infty)$; Range: $(-\infty,\infty)$

5. Use the graphs of $y = f(x)$ and $y = g(x)$ to evaluate $(g^{-1} \circ f^{-1})(-4)$.

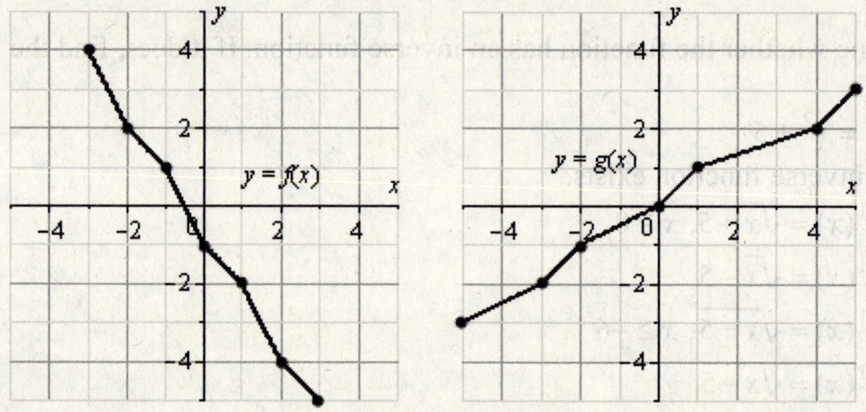

A) 4
B) 1.3
C) 0
D) –2
E) –2.5

6. Compare the graph of the following function with the graph of $f(x)=x^3$.

   $y=\left[5(x+10)\right]^3$

   A) vertical shift of 10 units up
   B) vertical shift of 10 units up
   horizontal shrink of $\frac{1}{5}$ units
   C) horizontal shift of 10 units to the left
   horizontal shrink of $\frac{1}{125}$ units
   D) horizontal shift of 10 units to the left
   horizontal stretch of $\frac{1}{5}$ units
   E) horizontal shift of 10 units to the left
   vertical shift of 5 units up

7. Find $f \circ g$.

   $f(x)=\left|x^2-6\right|$ $\qquad$ $g(x)=-9-x$

   A) $(f \circ g)(x)=\left|x^2+18x+75\right|$
   B) $(f \circ g)(x)=\left|x^2+75\right|$
   C) $(f \circ g)(x)=\left|-3-x^2\right|$
   D) $(f \circ g)(x)=\left|-15-x^2\right|$
   E) $(f \circ g)(x)=-9-\left|x^2-6\right|$

8. The average lengths L of cellular phone calls in minutes from 1999 to 2004 are shown in the table below.

Year	Average length, L (in minutes)
1999	2.38
2000	2.56
2001	2.74
2002	2.73
2003	2.87
2004	3.05

Use the *regression* feature of a graphing utility to find a linear model for the data. Let t represent the year, with $t = 9$ corresponding to 1999. Use the model to predict the average lengths of cellular phone calls for the year 2015. Round your answer to two decimal places.
A) 4.37 minutes
B) 8.74 minutes
C) 5.37 minutes
D) 3.37 minutes
E) 2.19 minutes

9. Decide whether the two functions shown in the graph below appear to be inverse functions of each other.

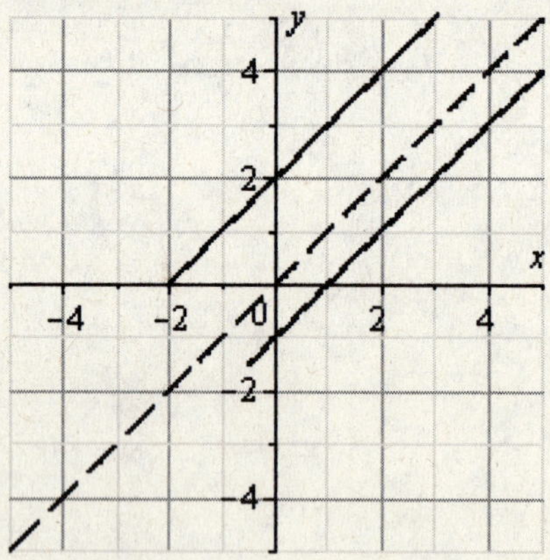

A) yes
B) no
C) not enough information

10. Use a graphing utility to graph the function and visually determine the intervals over which the function is increasing, decreasing, or constant.

    $f(x) = -x^3 + 3x + 1$

    A) increasing on $(-\infty, -1)$
    decreasing on $(-1, 1)$
    increasing on $(1, \infty)$
    B) decreasing on $(-\infty, 0)$
    increasing on $(0, \infty)$
    C) decreasing on $(-\infty, \infty)$
    D) increasing on $(-\infty, \infty)$
    E) decreasing on $(-\infty, -1)$
    increasing on $(-1, 1)$
    decreasing on $(1, \infty)$

11. Find the value(s) of x for which $f(x) = g(x)$.

    $f(x) = x^2 - 13x + 5 \qquad g(x) = -9x + 2$

    A) $5, 18, \frac{2}{9}$
    B) $5, -13, \frac{2}{9}$
    C) $3, 1$
    D) $-3, -1$
    E) $8, \frac{2}{9}$

12. Use function notation to write g in terms of $f(x)=x^3$.

$$g(x)=-\frac{1}{4}(x+9)^3$$

A) $g(x)=-\frac{1}{4}[f(x)]^3+9$

B) $g(x)=-\frac{1}{4}[f(x)+9]$

C) $g(x)=-[f(x)]^3+\frac{729}{4}$

D) $g(x)=-\frac{1}{4}[f(x)]^3+729$

E) $g(x)=-\frac{1}{4}[f(x+9)]$

13. Plot the points and find the slope of the line passing through the pair of points.
(3, 4), (–2, 4)

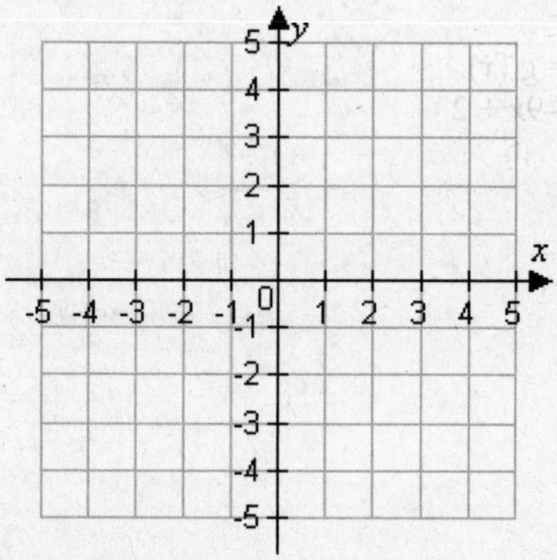

A) slope: 0
B) slope: 1
C) slope: –5
D) slope: $-\frac{1}{5}$
E) slope: undefined

14. Determine an equation that may represented by the graph shown below.

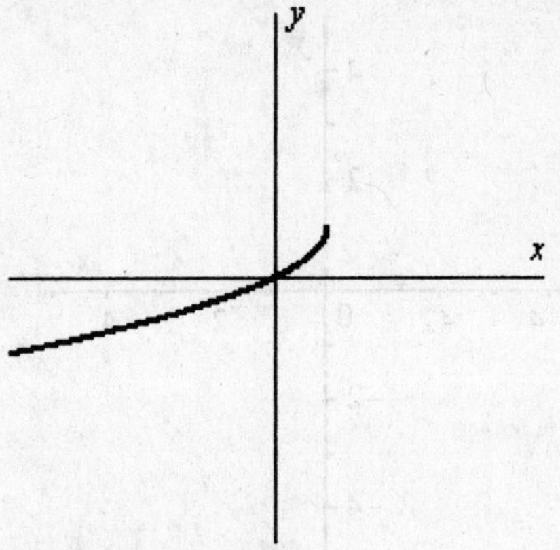

A) $f(x)=1-\sqrt{1-x}$
B) $f(x)=-1-\sqrt{1-x}$
C) $f(x)=-1+\sqrt{1-x}$
D) $f(x)=-1-\sqrt{1+x}$
E) $f(x)=-1+\sqrt{1+x}$

15. Use the graphs of f and g, shown below, to graph $h(x) = (f+g)(x)$.

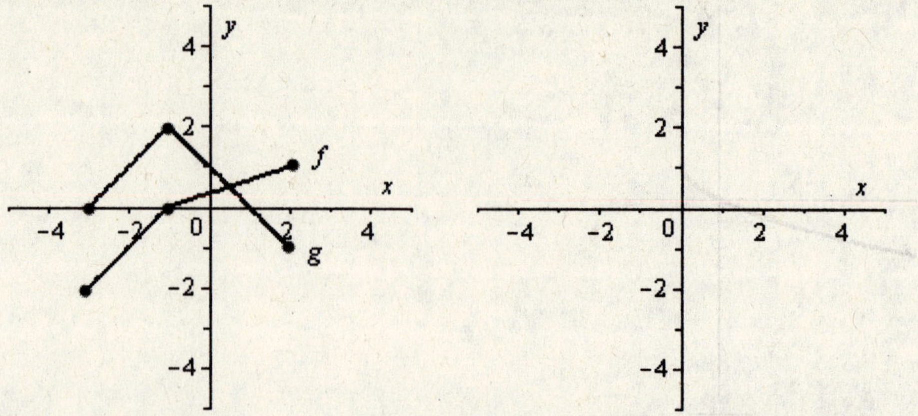

A)

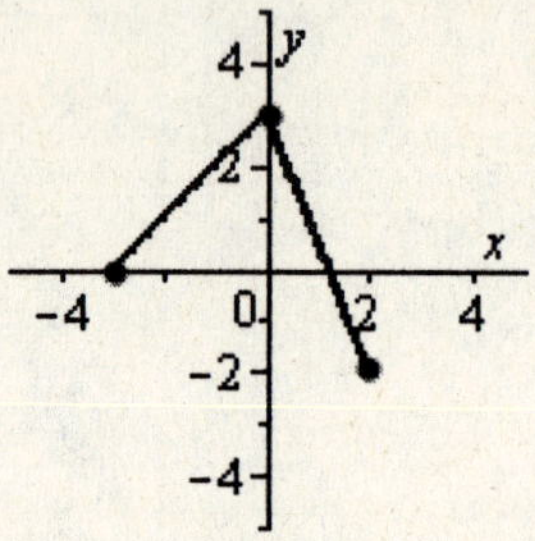

B)

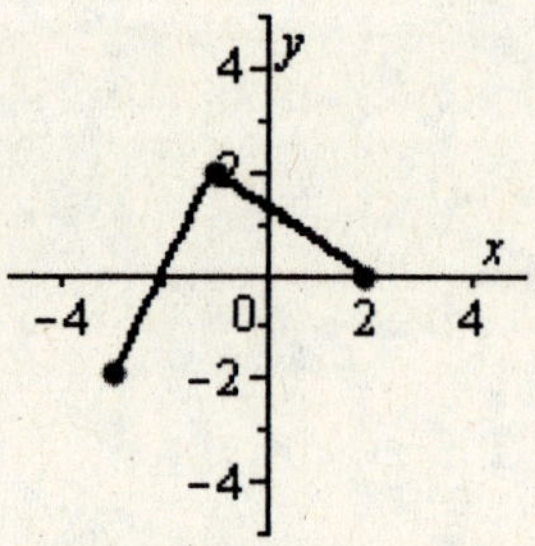

C)

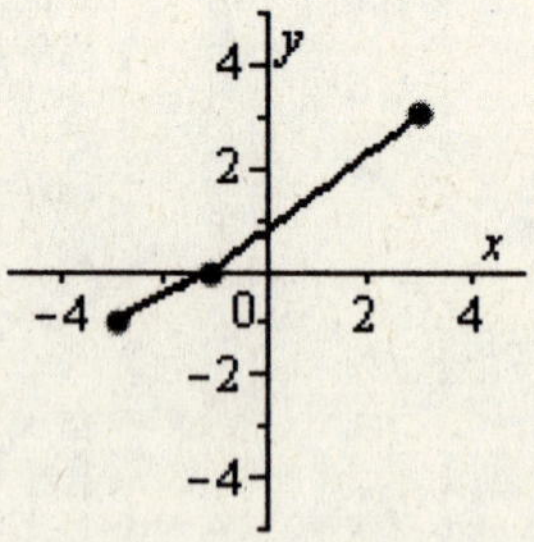

D)

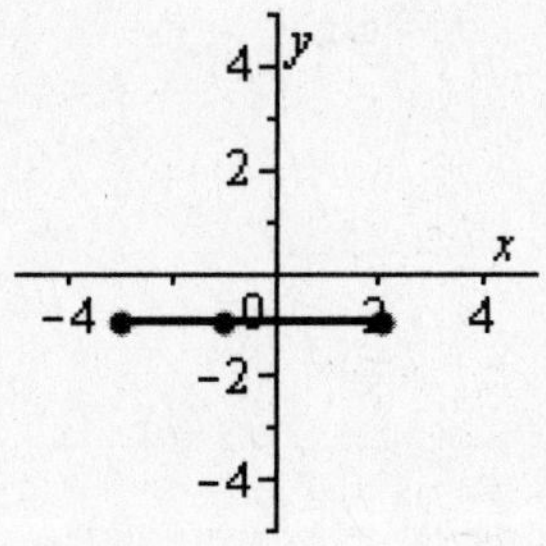

E)

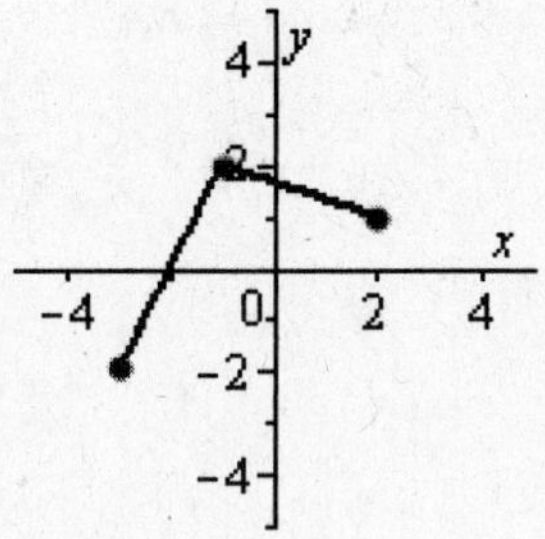

16. Evaluate the function at the specified value of the independent variable and simplify.
    $f(y) = 2y + 7$
    $f(-1.4)$
    A) $-2.8y + 14$
    B) -9.8
    C) 4.2
    D) $-1.4y + 7$
    E) $-1.4y - 7$

17. Estimate the slope of the line.

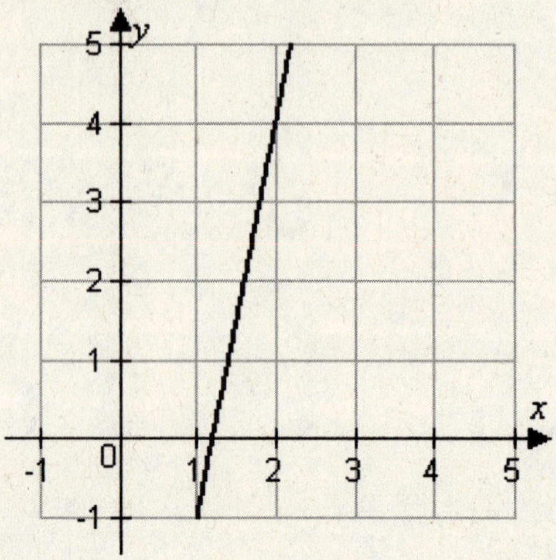

A) -5
B) 0
C) 5
D) $\frac{1}{5}$
E) $\frac{2}{5}$

18. Use the functions $f(x)=\frac{1}{125}x-5$ and $g(x)=x^3$ to find $(f\circ g)^{-1}$.

A) $(f\circ g)^{-1}=\frac{x^3+5}{5}$
B) $(f\circ g)^{-1}=\frac{x^3-625}{125}$
C) $(f\circ g)^{-1}=\frac{\sqrt[3]{x+5}}{5}$
D) $(f\circ g)^{-1}=5x+5$
E) $(f\circ g)^{-1}=5\sqrt[3]{x+5}$

19. Find all real values of x such that $f(x) = 0$.

$$f(x) = \frac{7x - 5}{7}$$

A) $\frac{5}{49}$

B) $\pm\frac{5}{49}$

C) $\pm\frac{5}{7}$

D) $\frac{5}{7}$

E) $-\frac{5}{7}$

20. Graph the function and determine the interval(s) for which $f(x) \geq 0$.

$$f(x) = -x^2 + 4x$$

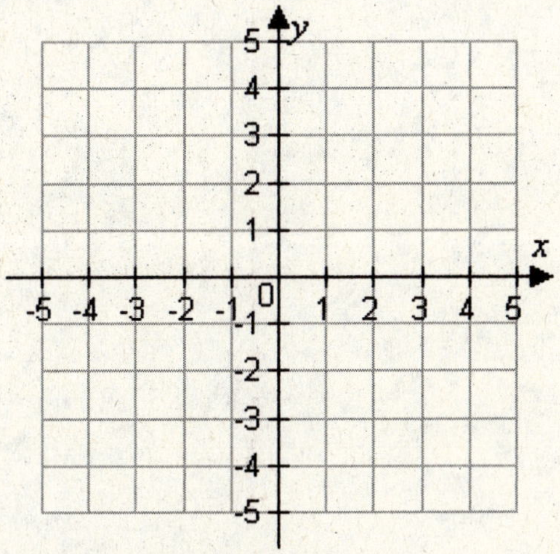

A) $(-\infty, 0] \cup [4, \infty)$

B) $[0, 4]$

C) $(0, 4)$

D) $(-\infty, 0) \cup (4, \infty)$

E) $\{4\}$

Answer Key

1. E
2. A
3. B
4. B
5. A
6. C
7. A
8. A
9. B
10. E
11. C
12. E
13. A
14. A
15. B
16. C
17. C
18. E
19. D
20. B

Name: ______________________ Date: __________

1. Determine an equation that may represented by the graph shown below.

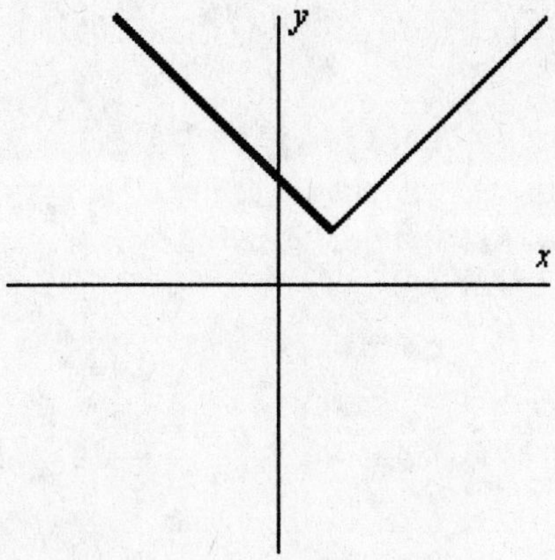

A) $f(x)=|x-1|-1$
B) $f(x)=-|x-1|+1$
C) $f(x)=|x-1|+1$
D) $f(x)=|x+1|+1$
E) $f(x)=|x+1|-1$

2. Find the inverse function of *f*.
$f(x)=x^5-1$
A) $f^{-1}(x)=-\sqrt[5]{x}-1$
B) $f^{-1}(x)=\sqrt[5]{x}-1$
C) $f^{-1}(x)=-\sqrt[5]{x-1}$
D) $f^{-1}(x)=\sqrt[5]{x+1}$
E) $f^{-1}(x)=\sqrt[5]{x}+1$

3. Find the domain of the function.
$g(w)=\dfrac{4w}{w+9}$
A) all real numbers $w\neq-9$
B) all real numbers $w\neq-9$, $w\neq0$
C) all real numbers
D) $w=-9$, $w=0$
E) $w=-9$

4. Decide whether the two functions shown in the graph below appear to be inverse functions of each other.

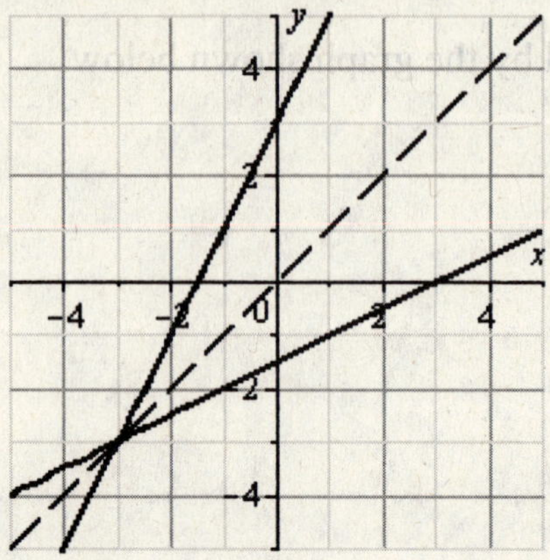

A) no
B) yes
C) not enough information

5. Determine an equation that may represented by the graph shown below.

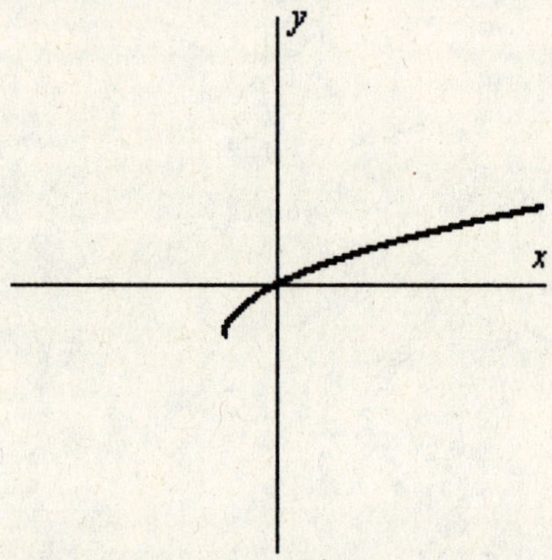

A) $f(x)=-1+\sqrt{1+x}$
B) $f(x)=1-\sqrt{1-x}$
C) $f(x)=-1-\sqrt{1-x}$
D) $f(x)=-1+\sqrt{1-x}$
E) $f(x)=-1-\sqrt{1+x}$

6. Which equation does not represent y as a function of x?
   A) $x = 6y - 9$
   B) $x = -5$
   C) $y = x + 5$
   D) $y = |-1 - x^2|$
   E) $y = \sqrt{-5 + 4x}$

7. Determine algebraically whether the following function is one-to-one.

   $f(x) = \dfrac{5x^2}{3x^2 + 6}$, where $x > 0$

   A)
   $$\begin{aligned} \frac{5a^2}{3a^2+6} &= \frac{5b^2}{3b^2+6} \\ \frac{5a^2}{3a^2} + \frac{5a^2}{6} &= \frac{5b^2}{3b^2} + \frac{5b^2}{6} \\ \frac{5}{3} + \frac{5a^2}{6} &= \frac{5}{3} + \frac{5b^2}{6} \\ \frac{30+5a^2}{18} &= \frac{30+5b^2}{18} \\ 30+5a^2 &= 30+5b^2 \\ 5a^2 &= 5b^2 \\ a^2 &= b^2 \\ \pm a &= \pm b \end{aligned}$$
   ; not one-to-one

   B)
   $$\begin{aligned} \frac{5a^2}{3a^2+6} &= \frac{5b^2}{3b^2+6} \\ \frac{5}{3+6} &= \frac{5}{3+6} \\ \frac{5}{6} &= \frac{3}{6} \\ a &= b \end{aligned}$$
   ; one-to-one

   C)
   $$\begin{aligned} \frac{5a^2}{3a^2+6} &= \frac{5b^2}{3b^2+6} \\ \frac{5a^2}{3a^2} &= \frac{5b^2}{3b^2} \\ \frac{5}{3} &= \frac{5}{3} \\ a &= b \end{aligned}$$
   ; one-to-one

D) $$\begin{aligned} \frac{5a^2}{3a^2+6} &= \frac{5b^2}{3b^2+6} \\ \frac{5a^2}{9a^2} &= \frac{5b^2}{9b^2} \\ \frac{5a}{9} &= \frac{5b}{9} \\ 5a &= 5b \\ a &= b \end{aligned}$$; one-to-one

E) $$\begin{aligned} \frac{5a^2}{3a^2+6} &= \frac{5b^2}{3b^2+6} \\ \frac{5a^2}{3a^2}+\frac{5a^2}{6} &= \frac{5b^2}{3b^2}+\frac{5b^2}{6} \\ \frac{5}{3}+\frac{5a^2}{6} &= \frac{5}{3}+\frac{5b^2}{6} \\ \frac{30+5a^2}{18} &= \frac{30+5b^2}{18} \\ 30+5a^2 &= 30+5b^2 \\ 5a^2 &= 5b^2 \\ a^2 &= b^2 \\ a &= b \end{aligned}$$; one-to-one

8. Find $f \circ g$.

$f(x) = x + 3$ $\qquad$ $g(x) = \dfrac{4}{x^2 - 9}$

A) $(f \circ g)(x) = \dfrac{4}{x^2}$

B) $(f \circ g)(x) = \dfrac{4}{x^2 + 6x}$

C) $(f \circ g)(x) = \dfrac{3x^2 + 1}{x^2 - 9}$

D) $(f \circ g)(x) = \dfrac{7}{x^2 - 9}$

E) $(f \circ g)(x) = \dfrac{3x^2 - 23}{x^2 - 9}$

9. Use function notation to write g in terms of $f(x)=\sqrt{x}$.

$$g(x)=-\frac{1}{3}\sqrt{x-8}+7$$

A) $g(x)=-f(x-8)+6$

B) $g(x)=-\frac{1}{3}f(x)-1$

C) $g(x)=-\frac{1}{3}f(x-8)+7$

D) $g(x)=f(x)+7$

E) $g(x)=f(x-8)-\frac{7}{3}$

10. Determine a piecewise-defined function for the graph shown below.

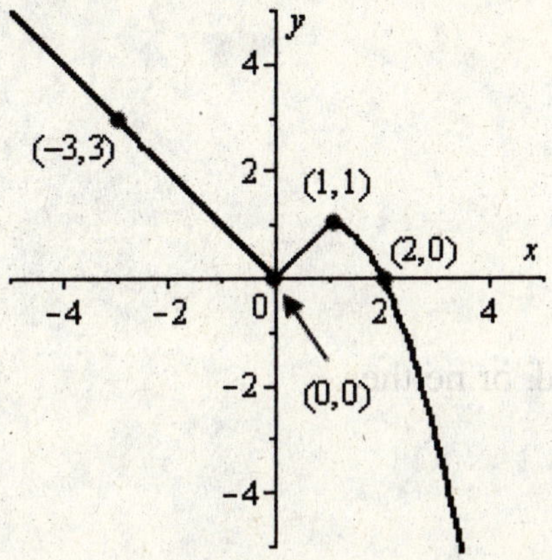

A) $f(x)=\begin{cases}|x|, & x\le 1\\ -(x-1)^2+1, & x>1\end{cases}$

B) $f(x)=\begin{cases}|x|, & x\le 0\\ -(x-1)^2+1, & x\le 0\end{cases}$

C) $f(x)=\begin{cases}|x|, & x\ge 1\\ -x^2, & x\le 1\end{cases}$

D) $f(x)=\begin{cases}|x|, & x\ge 0\\ -x^2, & x\le 1\end{cases}$

E) $f(x)=\begin{cases}|x|, & x\le 1\\ -(x-1)^2, & x>1\end{cases}$

11. Estimate the slope of the line.

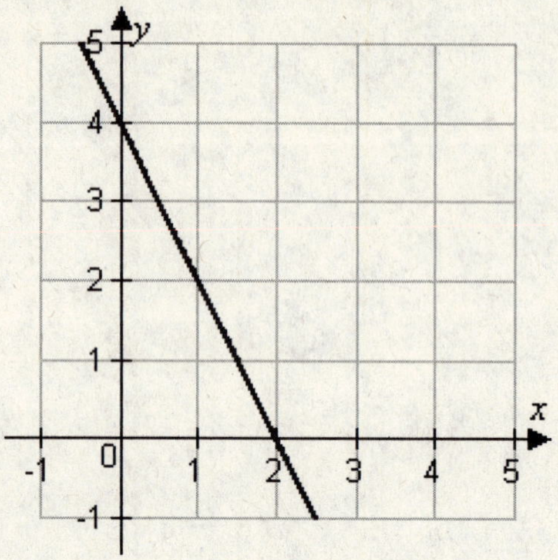

A) $-\frac{1}{2}$
B) 2
C) −2
D) $\frac{1}{2}$
E) −3

12. Determine whether the function is even, odd, or neither.
$f(x) = 4x^3 - 2x$

A) neither
B) even
C) odd

13. Find the slope and *y*-intercept of the equation of the line.
$y = -2x + 3$

A) slope: $-\frac{1}{2}$; *y*-intercept: 3
B) slope: $\frac{1}{3}$; *y*-intercept: −2
C) slope: −2; *y*-intercept: 3
D) slope: 3; *y*-intercept: −2
E) slope: −2; *y*-intercept: −3

14. Determine whether lines L_1 and L_2 passing through the pairs of points are parallel, perpendicular, or neither.
    $L_1: (-1, 1), (-1, -6)$
    $L_2: (3, -8), (24, -8)$
    A) parallel
    B) perpendicular
    C) neither

15. Show algebraically that the functions f and g shown below are inverse functions.
    $f(x) = \sqrt[3]{8x-7}, \quad g(x) = \dfrac{x^3+7}{8}$

A)
$$\begin{aligned} f(g(x)) &= \sqrt[3]{8\left(\frac{x^3+7}{8}\right)-7} \\ &= \sqrt[3]{\left(x^3+56\right)-56} \\ &= \sqrt[3]{x^3+56-56} \\ &= \sqrt[3]{x^3} \\ &= x \end{aligned} \qquad \begin{aligned} g(f(x)) &= \frac{\left(\sqrt[3]{8x-7}\right)^3+7}{8} \\ &= \frac{8x-7^3+7^3}{8} \\ &= \frac{8x}{8} \\ &= x \end{aligned}$$

B)
$$\begin{aligned} f(g(x)) &= \sqrt[3]{8\left(\frac{x^3+7}{8}\right)-7} \\ &= \sqrt[3]{\left(x^3+7\right)-7} \\ &= \sqrt[3]{x^3+7-7} \\ &= \sqrt[3]{x^3} \\ &= x \end{aligned} \qquad \begin{aligned} g(f(x)) &= \frac{\left(\sqrt[3]{8x-7}\right)^3+7}{8} \\ &= \frac{8x-7+7}{8} \\ &= \frac{8x}{8} \\ &= x \end{aligned}$$

C)
$$\begin{aligned} f(g(x)) &= \sqrt[3]{8\left(\frac{x^3+7}{8}\right)-7} \\ &= \sqrt[3]{\left(\frac{8x^3+7}{8}\right)-7} \\ &= \sqrt[3]{x^3+7-7} \\ &= \sqrt[3]{x^3} \\ &= x \end{aligned} \qquad \begin{aligned} g(f(x)) &= \frac{\left(\sqrt[3]{8x-7}\right)^3+7}{8} \\ &= \frac{8^3x-7+7}{8^3} \\ &= \frac{8^3x}{8^3} \\ &= x \end{aligned}$$

D) $f(g(x)) = \sqrt[3]{8\left(\dfrac{x^3+7}{8}\right)-7}$

$= \sqrt[3]{\left(8x^3+56\right)-56}$

$= \sqrt[3]{8x^3+56-56}$

$= \sqrt[3]{8x^3}$

$= x$

$g(f(x)) = \dfrac{\left(\sqrt[3]{8x-7}\right)^3+7}{8}$

$= \dfrac{8^3x-7^3+7^3}{8^3}$

$= \dfrac{8^3x}{8^3}$

$= x$

E) $f(g(x)) = \sqrt[3]{8\left(\dfrac{x^3+7}{8}\right)-7}$

$= \sqrt[3]{\left(x^3+\dfrac{7}{8}\right)-7}$

$= \sqrt[3]{x^3+\dfrac{0}{8}}$

$= \sqrt[3]{x^3}$

$= x$

$g(f(x)) = \dfrac{\left(\sqrt[3]{8x-7}\right)^3+7}{8}$

$= \dfrac{24x-21+21}{24}$

$= \dfrac{24x}{24}$

$= x$

16. Find all real values of x such that $f(x) = 0$.

$f(x)=\dfrac{-2x+5}{5}$

A) $\dfrac{1}{2}$

B) $\pm\dfrac{1}{2}$

C) $\pm\dfrac{5}{2}$

D) $\dfrac{5}{2}$

E) $-\dfrac{5}{2}$

17. Compare the graph of the following function with the graph of $f(x)=|x|$.

$y=\left|\frac{3}{4}x\right|$

A) vertical shift of $\frac{3}{4}$ units up

B) horizontal stretch of $\frac{4}{3}$ units

C) vertical shrink of $\frac{3}{4}$ units

D) horizontal shrink of $\frac{3}{4}$ units

vertical shift of $\frac{4}{3}$ units

E) horizontal shrink of $\frac{3}{4}$ units

18. Find the domain of the function.

$g(x)=\sqrt{25-x^2}$

A) $-5\le x\le 5$
B) $x\le -5$ or $x\ge 5$
C) $x\ge 0$
D) $x\le 5$
E) all real numbers

19. Use the functions $f(x)=x+4$ and $g(x)=5x-7$ to find $(g\circ f)^{-1}$.

A) $(g\circ f)^{-1}=\dfrac{5x+11}{4}$

B) $(g\circ f)^{-1}=5x-42$

C) $(g\circ f)^{-1}=\dfrac{x-13}{5}$

D) $(g\circ f)^{-1}=\dfrac{-7x-7}{5}$

E) $(g\circ f)^{-1}=5x+13$

20. Find the value(s) of x for which $f(x) = g(x)$.

$f(x) = x^2 - 11x - 36$ $\qquad$ $g(x) = -7x - 4$

A) $-36, -25, -\frac{4}{7}$

B) $-36, -11, -\frac{4}{7}$

C) $8, -4$

D) $-8, 4$

E) $47, -\frac{4}{7}$

Answer Key

1. C
2. D
3. A
4. B
5. A
6. B
7. E
8. E
9. C
10. A
11. C
12. C
13. C
14. B
15. B
16. D
17. B
18. A
19. C
20. C

Name: ______________________________ Date: _______________

1. Determine an equation that may represented by the graph shown below.

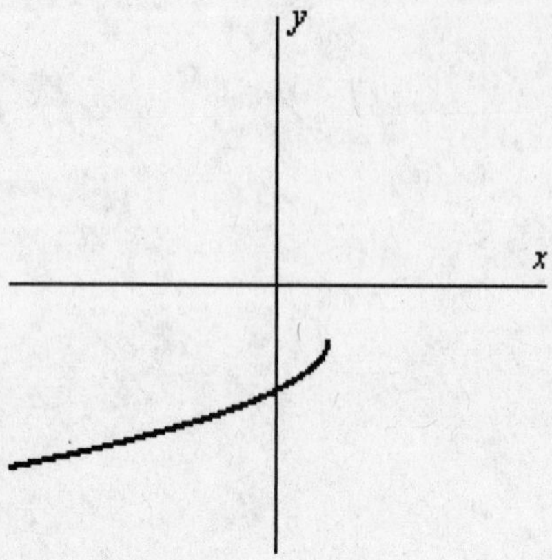

A) $f(x) = -1 - \sqrt{1-x}$
B) $f(x) = -1 + \sqrt{1-x}$
C) $f(x) = -1 - \sqrt{1+x}$
D) $f(x) = -1 + \sqrt{1+x}$
E) $f(x) = 1 - \sqrt{1-x}$

2. The scatter plots of different data are shown below. Determine whether there is a positive correlation, negative correlation, or no discernible correlation between the variables.

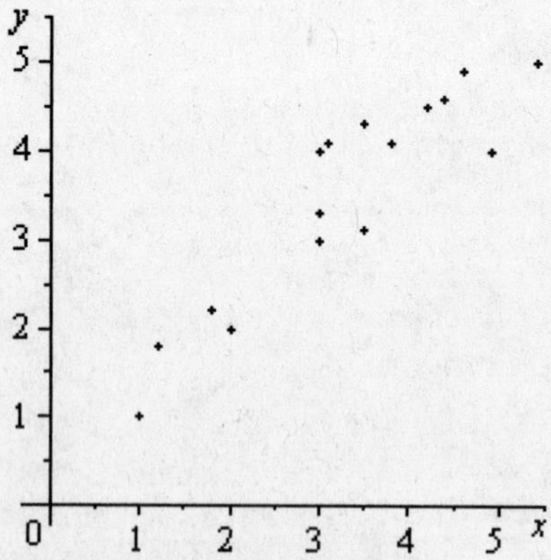

A) positive correlation
B) negative correlation
C) no discernible correlation

3. Does the table describe a function?

Input value	–6	–3	0	3	6
Output value	11	11	11	11	11

A) yes
B) no

4. Find the domain of the function.

$$g(w) = \frac{-7w}{w-5}$$

A) all real numbers $w \neq 5$
B) all real numbers $w \neq 5,\ w \neq 0$
C) all real numbers
D) $w = 5,\ w = 0$
E) $w = 5$

5. Determine the domain and range of the inverse function f^{-1} of the following function f.

$$f(x) = -|x+8| - 3, \text{ where } x > -8$$

A) Domain: $[-8, \infty)$; Range: $[-3, \infty)$
B) Domain: $(-\infty, -3]$; Range: $[-8, \infty)$
C) Domain: $[-8, -3]$; Range: $[-8, \infty)$
D) Domain: $(-\infty, -8]$; Range: $[3, \infty)$
E) Domain: $(-\infty, \infty)$; Range: $(-\infty, \infty)$

6. Use function notation to write g in terms of $f(x) = x^3$.

$$g(x) = -\frac{1}{2}(x+9)^3$$

A) $g(x) = -\frac{1}{2}[f(x)]^3 + 9$
B) $g(x) = -\frac{1}{2}[f(x)+9]$
C) $g(x) = -[f(x)]^3 + \frac{729}{2}$
D) $g(x) = -\frac{1}{2}[f(x)]^3 + 729$
E) $g(x) = -\frac{1}{2}[f(x+9)]$

7. Evaluate the indicated function for $f(x) = x^2 - 1$ and $g(x) = x - 6$.

$(fg)(-2)$

A) –24
B) 40
C) –2
D) 12
E) 24

8. If f is an even function, determine if g is even, odd, or neither.

$g(x) = f(x+4)$

A) even
B) odd
C) cannot be determined
D) neither

9. Plot the points and find the slope of the line passing through the pair of points.
   (1, 0), (5, 3)

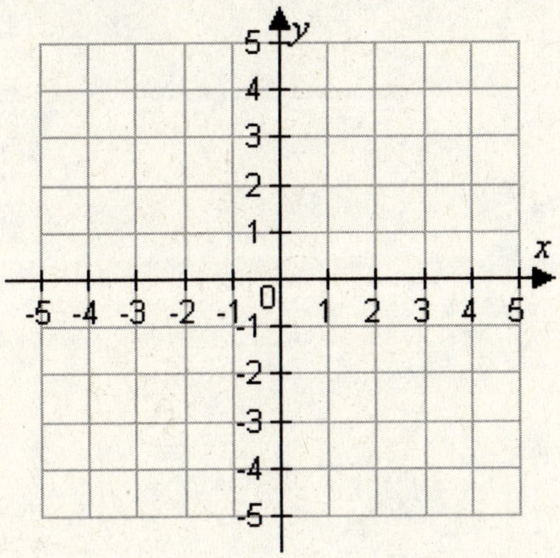

A) slope: $\frac{4}{3}$

B) slope: $-\frac{4}{3}$

C) slope: $\frac{1}{2}$

D) slope: $\frac{3}{4}$

E) slope: $-\frac{3}{4}$

10. Compare the graph of the following function with the graph of $f(x) = x^3$.
    $y = [5(x-2)]^3$

A) vertical shift of 2 units down

B) vertical shift of 2 units down
   horizontal shrink of $\frac{1}{5}$ units

C) horizontal shift of 2 units to the right
   horizontal shrink of $\frac{1}{125}$ units

D) horizontal shift of 2 units to the right
   horizontal stretch of $\frac{1}{5}$ units

E) horizontal shift of 2 units to the right
   vertical shift of 5 units down

11. Find the slope-intercept form of the line passing through the points.
    (–4, –2), (–1, 7)
    A) $y = 3x + 2$
    B) $y = 3x + 10$
    C) $y = \frac{1}{3}x - \frac{2}{3}$
    D) $y = -\frac{1}{3}x - \frac{10}{3}$
    E) $y = -3x - 14$

12. Match the graph of the function shown below with the graph of its inverse function

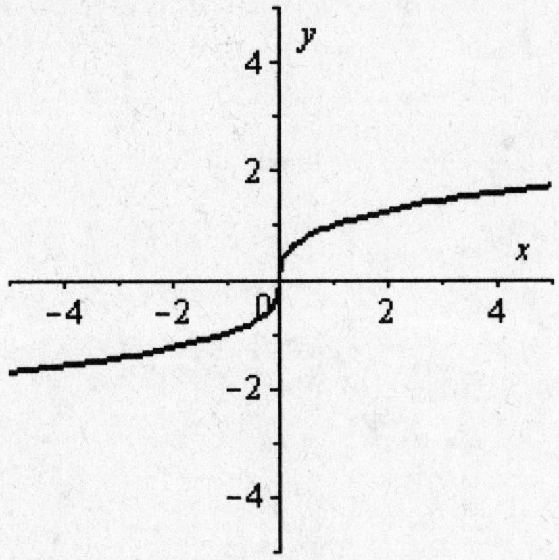

A)

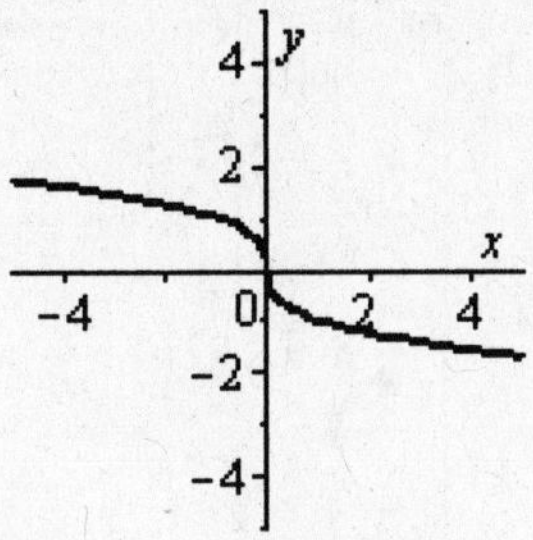

B)

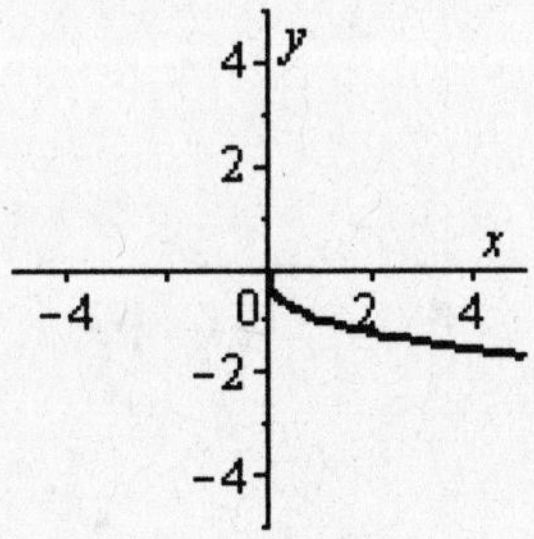

C)

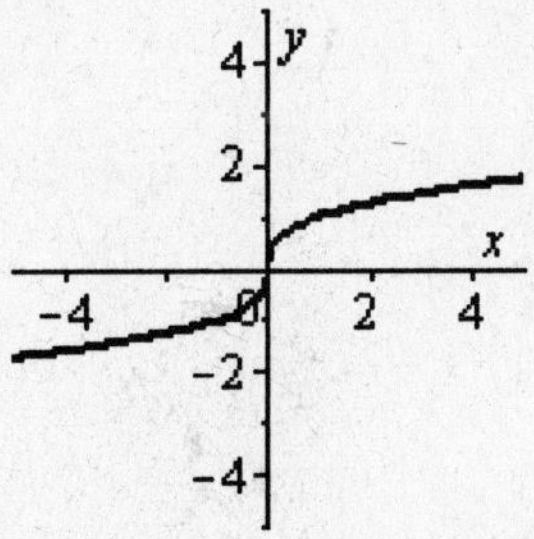

D)

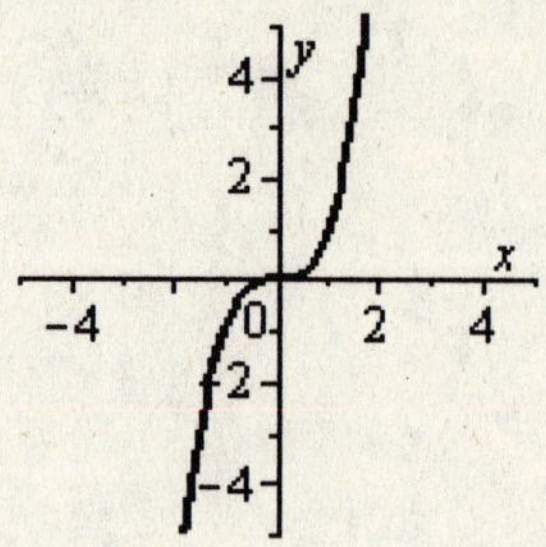

E)

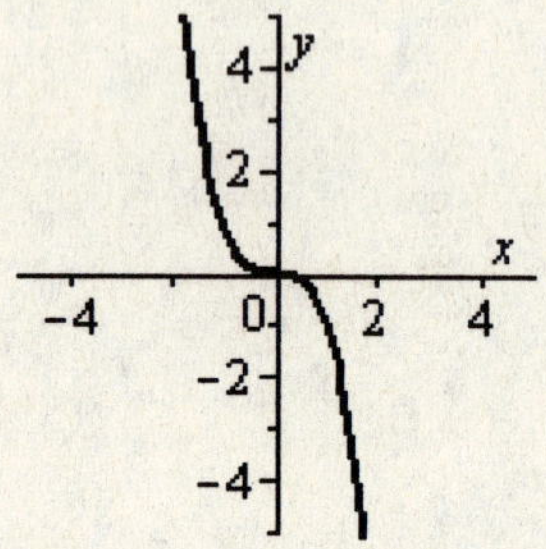

13. Use the graph of f to sketch the graph of the function indicated below.

$$y = \frac{1}{2}f(x)$$

f

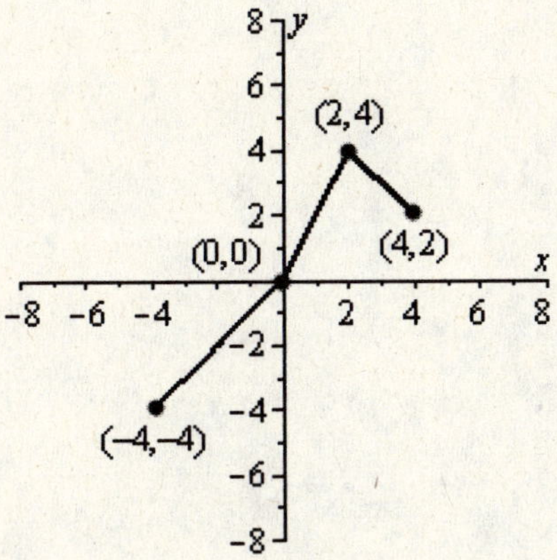

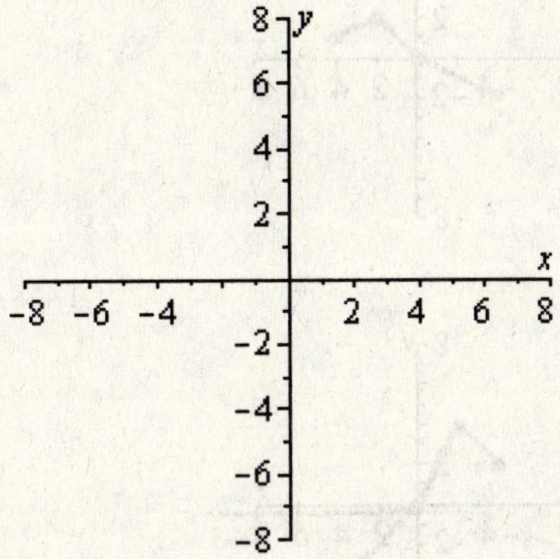

A)

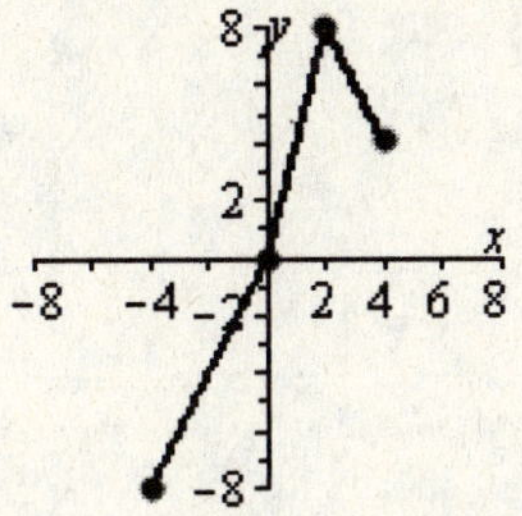

B)

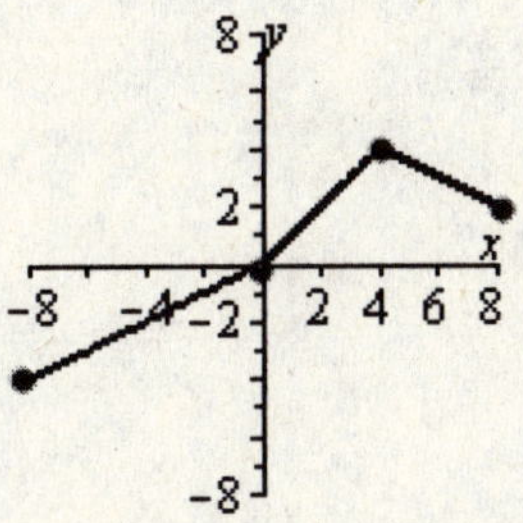

C)

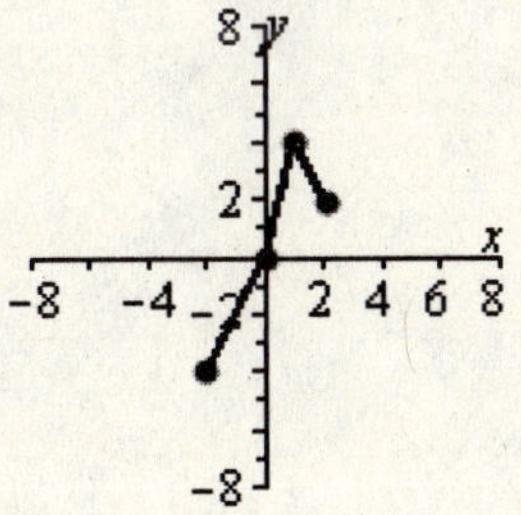

D)

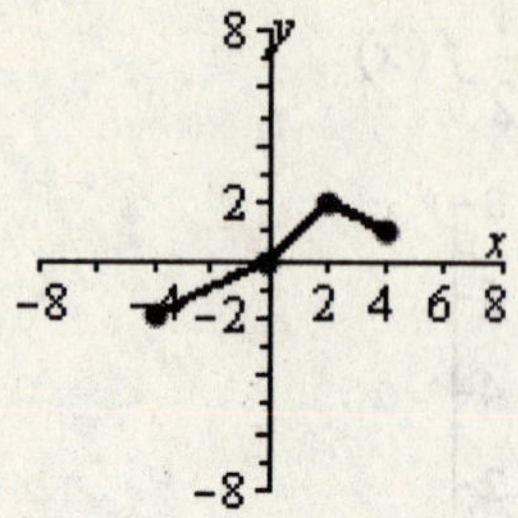

E)

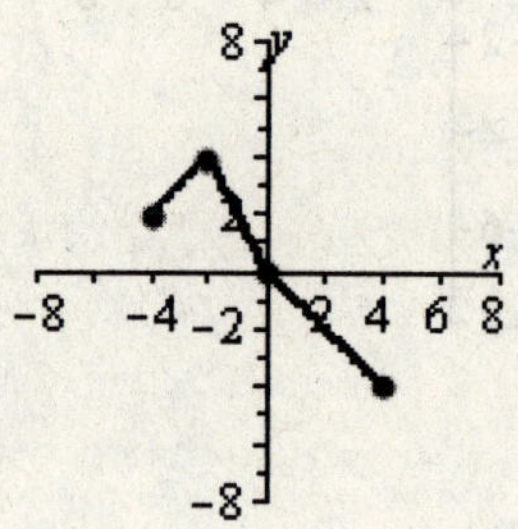

14. Compare the graph of the following function with the graph of $f(x) = |x|$.

$$y = \left|\frac{7}{9}x\right|$$

A) vertical shift of $\frac{7}{9}$ units up
B) horizontal stretch of $\frac{9}{7}$ units
C) vertical shrink of $\frac{7}{9}$ units
D) horizontal shrink of $\frac{7}{9}$ units

vertical shift of $\frac{9}{7}$ units

E) horizontal shrink of $\frac{7}{9}$ units

15. Write the slope-intercept form of the equation of the line through the given point parallel to the given line.
point: (3, –4) line: $28x + 7y = -4$
A) $y = -\frac{1}{28}x - \frac{109}{28}$
B) $y = \frac{1}{4}x - \frac{19}{4}$
C) $y = 28x + 80$
D) $y = -4x + 8$
E) $y = -4x - 13$

16. Does the table describe a function?

Input value	5	10	13	10	5
Output value	–13	–9	0	9	13

A) yes
B) no

17. Show algebraically that the functions f and g shown below are inverse functions.

$$f(x)=-\frac{5}{7}x-3,\quad g(x)=-\frac{7x+21}{5}$$

A)
$$\begin{aligned} f(g(x)) &= -\frac{5}{7}\left(\frac{7x+21}{5}\right)-3 \\ &= \left(\frac{7x+21}{7}\right)-3 \\ &= (x+3)-3 \\ &= x+3-3 \\ &= x \end{aligned}$$

$$\begin{aligned} g(f(x)) &= -\frac{7\left(-\frac{5}{7}x-3\right)+21}{5} \\ &= -\frac{(-5x-21)+21}{5} \\ &= \frac{-5x-21+21}{5} \\ &= \frac{5x}{5} \\ &= x \end{aligned}$$

B)
$$\begin{aligned} f(g(x)) &= -\frac{5}{7}\left(-\frac{7x+21}{5}\right)-21 \\ &= \left(\frac{35x+21}{35}\right)-21 \\ &= (x+21)-21 \\ &= x+21-21 \\ &= x \end{aligned}$$

$$\begin{aligned} g(f(x)) &= -\frac{7\left(-\frac{5}{7}x-3\right)+21}{5} \\ &= -\frac{(-5x-3)+21}{5} \\ &= \frac{5x+3-21}{5} \\ &= \frac{5x}{5} \\ &= x \end{aligned}$$

C)
$$\begin{aligned} f(g(x)) &= -\frac{5}{7}\left(-\frac{7x+3}{5}\right)-3 \\ &= \left(\frac{35x+3}{35}\right)-3 \\ &= (x+3)-3 \\ &= x+3-3 \\ &= x \end{aligned}$$

$$\begin{aligned} g(f(x)) &= -\frac{7\left(-\frac{5}{7}x-3\right)+21}{5} \\ &= -\frac{(-5x-3)+3}{5} \\ &= \frac{5x+3-3}{5} \\ &= \frac{5x}{5} \\ &= x \end{aligned}$$

D) $f(g(x)) = -\frac{5}{7}\left(-\frac{7x+21}{5}\right)-3$
$= \left(\frac{7x+21}{7}\right)-3$
$= (x+3)-3$
$= x+3-3$
$= x$

$g(f(x)) = -\frac{7\left(-\frac{5}{7}x-3\right)+21}{5}$
$= -\frac{(-5x-21)+21}{5}$
$= \frac{5x+21-21}{5}$
$= \frac{5x}{5}$
$= x$

E) $f(g(x)) = -\frac{7}{5}\left(-\frac{5x+15}{7}\right)-3$
$= \left(\frac{5x+15}{5}\right)-3$
$= (x+3)-3$
$= x+3-3$
$= x$

$g(f(x)) = -\frac{7\left(-\frac{5}{7}x-3\right)+21}{5}$
$= -\frac{(-5x-3)+21}{35}$
$= \frac{5x+3-21}{35}$
$= \frac{35x}{35}$
$= x$

18. Find the domain of the function.

$f(t)=\sqrt{64-t^2}$

A) $-8 \le t \le 8$
B) $t \le -8$ or $t \ge 8$
C) $t \ge 0$
D) $t \le 8$
E) all real numbers

19. Find the inverse function of f.

    $f(x) = x^9 - 2$

    A) $f^{-1}(x) = -\sqrt[9]{x} - 2$
    B) $f^{-1}(x) = \sqrt[9]{x} - 2$
    C) $f^{-1}(x) = -\sqrt[9]{x-2}$
    D) $f^{-1}(x) = \sqrt[9]{x+2}$
    E) $f^{-1}(x) = \sqrt[9]{x} + 2$

20. Find $f \circ g$.

    $f(x) = -4x + 3$ $\qquad$ $g(x) = x + 7$

    A) $(f \circ g)(x) = -4x - 25$
    B) $(f \circ g)(x) = -4x + 10$
    C) $(f \circ g)(x) = -4x^2 - 25x + 21$
    D) $(f \circ g)(x) = -5x - 4$
    E) $(f \circ g)(x) = -5x + 10$

Answer Key

1. A
2. A
3. A
4. A
5. B
6. E
7. A
8. C
9. D
10. C
11. B
12. D
13. D
14. B
15. D
16. B
17. D
18. A
19. D
20. A

Name: ___________________________ Date: _____________

1. Hooke's Law states that the force F required to compress or stretch a spring (within its elastic limits) is proportional to the distance d that the spring is compressed or stretched from its original length. That is, $F = kd$, where k is the measure of the stiffness of the spring and is called the *spring constant*. The table below shows the elongation d in centimeters of a spring when a force of F kilograms is applied.

Force, *F*	**Elongation, *d***
20	1.4
40	2.5
60	4.0
80	5.3
100	6.6

Find the equation of the line that seems to best fit the data. Use the model to estimate the elongation of the spring when a force of 55 kilograms is applied. Round your answer to one decimal place.
A) 7.2 centimeters
B) 5.4 centimeters
C) 1.8 centimeters
D) 3.6 centimeters
E) 2.7 centimeters

2. If f is an even function, determine if g is even, odd, or neither.
$g(x) = -f(x+3)$
A) even
B) odd
C) cannot be determined
D) neither

3. Given $f(x)=\dfrac{10}{x^2-9}$ and $g(x)=x+3$ determine the domain of $f\circ g$.

A) $(-\infty,-3)\cup(3,\infty)$
B) $(-\infty,-6)\cup(-6,0)\cup(0,\infty)$
C) $\left(-\infty,-\dfrac{10}{3}\right)\cup\left(\dfrac{10}{3},\infty\right)$
D) $(-\infty,-3)\cup(-3,3)\cup(3,\infty)$
E) $(-\infty,\infty)$

4. Determine an equation that may represented by the graph shown below.

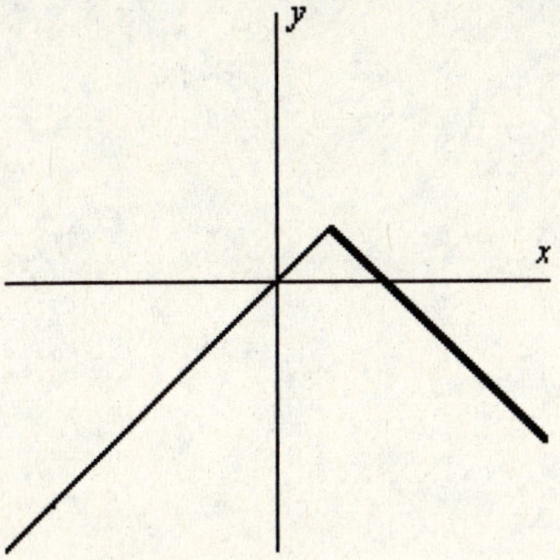

A) $f(x)=|x+1|+1$
B) $f(x)=|x+1|-1$
C) $f(x)=-|x-1|+1$
D) $f(x)=|x-1|+1$
E) $f(x)=|x-1|-1$

5. Find all real values of x such that $f(x) = 0$.

$f(x) = 49x^2 - 64$

A) $\pm\frac{7}{8}$

B) $\pm\frac{8}{7}$

C) $\pm\frac{64}{49}$

D) $-\frac{64}{49}$

E) $\frac{8}{7}$

6. Find $(f+g)(x)$.

$f(x) = -8x^2 + 5x - 2$

$g(x) = 4x^2 + 7x + 4$

A) $(f+g)(x) = -12x^4 - 2x^2 - 6$

B) $(f+g)(x) = -4x^4 + 12x^2 + 2$

C) $(f+g)(x) = -12x^2 - 2x - 6$

D) $(f+g)(x) = -4x^2 + 12x + 2$

E) $(f+g)(x) = 4x^2 - 12x - 2$

7. Find $f \circ g$.

$f(x) = x + 4 \qquad g(x) = \frac{3}{x^2 - 16}$

A) $(f \circ g)(x) = \frac{3}{x^2}$

B) $(f \circ g)(x) = \frac{3}{x^2 + 8x}$

C) $(f \circ g)(x) = \frac{4x^2 - 1}{x^2 - 16}$

D) $(f \circ g)(x) = \frac{7}{x^2 - 16}$

E) $(f \circ g)(x) = \frac{4x^2 - 61}{x^2 - 16}$

8. Graph the function and determine the interval(s) for which $f(x) \geq 0$.
$f(x) = -x^2 + 4x$

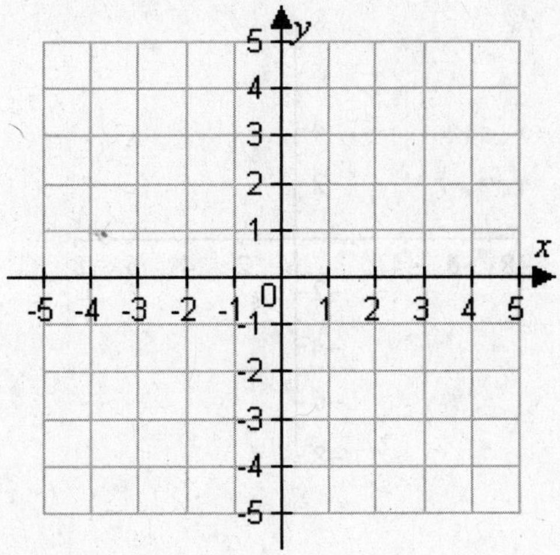

A) $(-\infty, 0] \cup [4, \infty)$
B) $[0, 4]$
C) $(0, 4)$
D) $(-\infty, 0) \cup (4, \infty)$
E) $\{4\}$

9. Restrict the domain of the following function f so that the function is one-to-one and has an inverse function.
$f(x) = -|x - 4| + 2$

A) $[-4, \infty)$
B) $[2, 4]$
C) $[4, \infty)$
D) $[-2, 4]$
E) $(-\infty, 2]$

10. Use the graph of f to sketch the graph of the function indicated below.

f $\qquad\qquad\qquad\qquad$ $y = 2f(x+2)$

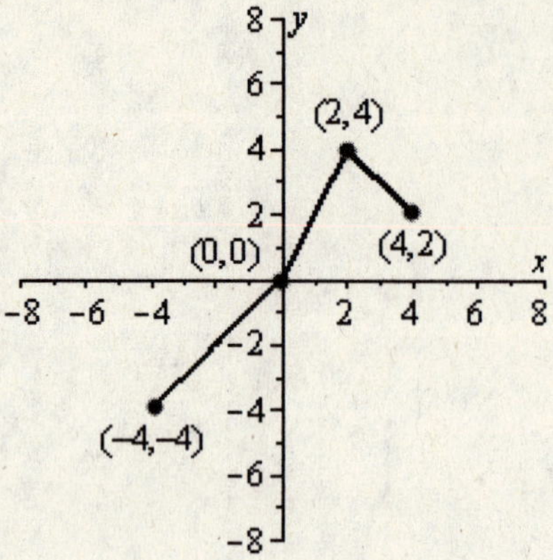

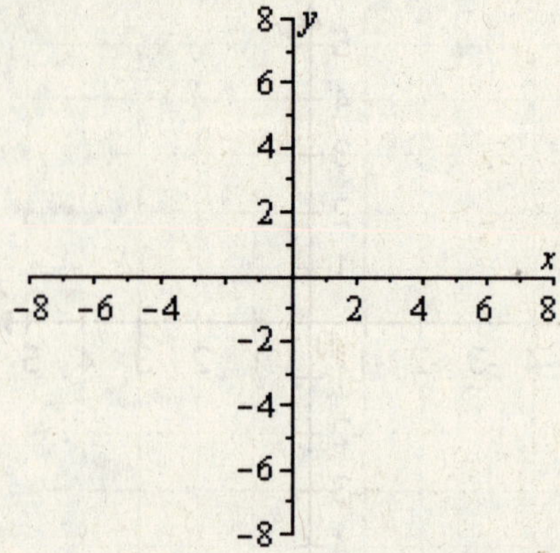

A)

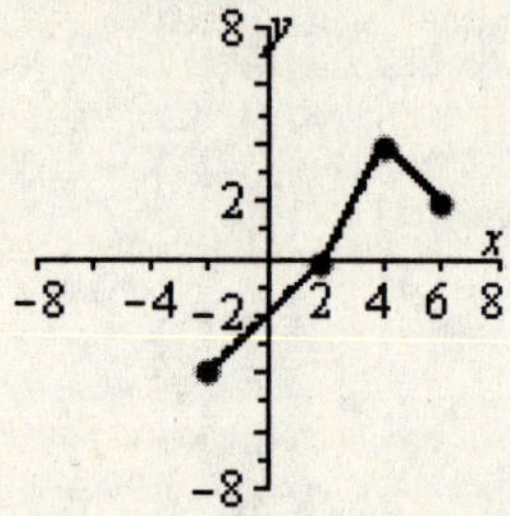

B)

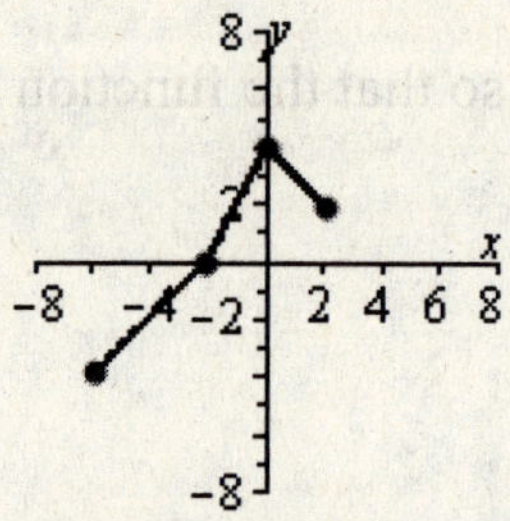

C)

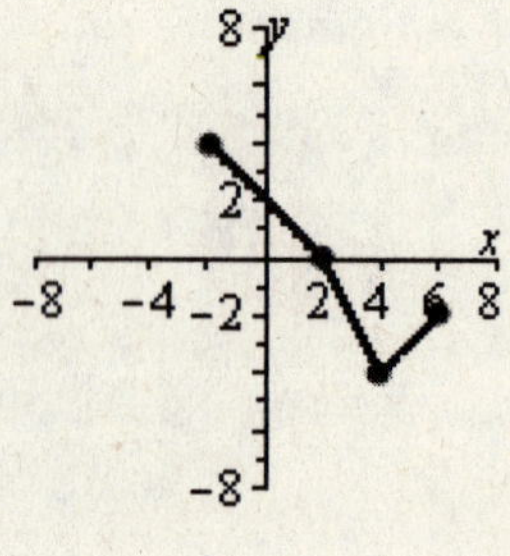

D)

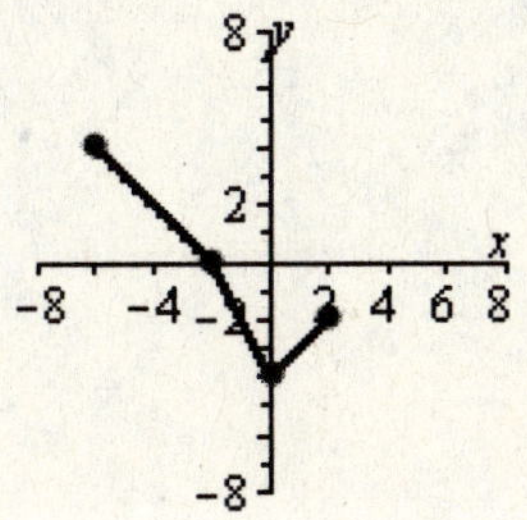

E)

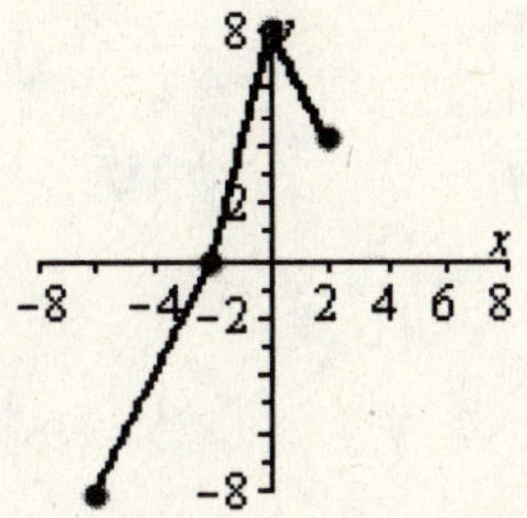

11. Algebraically determine whether the function below is even, odd, or neither.
    $f(s) = 8s^{7/6}$
    A) even
    B) odd
    C) cannot be determined
    D) neither

12. Compare the graph of the following function with the graph of $f(x) = \sqrt{x}$.
    $y = \sqrt{-x+4}$
    A) First a vertical shift of 4 units up then a reflection in the y-axis.
    B) First a horizontal shift of 4 units to the left then a reflection in the y-axis.
    C) First a vertical shift of 4 units up then a reflection in the x-axis.
    D) First a horizontal shift of 4 units to the left, then a vertical shift of 4 units up and then a reflection in the y-axis.
    E) First a horizontal shift of 4 units to the left then a reflection in the x-axis.

13. Find the domain of the function.

$$g(p) = \sqrt{4 - p^2}$$

A) $-2 \le p \le 2$
B) $p \le -2$ or $p \ge 2$
C) $p \ge 0$
D) $p \le 2$
E) all real numbers

14. Use the graphs of f and g, shown below, to graph $h(x) = (f + g)(x)$.

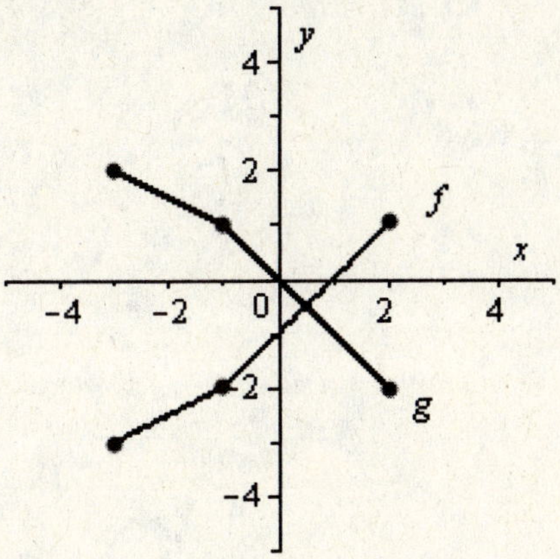

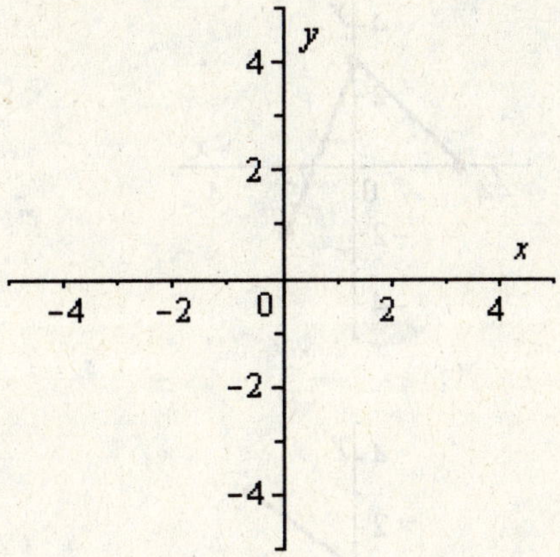

A)

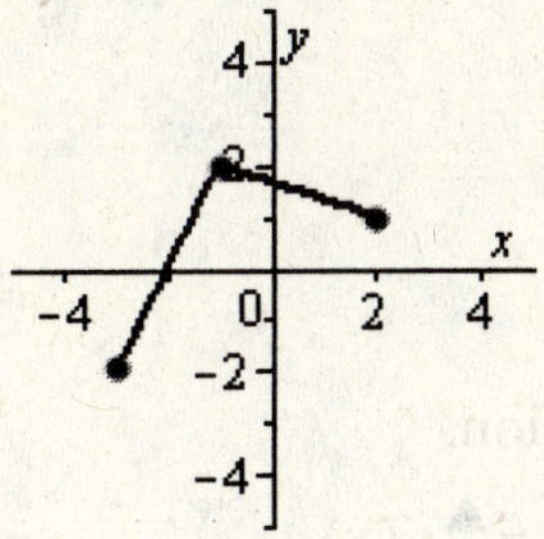

B)

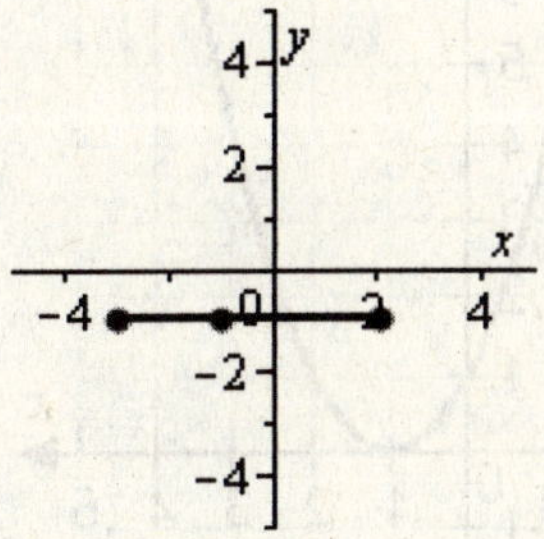

C)

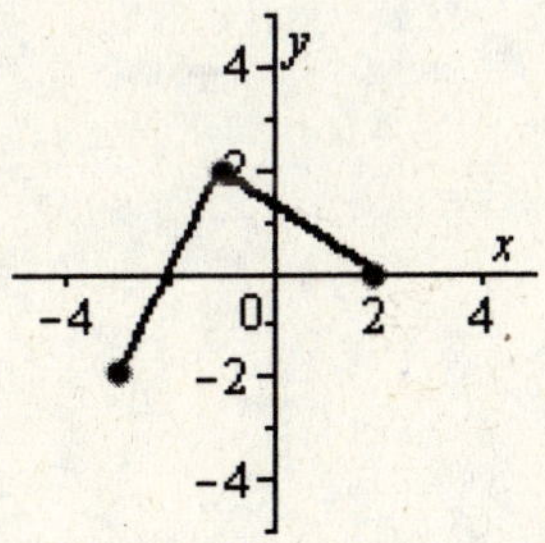

D)

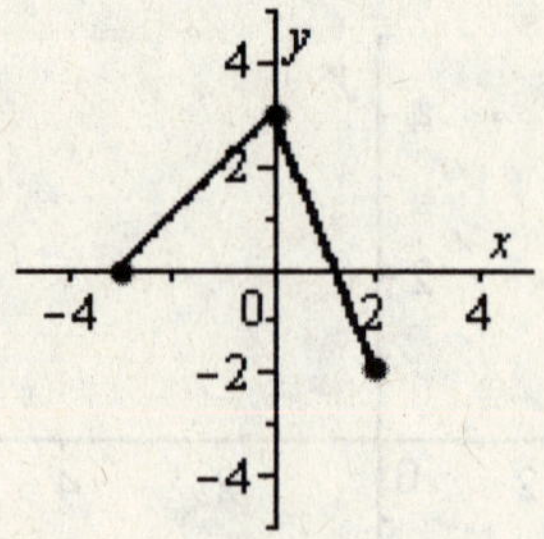

E)

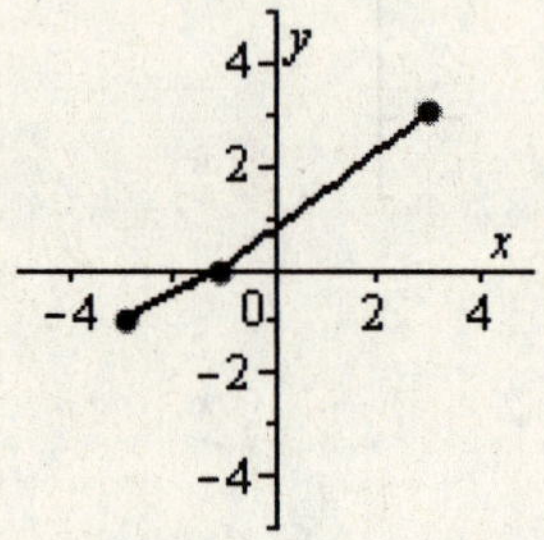

15. Use the graphs of f and g to evaluate the function.

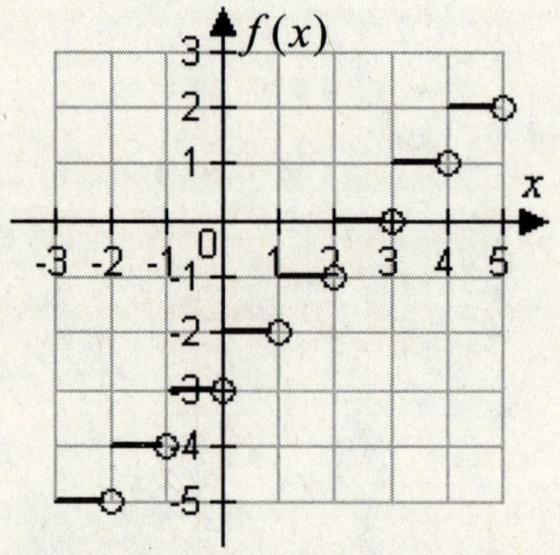

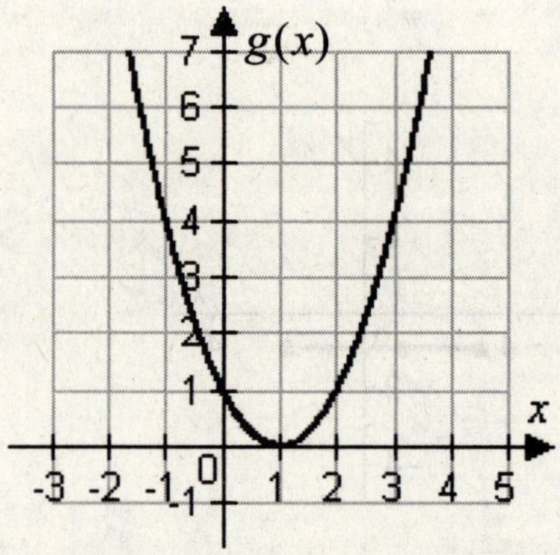

$(f \circ g)(1)$

A) 9
B) –1
C) 0
D) –4
E) –2

16. Find the slope and y-intercept of the equation of the line.
$y = -2x - 9$

A) slope: $-\frac{1}{2}$; y-intercept: –9
B) slope: $-\frac{1}{9}$; y-intercept: –2
C) slope: –2; y-intercept: –9
D) slope: –9; y-intercept: –2
E) slope: –2; y-intercept: 9

17. Use the graph of the function to find the domain and range of f.

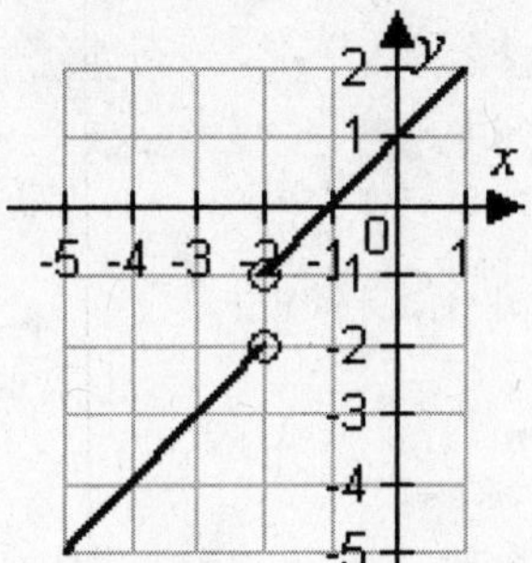

A) domain : all real numbers
range : $(-\infty,-2)\cup(-1,\infty)$
B) domain : all real numbers
range : all real numbers
C) domain : $(-\infty,-2)\cup(-2,\infty)$
range : $(-\infty,-2)\cup(-1,\infty)$
D) domain : $(-\infty,-2)\cup(-1,\infty)$
range : $(-\infty,-2)\cup(-2,\infty)$
E) Domain: all real numbers
Range: $(-\infty,-2]\cup[-1,\infty)$

18. Given that $f(x)=\sqrt[4]{x-4}$ and $g(x)=x^4+4$ determine the value of the following (if possible).

$$(f\circ g)(0)$$

A) 0
B) 2
C) 4
D) x^4-16
E) not possible

19. Find the inverse function of $f(x)=8x+3$

A) $g(x)=\dfrac{x-3}{8}$
B) $g(x)=3x+8$
C) $g(x)=\dfrac{x+3}{8}$
D) $g(x)=\dfrac{x}{3}$
E) $g(x)=\dfrac{1}{8}x-3$

20. Show algebraically that the functions f and g shown below are inverse functions.

$$f(x)=\frac{2}{2+x},\ x\geq 0,\quad g(x)=\frac{2-2x}{x},\ 0<x\leq 1$$

A)
$$\begin{aligned} f(g(x)) &= \frac{2}{2+\left(\frac{2-2x}{x}\right)} \\ &= \frac{2}{2+\left(\frac{1}{x}\right)} \\ &= \frac{1}{\left(\frac{1}{x}\right)} \\ &= 1\cdot\frac{x}{1} \\ &= x \end{aligned}$$

$$\begin{aligned} g(f(x)) &= \frac{2-2\left(\frac{2}{2+x}\right)}{\left(\frac{2}{2+x}\right)} \\ &= \frac{0-\left(\frac{2}{2+x}\right)}{\left(\frac{2}{2+x}\right)} \\ &= \frac{\frac{-2}{2+x}}{\left(\frac{2}{2+x}\right)} \\ &= \left(\frac{-2}{2+x}\right)\left(\frac{2+x}{2}\right) \\ &= \frac{2x+2}{2+x} \\ &= x \end{aligned}$$

B)
$$\begin{aligned} f(g(x)) &= \frac{2}{2+\left(\frac{2-2x}{x}\right)} \\ &= \frac{1}{1+\frac{2-2x}{x}} \\ &= \frac{1}{\left(\frac{0}{x}\right)} \\ &= x \end{aligned}$$

$$\begin{aligned} g(f(x)) &= \frac{2-2\left(\frac{2}{2+x}\right)}{\left(\frac{2}{2+x}\right)} \\ &= \frac{2-\left(\frac{4}{2+x}\right)}{\left(\frac{2}{2+x}\right)} \\ &= \frac{\frac{4+2x-4}{2+x}}{\left(\frac{2}{2+x}\right)} \\ &= \frac{\frac{2x}{2+x}}{\left(\frac{x}{2+x}\right)} \\ &= \frac{2x}{x} \\ &= x \end{aligned}$$

C) $f(g(x)) = \dfrac{2}{2+\left(\dfrac{2-2x}{x}\right)}$

$= \dfrac{4}{\dfrac{2-2x}{x}}$

$= \dfrac{2}{\left(\dfrac{2x}{x}\right)}$

$= 2\cdot\dfrac{x}{2}$

$= x$

$g(f(x)) = \dfrac{2-2\left(\dfrac{2}{2+x}\right)}{\left(\dfrac{2}{2+x}\right)}$

$= \dfrac{\left(\dfrac{2}{2+2x}\right)}{\left(\dfrac{2}{2+x}\right)}$

$= \left(\dfrac{2}{2+2x}\right)\left(\dfrac{2+x}{2}\right)$

$= \dfrac{2+x}{2+2x}$

$= \dfrac{x}{2x}$

$= x$

D) $f(g(x)) = \dfrac{2}{2+\left(\dfrac{2-2x}{x}\right)}$

$= \dfrac{2}{\dfrac{2x+2-2x}{x}}$

$= \dfrac{2-2x}{\left(\dfrac{2}{x}\right)}$

$= \dfrac{1-x}{1}$

$= x$

$g(f(x)) = \dfrac{2-2\left(\dfrac{2}{2+x}\right)}{\left(\dfrac{2}{2+x}\right)}$

$= \dfrac{\left(\dfrac{4}{2+x}\right)}{\left(\dfrac{2}{2+x}\right)}$

$= \left(\dfrac{4}{2+x}\right)\left(\dfrac{2+x}{2}\right)$

$= \dfrac{2(2+x)}{2+x}$

$= \dfrac{2x}{2}$

$= x$

E) $f(g(x)) = \dfrac{2}{2+\left(\dfrac{2-2x}{x}\right)}$

$= \dfrac{2}{\dfrac{2x+2-2x}{x}}$

$= \dfrac{2}{\left(\dfrac{2}{x}\right)}$

$= 2\cdot\dfrac{x}{2}$

$= x$

$g(f(x)) = \dfrac{2-2\left(\dfrac{2}{2+x}\right)}{\left(\dfrac{2}{2+x}\right)}$

$= \dfrac{2-\left(\dfrac{4}{2+x}\right)}{\left(\dfrac{2}{2+x}\right)}$

$= \dfrac{\dfrac{4+2x-4}{2+x}}{\left(\dfrac{2}{2+x}\right)}$

$= \dfrac{\dfrac{2x}{2+x}}{\left(\dfrac{2}{2+x}\right)}$

$= \dfrac{2x}{2}$

$= x$

Answer Key

1. D
2. C
3. B
4. C
5. B
6. D
7. E
8. B
9. C
10. E
11. D
12. B
13. A
14. B
15. E
16. C
17. C
18. A
19. A
20. E

Name: ______________________ Date: ___________

1. Use long division to divide.
   $$\left(x^4 - x^2 - 5\right) \div \left(x^2 + 4x - 1\right)$$
   A) $x^2 - 4x + 4$
   B) $x^2 + 4x - 4$
   C) $x^2 - 4x + 16 + \dfrac{-68x + 11}{x^2 + 4x - 1}$
   D) $x^2 + 4x - 4 + \dfrac{-5x - 1}{x^2 + 4x - 1}$
   E) $x^2 - 4x + 4 - \dfrac{4}{x^2 - 4x + 4}$

2. Write $f(x) = x^4 - 12x^3 + 59x^2 - 138x + 130$ as a product of linear factors.
   A) $(x-3-i)(x-3+2i)(x-3-2i)(x-2+i)$
   B) $(x-3-i)(x-3+i)(x-2-i)(x-2+i)$
   C) $(x-3-i)(x-3+i)(x+3-2i)(x-2+i)$
   D) $(x-3+i)(x-3-i)(x-2+3i)(x-2-3i)$
   E) $(x-3+i)(x-3-i)(x-3+2i)(x-3-2i)$

3. Find the zeros of the function below algebraically, if any exist.
   $$f(x) = x^5 - 5x^3 + 4x$$
   A) –4, –1, 1, 4
   B) –4, –2, 2, 4
   C) –2, –1, 0, 1, 2
   D) –4, –2, 0, 2, 4
   E) No zeros exist.

4. Find two positive real numbers whose product is a maximum and whose sum of the first number and four times the second is 200.
   A) 160, 10
   B) 116, 21
   C) 108, 23
   D) 100, 25
   E) 76, 31

5. Determine the equations of any horizontal and vertical asymptotes of

$$f(x)=\frac{x^2}{x^2+16}.$$

A) horizontal: $y=4$; vertical: $x=-4$
B) horizontal: $x=1$; vertical: none
C) horizontal: $y=-4$; vertical: $x=1$
D) horizontal: $y=1$; vertical: none
E) horizontal: none; vertical: none

6. Find a polynomial function with following characteristics.

Degree: 4
Zero: -1, multiplicity: 2
Zero: -3, multiplicity: 2
Falls to the left,
Falls to the right
Absolute value of the leading coefficient is one

A) $y=x^4-4x^3+22x^2+24x+3$
B) $y=-x^4-4x^3+12x^2+9$
C) $y=x^4-6x^3-18x^2+10x+3$
D) $y=-x^4-8x^3-22x^2-24x-9$
E) $y=-x^4-8x^3-24x+9$

7. A polynomial function f has degree 3, the zeros below, and a solution point of $f(-3)=-4$. Write f in completely factored form.

$-4,\ -3+2i$

A) $f(x)=(x+3)(x+4-2i)(x+4+2i)$
B) $f(x)=-(x+4)(x+2-3i)(x+2+3i)$
C) $f(x)=(x+4)(x+2-3i)(x+2+3i)$
D) $f(x)=-(x+4)(x+3-2i)(x+3+2i)$
E) $f(x)=(x+4)(x+3-2i)(x+3+2i)$

8. The interest rates that banks charge to borrow money fluctuate with the economy. The interest rate charged by a bank in a certain country is given in the table below. Let t represent the year, with $t = 0$ corresponding to 1986. Use the *regression* feature of a graphing utility to find a quadratic model of the form $y = at^2 + bt + c$ for the data.

Year t	**Percent** y
1986	12.4
1988	9.7
1990	7.3
1992	6.3
1994	9.7
1996	11.6

A) $y = -2.13t^2 + 12.61t + 0.21$
B) $y = 12.61t^2 + 0.21t - 2.13$
C) $y = 0.21t^2 - 2.13t + 12.61$
D) $y = 0.17t^2 - 2.58t + 10.59$
E) $y = 0.25t^2 - 1.73t + 14.37$

9. Find the zeros of the function below algebraically, if any exist.
$f(x) = 25x^3 - 60x^2 + 36x$

A) $-\frac{6}{5}$ and 0
B) 0 and $\frac{6}{5}$
C) $-\frac{6}{5}$, 0, and $\frac{6}{5}$
D) $-\frac{6}{5}$ and $\frac{6}{5}$
E) No zeros exist.

10. Determine the zeros (if any) of the rational function $f(x) = \dfrac{x^2 - 64}{x + 5}$.

A) $x = -5$
B) $x = -\frac{8}{5}, x = \frac{8}{5}$
C) $x = -64, x = 64$
D) $x = -8, x = 8$
E) no zeros

11. The graph of the function

$$f(x) = \frac{3x}{x+1}$$

is shown below. Determine the domain.

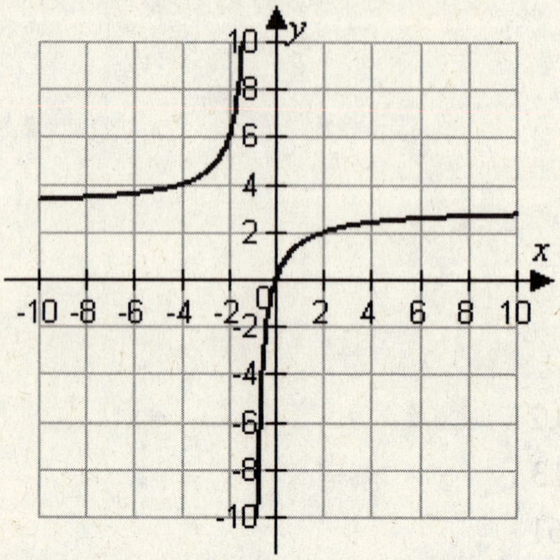

A) Domain: all real numbers except $x = -1$
B) Domain: all real numbers except $x = -1$ and $x = 0$
C) Domain: all real numbers except $x = -1$ and $x = 3$
D) Domain: all real numbers except $x = 3$
E) Domain: all real numbers except $x = 0$

12. Which of the given graphs is the graph of the polynomial function below?

$$h(x) = x^5 - \frac{3}{2}x^3 - \frac{1}{2}x$$

Graph 1 :

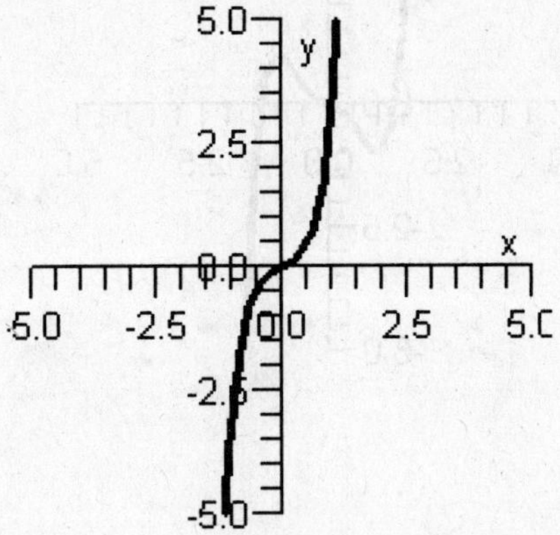

Graph 2 :

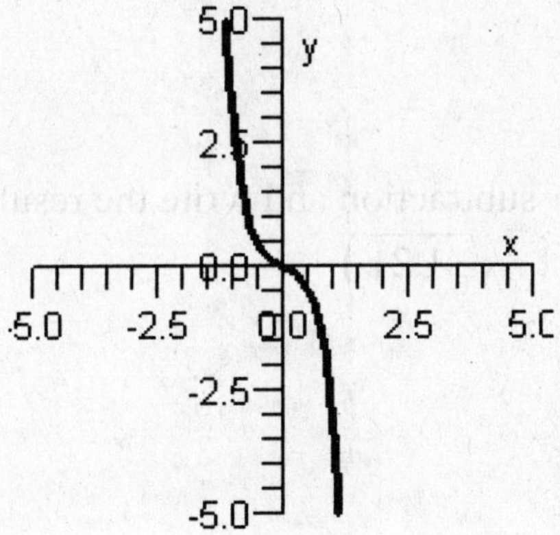

Graph 3 :

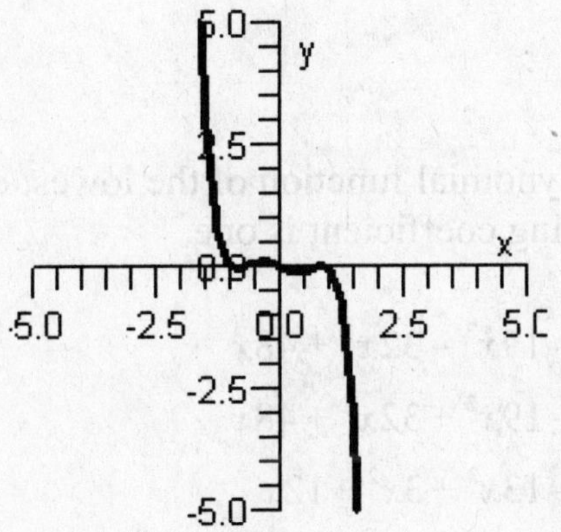

Graph 4 :

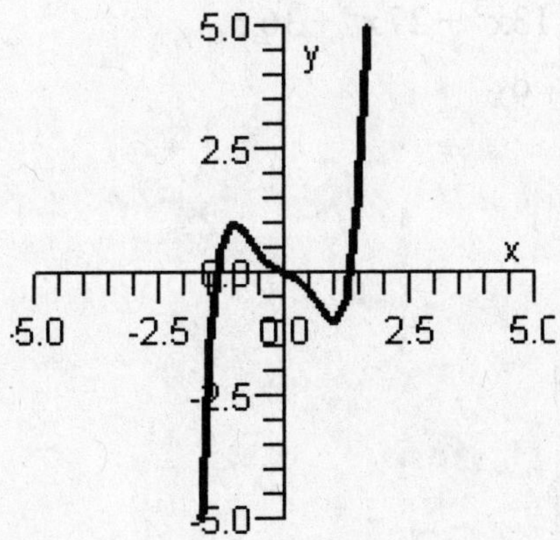

Graph 5 :

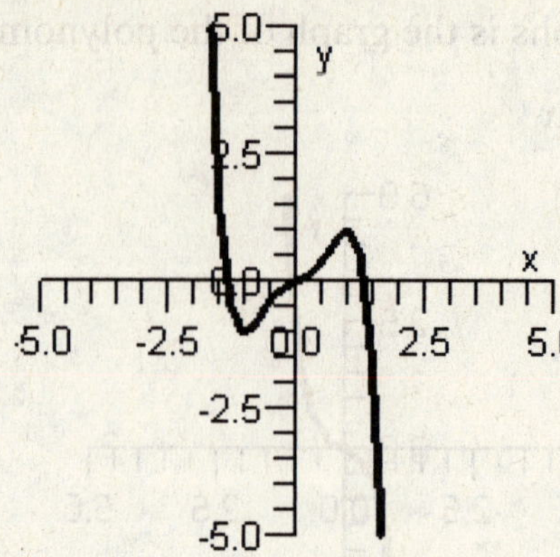

A) Graph 2
B) Graph 5
C) Graph 4
D) Graph 1
E) Graph 3

13. Perform the addition or subtraction and write the result in standard form.

$$-(7.2-12.3\,i)-\left(8.1-\sqrt{-1.21}\right)$$

A) $-15.3+13.4i$
B) $0.9+13.4i$
C) $-0.9+11.2i$
D) $-15.3+11.2i$
E) $15.3+13.4i$

14. Find a fifth degree polynomial function of the lowest degree that has the zeros below and whose leading coefficient is one.

$-3,\ -1,\ 0,\ 1,\ 3$

A) $f(x)=x^5+7x^4-19x^3-32x^2+48x$
B) $f(x)=x^5+7x^4-19x^3+32x^2+48x$
C) $f(x)=x^5+4x^4-13x^3+3x^2+12x$
D) $f(x)=x^5+5x^4-13x^3+27x^2+36x$
E) $f(x)=x^5-10x^3+9x$

15. Find the zeros of the function below algebraically, if any exist.

$f(x) = x^6 - 9x^3 + 8$

A) 1 and 2
B) –4 and 1
C) –4 and 2
D) –4, –1, 1, and 4
E) –4, –2, 2, and 4

16. Identify any horizontal and vertical asymptotes of the function below.

$f(x) = \dfrac{2x-8}{|x|+6}$

A) vertical asymptotes: $x = -2$ and $x = 2$; horizontal asymptotes: $y = -6$ and $y = 6$
B) vertical asymptotes: $x = -2$ and $x = 2$; horizontal asymptotes: none
C) vertical asymptotes: $x = -6$ and $x = 6$; horizontal asymptotes: none
D) vertical asymptotes: none; horizontal asymptotes: $y = -2$ and $y = 2$
E) vertical asymptotes: $x = -6$ and $x = 6$; horizontal asymptotes: $y = -2$ and $y = 2$

17. Find all the rational zeros of the function

$f(x) = -2x^5 - 11x^4 - 19x^3 - 17x^2 - 17x - 6$.

A) $x = \frac{1}{2}, -3, -1$
B) $x = -\frac{2}{3}, 1, -2$
C) $x = -\frac{1}{2}, \frac{3}{2}, -2$
D) $x = -\frac{1}{2}, \frac{3}{2}$
E) $x = -\frac{1}{2}, -3, -2$

18. Find real numbers a and b such that the equation $a+bi=-10+10i$ is true.
 A) $a=10,\ b=-10$
 B) $a=-10,\ b=-10$
 C) $a=10,\ b=10$
 D) $a=-10,\ b=10$
 E) $a=-20,\ b=0$

19. Use long division to divide.

$$\left(x^3+3x^2+x+3\right)\div\left(x+3\right)$$

 A) x^2+3
 B) $x^2+6x+17-\dfrac{53}{x+3}$
 C) $x^2+6x+19+\dfrac{48}{x+3}$
 D) $x^2+6x+17$
 E) x^2+1

20. Find two positive real numbers whose product is a maximum and whose sum is 146.
 A) 71, 75
 B) 73, 73
 C) 78, 68
 D) 82, 64
 E) 61, 85

Answer Key

1. C
2. E
3. C
4. D
5. D
6. D
7. D
8. C
9. B
10. D
11. A
12. C
13. A
14. E
15. A
16. D
17. E
18. D
19. E
20. B

Name: ___________________________ Date: _____________

1. Describe the right-hand and the left-hand behavior of the graph of $t(x) = -\frac{4}{7}\left(x^3 + 5x^2 + 8x + 1\right)$.
   A) Because the degree is odd and the leading coefficient is positive, the graph falls to the left and falls to the right.
   B) Because the degree is odd and the leading coefficient is negative, the graph rises to the left and falls to the right.
   C) Because the degree is odd and the leading coefficient is negative, the graph falls to the left and rises to the right.
   D) Because the degree is odd and the leading coefficient is positive, the graph rises to the left and rises to the right.
   E) Because the degree is even and the leading coefficient is negative, the graph rises to the left and falls to the right.

2. If $x = \frac{2}{5}$ is a root of $25x^3 - 70x^2 + 44x - 8 = 0$, use synthetic division to factor the polynomial completely and list all real solutions of the equation.
   A) $(5x-2)(5x+2)(x-2)$; $x = \frac{2}{5}, -\frac{2}{5}, 2$
   B) $(5x+2)^2(x-2)$; $x = -\frac{2}{5}, 2$
   C) $(5x-2)(x-2)^2$; $x = \frac{2}{5}, 2$
   D) $(5x+2)(x-2)^2$; $x = -\frac{2}{5}, 2$
   E) $(5x-2)^2(x-2)$; $x = \frac{2}{5}, 2$

3. Simplify $(3-6i)^2 - (3+6i)^2$ and write the answer in standard form.
   A) 0
   B) $-72i$
   C) $18-72i$
   D) $18+72i$
   E) $6-24i$

4. Determine the domain of $f(x)=\dfrac{6x+6}{x^2-6x}$.

A) all real numbers except $x=-1$, $x=0$, and $x=6$
B) all real numbers except $x=0$ and $x=6$
C) all real numbers except $x=-6$ and $x=-1$
D) all real numbers except $x=6$
E) all real numbers

5. Suppose the IQ scores (y, rounded to the nearest 10) for a group of people are summarized in the table below. Use the *regression* feature of a graphing utility to find a quadratic function of the form $y=ax^2+bx+c$ for the data.

IQ Score y	Number of People x
70	50
80	76
90	89
100	93
110	74
120	53
130	16

A) $y=-0.04x^2+15.08x-411.58$
B) $y=-0.06x^2+12.06x-484.21$
C) $y=-0.08x^2+10.98x-508.43$
D) $y=-0.07x^2+13.63x-460$
E) $y=-0.09x^2+8.56x-556.85$

6. Simplify f below and find any vertical asymptotes of f.

$$f(x)=\frac{x^2-25}{x+5}$$

A) $f(x)=x+5,\ x\neq 5$; vertical asymptotes: none
B) $f(x)=x-5,\ x\neq -5$; vertical asymptotes: none
C) $f(x)=x-5,\ x\neq 5$; vertical asymptotes: none
D) $f(x)=x+5,\ x\neq -5$; vertical asymptotes: $x=-5$
E) $f(x)=x-5,\ x\neq 5$; vertical asymptotes: $x=5$

7. Find the quadratic function *f* whose graph intersects the *x*-axis at $(2,0)$ and $(3,0)$ and the y-axis at $(0,-18)$.
   A) $f(x)=3x^2+3x+9$
   B) $f(x)=-3x^2+15x-18$
   C) $f(x)=-3x^2-3x+6$
   D) $f(x)=3x^2-3x-18$
   E) $f(x)=3x^2-15x-18$

8. Using the factors $(-5x+2)$ and $(x-1)$, find the remaining factor(s) of $f(x)=10x^4+31x^3-84x^2+53x-10$ and write the polynomial in fully factored form.
   A) $f(x)=(-5x+2)(-5x+2)(2x-1)(x-1)$
   B) $f(x)=(-5x+2)(-x-5)(2x-1)(x-1)$
   C) $f(x)=(-5x+2)^2(2x-1)(x+1)$
   D) $f(x)=(-5x+2)(-x+5)^2(x+1)$
   E) $f(x)=(-5x+2)^2(x-1)^2$

9. Simplify $\dfrac{4+3i}{5+2i}$ and write the answer in standard form.
   A) $-\dfrac{26}{29}+\dfrac{7}{29}i$
   B) $\dfrac{26}{29}-\dfrac{7}{29}i$
   C) $\dfrac{26}{29}+\dfrac{7}{29}i$
   D) $\dfrac{7}{29}+\dfrac{26}{29}i$
   E) $\dfrac{7}{29}-\dfrac{26}{29}i$

10. Write the complex conjugate of the complex number $-5-\sqrt{10}\,i$.
    A) $5-\sqrt{10}\,i$
    B) $-5-\sqrt{-10}\,i$
    C) $5-\sqrt{-10}\,i$
    D) $-5+\sqrt{10}\,i$
    E) $5+\sqrt{10}\,i$

11. Determine the value that $f(x)=\dfrac{4x-6}{x^2-7}$ approaches as x increases and decreases in magnitude without bound.
    A) 8
    B) 6
    C) 4
    D) 2
    E) 0

12. Find all real zeros of the polynomial $f(x)=x^4+13x^3+40x^2$ and determine the mutiplicity of each.
    A) $x=0$, multiplicity 2; $x=-8$, multiplicity 1; $x=-5$, multiplicity 1
    B) $x=8$, multiplicity 2; $x=5$, multiplicity 2
    C) $x=0$, multiplicity 2; $x=8$, multiplicity 1; $x=5$, multiplicity 1
    D) $x=-8$, multiplicity 2; $x=-5$, multiplicity 2
    E) $x=0$, multiplicity 1; $x=8$, multiplicity 1; $x=-8$, multiplicity 1; $x=5$, multiplicity 1

13. Given $3+i$ is a root, determine all other roots of $f(x)=x^4-10x^3+42x^2-88x+80$.
    A) $x=3+i,\ 2\pm 2i,\ 2-i$
    B) $x=3-i,\ 2\pm i$
    C) $x=3-i,\ 2-2i,\ 2+i$
    D) $x=3-i,\ -2\pm 2i$
    E) $x=3-i,\ 2\pm 2i$

14. Determine the zeros (if any) of the rational function $f(x)=\dfrac{x^2-9}{x-2}$.

A) $x=2$
B) $x=\dfrac{3}{2}, x=-\dfrac{3}{2}$
C) $x=-9, x=9$
D) $x=-3, x=3$
E) no zeros

15. Determine the zeros (if any) of the rational function $g(x)=\dfrac{x^3-1}{x^2+5}$.

A) $x=-1, x=1$
B) $x=1$
C) $x=-\sqrt{5}, x=\sqrt{5}, x=1$
D) $x=-\sqrt{5}, x=\sqrt{5}, x=-1, x=1$
E) no zeros

16. Match the equation with its graph.

$$f(x) = \frac{x^4 - 17x^2 + 16}{20}$$

A)

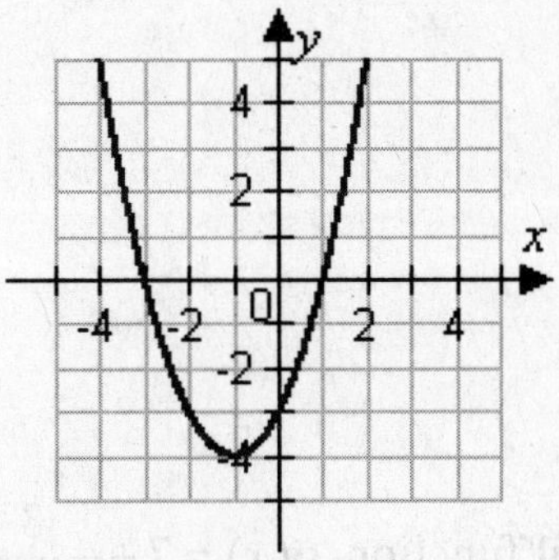

B)

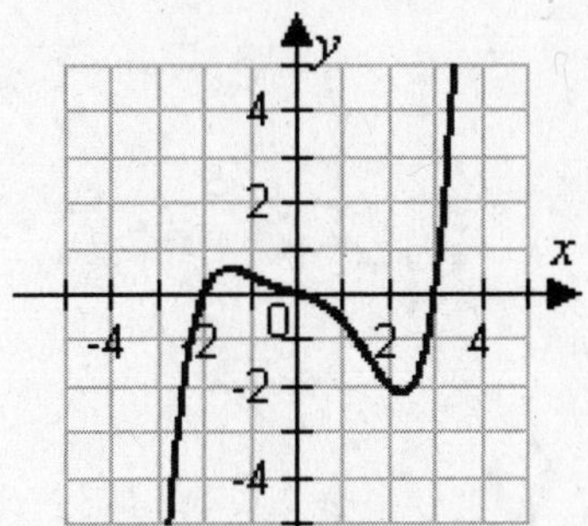

C)

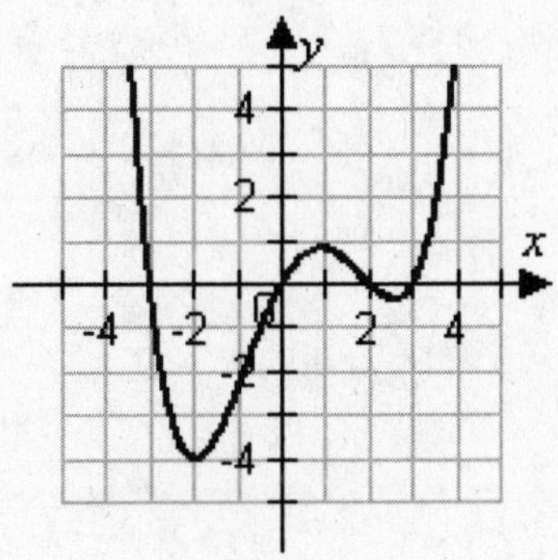

D)

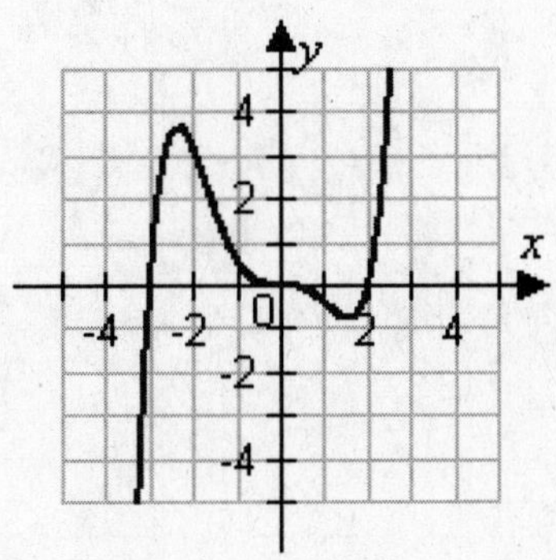

E)

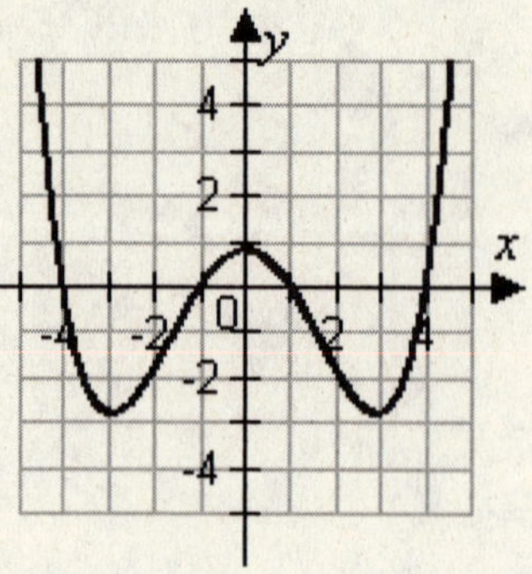

17. Determine the zeros (if any) of the rational function $g(x) = 7 + \dfrac{3}{x^2 + 7}$.

A) $x = -\sqrt{7}, x = \sqrt{7}$
B) $x = -3$
C) $x = -\dfrac{3}{7}, x = \dfrac{3}{7}$
D) $x = -7, x = 7$
E) no zeros

18. Find all zeros of the function $f(x) = (x+6)(x+3i)(x-3i)$.

A) $x = 6, -3i, 3i$
B) $x = -6, 3i$
C) $x = -6, -3, 3$
D) $x = -6, -3i, 3i$
E) $x = -6$

19. Find the zeros of the function below algebraically, if any exist.

$$f(x) = 2x^4 + 10x^2 + 12$$

A) $-\sqrt{3},\ -\sqrt{2},\ \sqrt{2}$, and $\sqrt{3}$
B) $-\sqrt{3},\ 0$, and $\sqrt{3}$
C) $-\sqrt{3}$ and $\sqrt{3}$
D) $-\sqrt{2}$ and $\sqrt{2}$
E) No zeros exist.

20. Simplify $\dfrac{2+5i}{3i}$ and write the answer in standard form.

A) $-\dfrac{5}{3}-\dfrac{2i}{3}$
B) $\dfrac{5}{3}-\dfrac{2i}{3}$
C) $\dfrac{5}{3}+\dfrac{2i}{3}$
D) $\dfrac{2}{3}+\dfrac{5i}{3}$
E) $-\dfrac{2}{3}+\dfrac{5i}{3}$

Answer Key

1. B
2. E
3. B
4. B
5. B
6. B
7. B
8. B
9. C
10. D
11. E
12. A
13. E
14. D
15. B
16. E
17. E
18. D
19. E
20. B

Name: ______________________________ Date: _______________

1. Find two positive real numbers whose product is a maximum and whose sum is 146.
   A) 71, 75
   B) 73, 73
   C) 78, 68
   D) 82, 64
   E) 61, 85

2. Write the complex conjugate of the following complex number and then multiply the number by the complex conjugate. Write the result in standard form.
   $1+\sqrt{-20}$
   A) $1-20i$; 19
   B) $1-5\sqrt{2}i$; 21
   C) $-1-2\sqrt{5}i$; 21
   D) $-1-2\sqrt{5}i$; 19
   E) $1-2\sqrt{5}i$; 21

3. Use synthetic division to divide.
   $$\left(2x^3-5x^2-22x-15\right)\div\left(x-5\right)$$
   A) $2x^2-3x-5$
   B) $2x^2+5x+3$
   C) $2x^2-2x-15$
   D) $2x^2-7x+6$
   E) $2x^2+5x+2$

4. Sketch the graph of the rational function below.

$$f(x) = \frac{x^2 - 1}{x}$$

A)

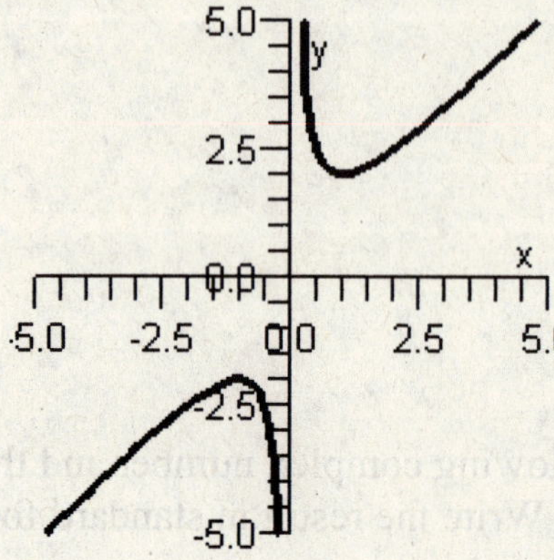

B)

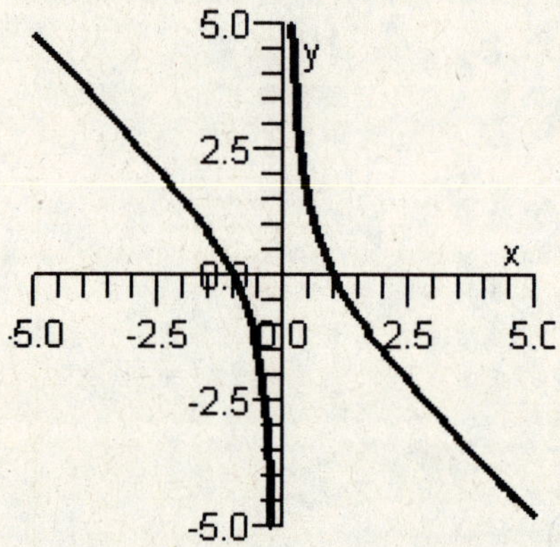

C)

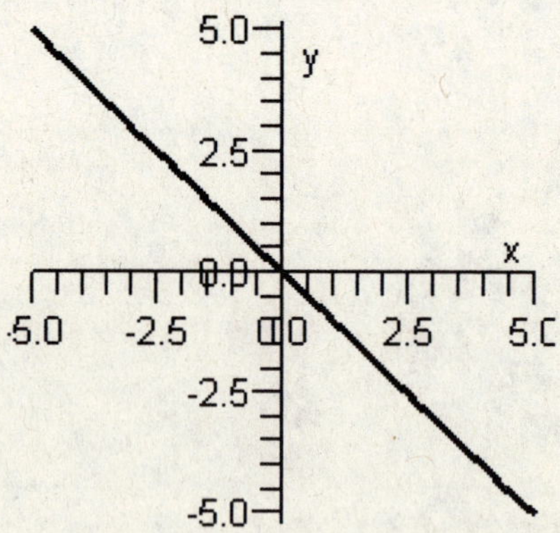

D)

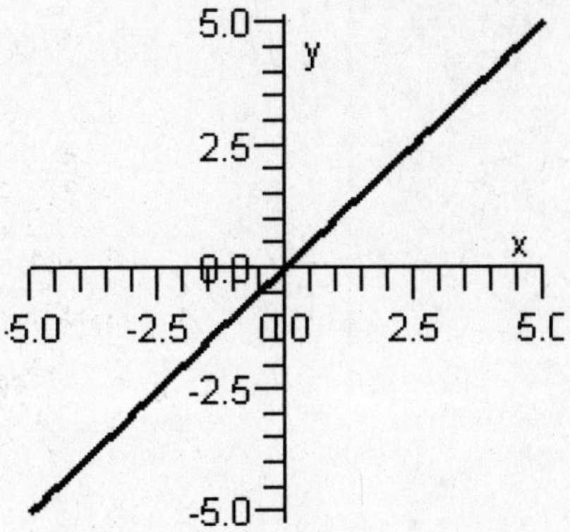

E)

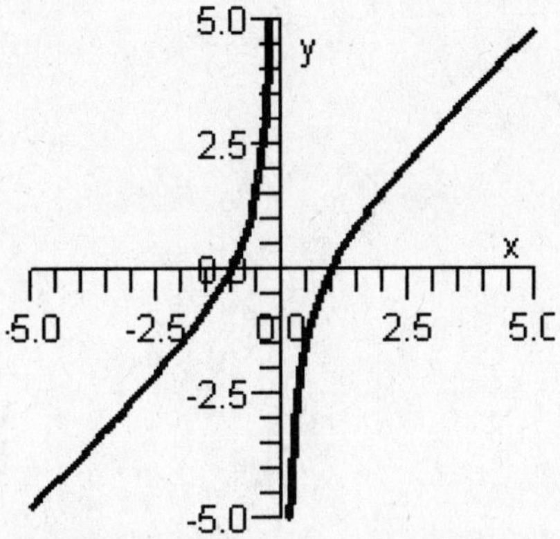

5. Use long division to divide.

$$\left(2x^2+11x+12\right)\div\left(x+4\right)$$

A) $2x+19+\dfrac{88}{x+4}$

B) $2x+3$

C) $2x+19+\dfrac{22}{x+4}$

D) $2x+22$

E) $-2x-3$

6. Sketch the graph of the quadratic function below.

$h(x) = -x^2 - 4x - 1$

A)

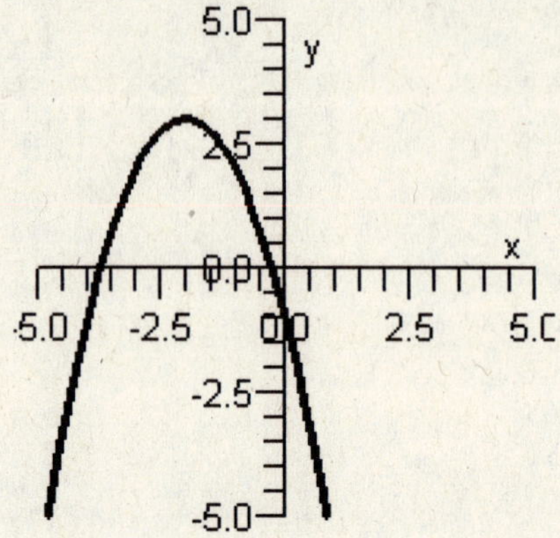

B)

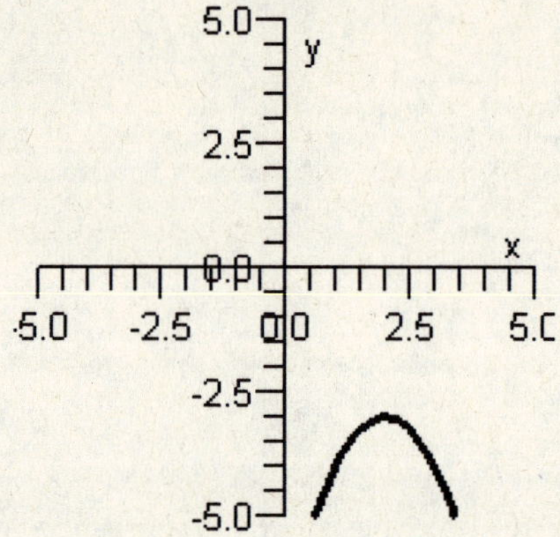

C)

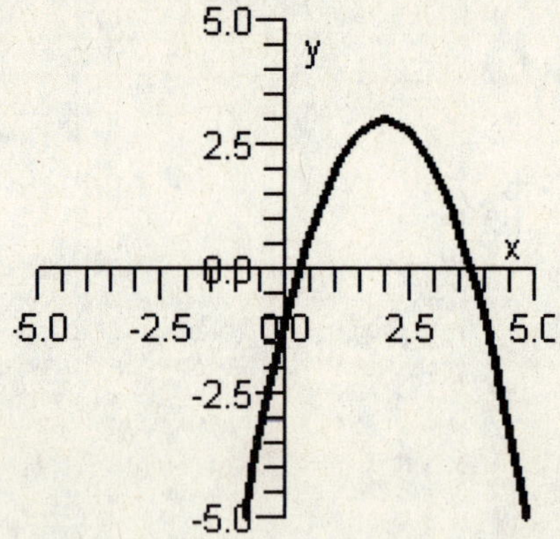

D)

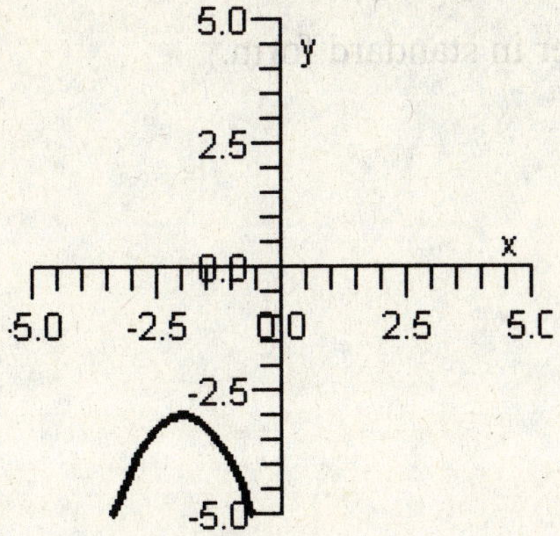

E)

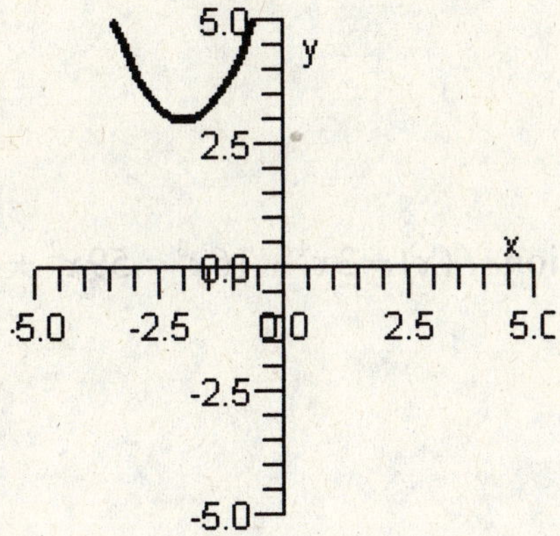

7. If $x = 4$ is a root of $x^3 + 5x^2 - 16x - 80 = 0$, use synthetic division to factor the polynomial completely and list all real solutions of the equation.
   A) $(x-5)(x-4)(x+4)$; $x = 5, 4, -4$
   B) $(x+5)(x-4)(x+4)$; $x = -5, 4, -4$
   C) $(x+5)(x-4)^2$; $x = -5, 4$
   D) $(x+5)^2(x-4)$; $x = -5, 4$
   E) $(x+5)(x-5)(x+4)$; $x = -5, 5, -4$

8. Simplify $\dfrac{-1-5i}{7i}$ and write the answer in standard form.

A) $\dfrac{5}{7}+\dfrac{i}{7}$

B) $-\dfrac{5}{7}+\dfrac{i}{7}$

C) $-\dfrac{5}{7}-\dfrac{i}{7}$

D) $-\dfrac{1}{7}-\dfrac{5i}{7}$

E) $\dfrac{1}{7}-\dfrac{5i}{7}$

9. Find all the rational zeros of the function $f(x)=3x^4-16x^3-59x^2+400x-400$.

A) $x=-4,5,-5,-\dfrac{3}{4}$

B) $x=3,-20,5$

C) $x=4,5,-5,\dfrac{4}{3}$

D) $x=-\dfrac{4}{5},\dfrac{5}{3},\dfrac{4}{3},-5$

E) $x=3,-20,\dfrac{5}{3},\dfrac{4}{3}$

10. Use long division to divide.

$$\left(x^3+27\right)\div\left(x+3\right)$$

A) x^2-3x+9

B) x^2-9

C) x^2+3x-9

D) x^2+9

E) $x^2-9+\dfrac{3}{x+3}$

11. Find all real zeros of the polynomial $f(x)=x^3+3x^2-49x-147$ and determine the mutiplicity of each.
   A) $x=7$, multiplicity 2; $x=-3$, multiplicity 1
   B) $x=7$, multiplicity 1; $x=-7$, multiplicity 1; $x=-3$, multiplicity 1
   C) $x=-3$, multiplicity 2; $x=-7$, multiplicity 1
   D) $x=-7$, multiplicity 1; $x=3$, multiplicity 1; $x=-3$, multiplicity 1
   E) $x=-3$, multiplicity 3

12. Use long division to divide.
   $(x^3+4x-1)\div(x+2)$
   A) $x^2-2x+8-\dfrac{17}{x+2}$
   B) $x^2+2x+8-\dfrac{15}{x+2}$
   C) $x^2+2-\dfrac{3}{x+2}$
   D) $x^2-2+\dfrac{3}{x+2}$
   E) $x^2+2x-8+\dfrac{17}{x+2}$

13. Find real numbers a and b such that the equation $a+bi=10-12i$ is true.
   A) $a=-10,\ b=12$
   B) $a=10,\ b=12$
   C) $a=-10,\ b=-12$
   D) $a=10,\ b=-12$
   E) $a=22,\ b=-2$

14. Determine the x-intercept(s) of the quadratic function $f(x)=x^2-10x+26$.
   A) $(0,0),(4,0)$
   B) $(5,0),(10,0)$
   C) $(7,0),(2,0)$
   D) $(0,0),(2,0)$
   E) no x-intercept(s)

15. Determine the domain of $f(x)=\dfrac{3x+3}{x^2-3x}$.

A) all real numbers except $x=-1$, $x=0$, and $x=3$
B) all real numbers except $x=0$ and $x=3$
C) all real numbers except $x=-3$ and $x=-1$
D) all real numbers except $x=3$
E) all real numbers

16. Perform the following operation and write the result in standard form.

$$\frac{9i}{9+i}+\frac{2}{9-i}$$

A) $\dfrac{9}{40}+\dfrac{83}{80}i$
B) $\dfrac{27}{10}+\dfrac{83}{10}i$
C) $\dfrac{1}{41}+\dfrac{9}{82}i$
D) $\dfrac{27}{82}+\dfrac{83}{82}i$
E) $\dfrac{27}{8}+\dfrac{83}{8}i$

17. The interest rates that banks charge to borrow money fluctuate with the economy. The interest rate charged by a bank in a certain country is given in the table below. Let t represent the year, with $t = 0$ corresponding to 1986. Use the *regression* feature of a graphing utility to find a quadratic model of the form $y = at^2 + bt + c$ for the data.

Year t	Percent y
1986	12.1
1988	10.1
1990	6.8
1992	6.6
1994	8.6
1996	12.0

A) $y = -2.23t^2 + 12.61t + 0.22$
B) $y = 12.61t^2 + 0.22t - 2.23$
C) $y = 0.22t^2 - 2.23t + 12.61$
D) $y = 0.17t^2 - 2.69t + 10.59$
E) $y = 0.26t^2 - 1.8t + 14.37$

18. Which of the following is the graph of the given equation?

$$f(x) = \frac{2 - x}{x + 3}$$

A)

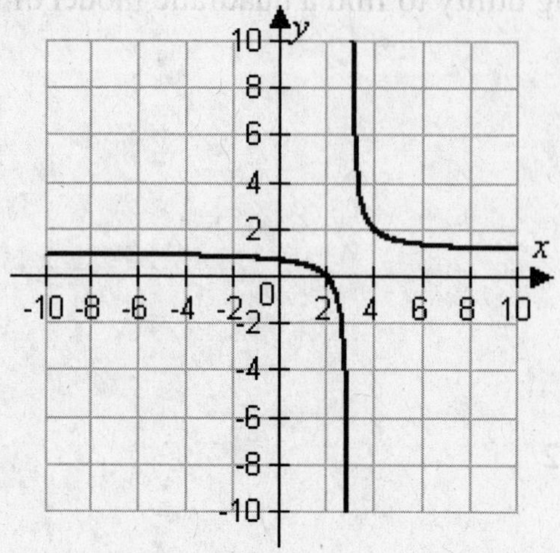

B)

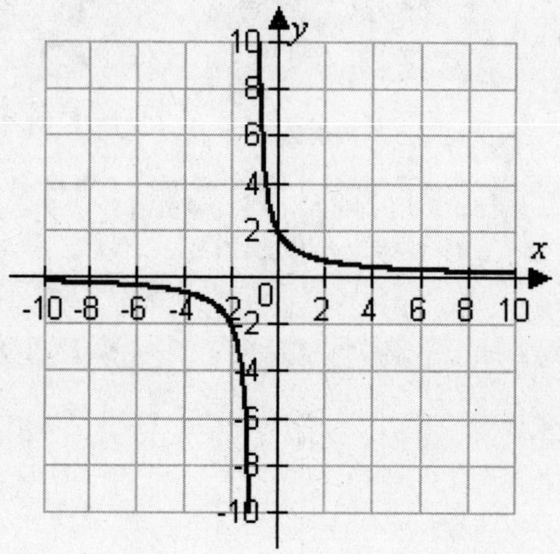

C)

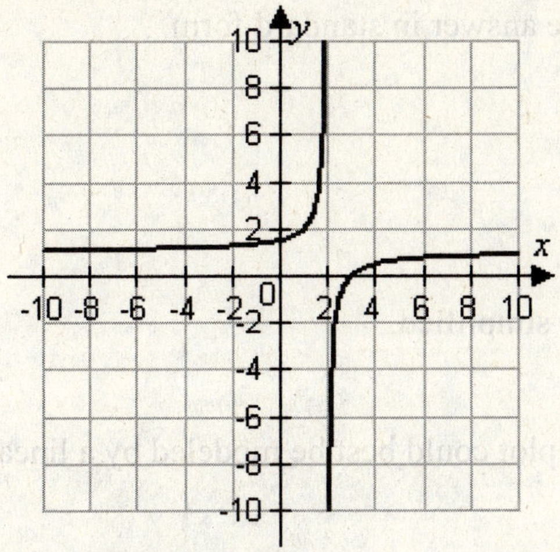
y
x
10
8
6
4
2
-2
-4
-6
-8
-10
-10 -8 -6 -4 -2 0 2 4 6 8 10

D)

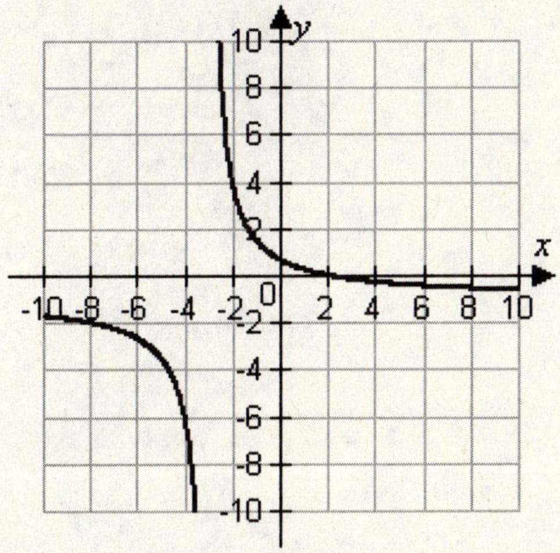
y
x
10
8
6
4
2
-2
-4
-6
-8
-10
-10 -8 -6 -4 -2 0 2 4 6 8 10

E)

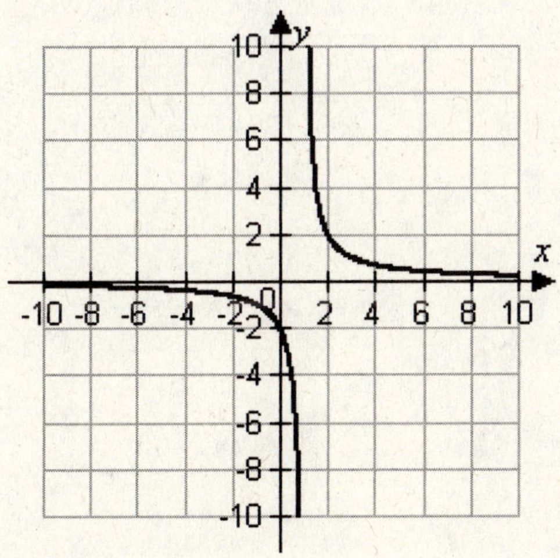
y
x
10
8
6
4
2
-2
-4
-6
-8
-10
-10 -8 -6 -4 -2 0 2 4 6 8 10

19. Simplify $\left(\sqrt{-2}\right)^{11}$ and write the answer in standard form.
   A) $32\sqrt{2}\,i$
   B) $-32\sqrt{2}\,i$
   C) $1024\sqrt{2}\,i$
   D) $-32\sqrt{2}$
   E) The expression cannot be simplified.

20. Determine whether the scatter plot could best be modeled by a linear model, a quadratic model, or neither.

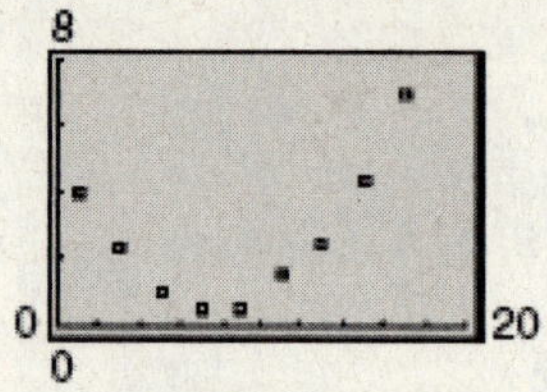

   A) linear model
   B) quadratic model
   C) neither

Answer Key

1. B
2. E
3. B
4. E
5. B
6. A
7. B
8. B
9. C
10. A
11. B
12. A
13. D
14. E
15. B
16. D
17. C
18. D
19. B
20. B

Name: ______________________________ Date: _______________

1. Use the *regression* feature of a graphing utility to find a quadratic model for the data below.

x	y
−2	9.8
−1	4.1
0	3.3
1	6.6
2	13.8
3	24.1
4	39.5

A) $y = 1.91x^2 + 1.06x + 3.15$

B) $y = 1.81x^2 + 1.02x + 3.3$

C) $y = 2.21x^2 + 0.92x + 3.3$

D) $y = 2.11x^2 + 0.87x + 3.65$

E) $y = 2.01x^2 + 0.97x + 3.44$

2. Find a fifth degree polynomial function of the lowest degree that has the zeros below and whose leading coefficient is one.

$-4, \ -1, \ 0, \ 1, \ 4$

A) $f(x) = x^5 + 4x^4 - 13x^3 + 27x^2 + 36x$

B) $f(x) = x^5 + 4x^4 - 13x^3 - 27x^2 + 36x$

C) $f(x) = x^5 + 5x^4 - 19x^3 - 16x^2 + 48x$

D) $f(x) = x^5 + 7x^4 - 13x^3 - x^2 + 12x$

E) $f(x) = x^5 - 17x^3 + 16x$

3. The interest rates that banks charge to borrow money fluctuate with the economy. The interest rate charged by a bank in a certain country is given in the table below. Let t represent the year, with $t = 0$ corresponding to 1986. Use the *regression* feature of a graphing utility to find a quadratic model of the form $y = at^2 + bt + c$ for the data.

Year t	Percent y
1986	12.8
1988	10.0
1990	6.9
1992	5.7
1994	8.6
1996	12.7

A) $y = -2.7t^2 + 13.35t + 0.26$
B) $y = 13.35t^2 + 0.26t - 2.7$
C) $y = 0.26t^2 - 2.7t + 13.35$
D) $y = 0.21t^2 - 3.26t + 11.22$
E) $y = 0.32t^2 - 2.18t + 15.22$

4. Find all real zeros of the polynomial $f(x) = x^4 + 8x^3 + 12x^2$ and determine the mutiplicity of each.
A) $x = 0$, multiplicity 2; $x = -2$, multiplicity 1; $x = -6$, multiplicity 1
B) $x = 2$, multiplicity 2; $x = 6$, multiplicity 2
C) $x = 0$, multiplicity 2; $x = 2$, multiplicity 1; $x = 6$, multiplicity 1
D) $x = -2$, multiplicity 2; $x = -6$, multiplicity 2
E) $x = 0$, multiplicity 1; $x = 2$, multiplicity 1; $x = -2$, multiplicity 1; $x = 6$, multiplicity 1

5. Which of the given graphs is the graph of the polynomial function below?

$$h(x) = x^5 + \frac{3}{2}x^3 - \frac{1}{2}x$$

Graph 1 :

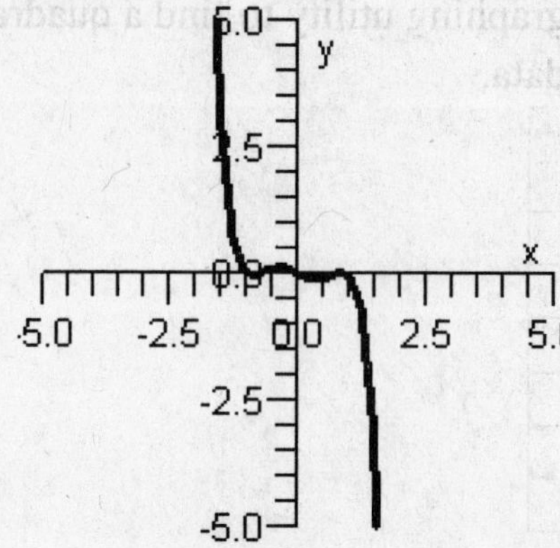

Graph 2 :

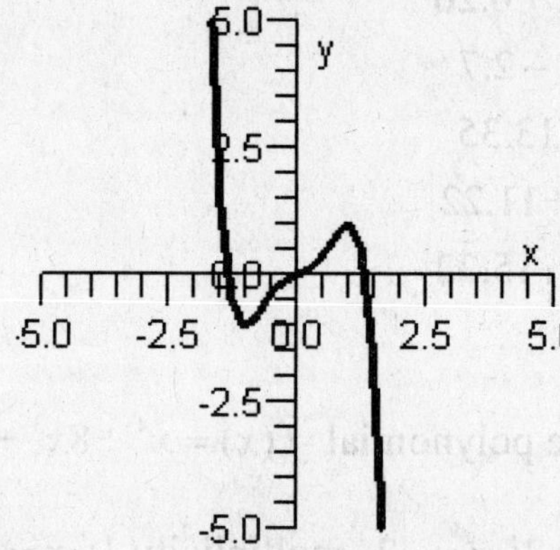

Graph 3 :

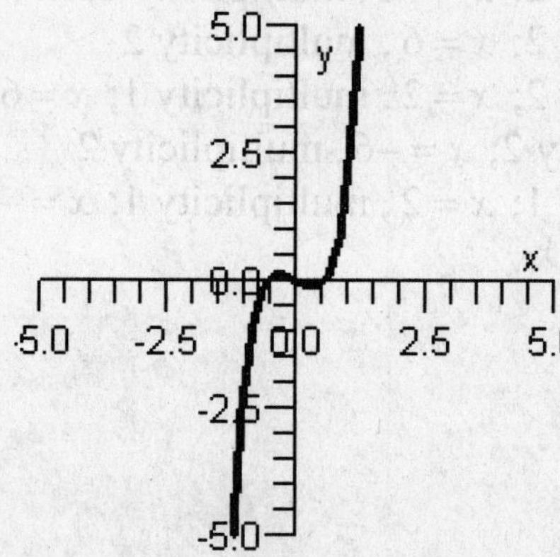

Graph 4 :

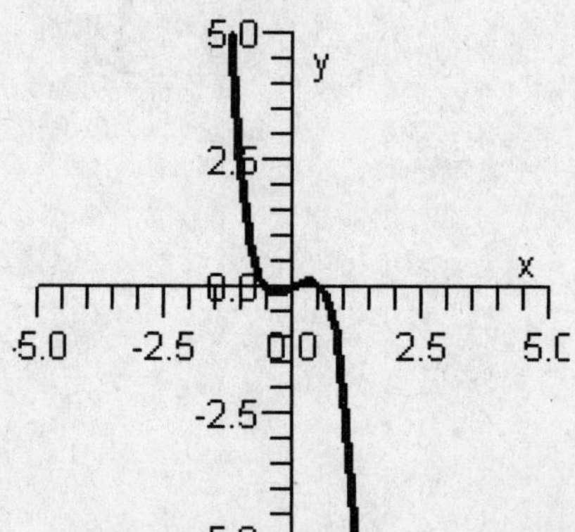

Graph 5 :

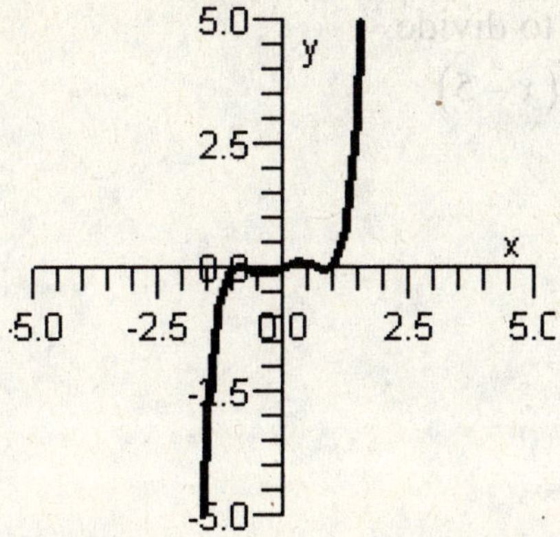

A) Graph 1
B) Graph 4
C) Graph 3
D) Graph 5
E) Graph 2

6. Determine the *x*-intercept(s) of the quadratic function $f(x)=x^2-12x+37$.
A) $(1,0),(9,0)$
B) $(6,0),(15,0)$
C) $(8,0),(3,0)$
D) $(1,0),(3,0)$
E) no *x*-intercept(s)

7. Find a polynomial function with following characteristics.

Degree: 4
Zero: 4, multiplicity: 2
Zero: –3, multiplicity: 2
Falls to the left,
Falls to the right
Absolute value of the leading coefficient is one

A) $y=x^4+x^3-23x^2+24x-12$
B) $y=-x^4+x^3-48x^2+9$
C) $y=x^4-6x^3-18x^2+25x+48$
D) $y=-x^4+2x^3+23x^2-24x-144$
E) $y=-x^4+2x^3-24x-36$

8. Use synthetic division to divide.

   $\left(x^3 - 75x + 250\right) \div (x - 5)$

   A) $x^2 + 5x - 50$
   B) $x^2 - 5x - 75$
   C) $x^2 + 10x + 25$
   D) $x^2 + 15x + 50$
   E) $x^2 + 25x - 10$

9. Find the zeros of the function below algebraically, if any exist.

   $f(x) = x^4 + 8x^2 + 15$

   A) $-\sqrt{5},\ -\sqrt{3},\ \sqrt{3}$, and $\sqrt{5}$
   B) $-\sqrt{3},\ 0$, and $\sqrt{3}$
   C) $-\sqrt{3}$ and $\sqrt{3}$
   D) $-\sqrt{5}$ and $\sqrt{5}$
   E) No zeros exist.

10. A polynomial function f has degree 3, the zeros below, and a solution point of $f(1) = -96$. Write f in completely factored form.

    $-2,\ -3 + 4i$

    A) $f(x) = (x+3)(x+2-4i)(x+2+4i)$
    B) $f(x) = -(x+2)(x-4-3i)(x-4+3i)$
    C) $f(x) = (x+2)(x-4-3i)(x-4+3i)$
    D) $f(x) = -(x+2)(x+3-4i)(x+3+4i)$
    E) $f(x) = (x+2)(x+3-4i)(x+3+4i)$

11. Describe the right-hand and the left-hand behavior of the graph of $q(x) = 7x^5 + x^3 + 7$.
    A) Because the degree is odd and the leading coefficient is positive, the graph falls to the left and falls to the right.
    B) Because the degree is odd and the leading coefficient is positive, the graph rises to the left and falls to the right.
    C) Because the degree is odd and the leading coefficient is positive, the graph falls to the left and rises to the right.
    D) Because the degree is odd and the leading coefficient is positive, the graph rises to the left and rises to the right.
    E) Because the degree is even and the leading coefficient is positive, the graph rises to the left and rises to the right.

12. Find the zeros of the function below algebraically, if any exist.
    $f(x) = x^5 - 9x^3 + 27x^2 - 243$
    A) –2 and 2
    B) –4 and 2
    C) –4 and 4
    D) –3 and 4
    E) –3 and 3

13. Determine the vertex of the graph of the quadratic function $f(x) = x^2 - x + \frac{5}{4}$.
    A) $\left(\frac{1}{2}, \frac{3}{2}\right)$
    B) $\left(-1, \frac{5}{4}\right)$
    C) $\left(-\frac{1}{2}, \frac{5}{4}\right)$
    D) $\left(-\frac{1}{4}, -\frac{3}{4}\right)$
    E) $\left(\frac{1}{2}, 1\right)$

14. Write the standard form of the equation of the parabola that has a vertex at $(3,8)$ and passes through the point $(5,-2)$.
    A) $f(x)=-\frac{1}{2}(x-3)^2+5$
    B) $f(x)=-\frac{5}{2}(x-3)^2+8$
    C) $f(x)=-\frac{10}{9}(x+3)^2+8$
    D) $f(x)=\frac{8}{5}(x-8)^2-2$
    E) $f(x)=\frac{8}{9}(x-8)^2-5$

15. Find two positive real numbers whose product is a maximum and whose sum is 116.
    A) 56, 60
    B) 58, 58
    C) 63, 53
    D) 67, 49
    E) 46, 70

16. Determine the zeros (if any) of the rational function $g(x)=7+\frac{2}{x^2+7}$.
    A) $x=-\sqrt{7}, x=\sqrt{7}$
    B) $x=-2$
    C) $x=-\frac{2}{7}, x=\frac{2}{7}$
    D) $x=-7, x=7$
    E) no zeros

17. Determine the domain of the function $f(x)=\frac{x^2+7x+12}{x^2+16}$.
    A) Domain: all real numbers except $x=-4$ and 3
    B) Domain: all real numbers except $x=-16$
    C) Domain: all real numbers except $x=-4$ and -3
    D) Domain: all real numbers except $x=4$ and 3
    E) Domain: all real numbers

18. Write the complex conjugate of the complex number $3-\sqrt{2}\,i$.

A) $-3-\sqrt{2}\,i$
B) $3-\sqrt{-2}\,i$
C) $-3-\sqrt{-2}\,i$
D) $3+\sqrt{2}\,i$
E) $-3+\sqrt{2}\,i$

19. Match the equation with its graph.

$f(x) = x^2 + 2x - 3$

A)

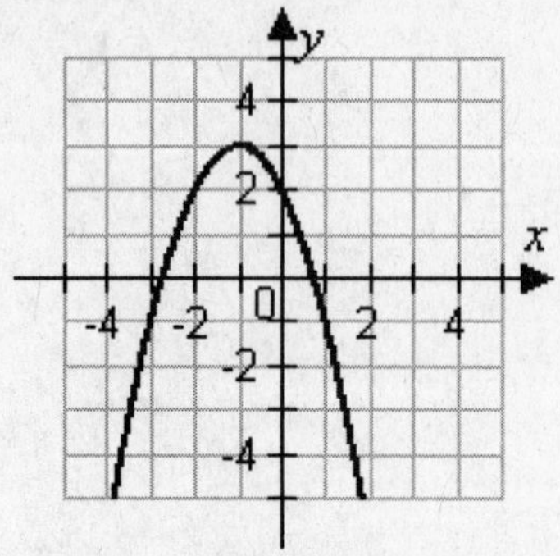

B)

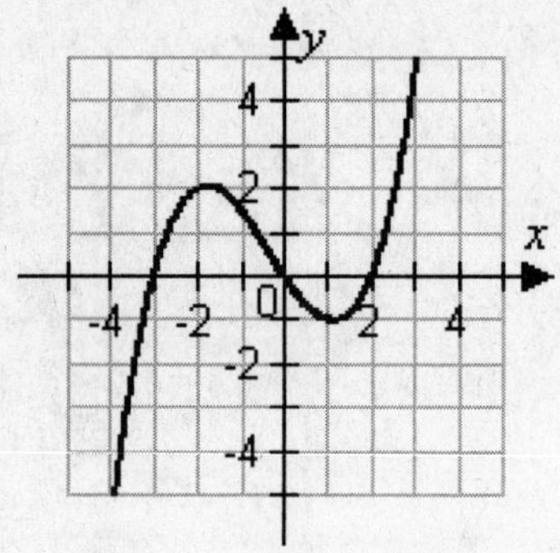

C)

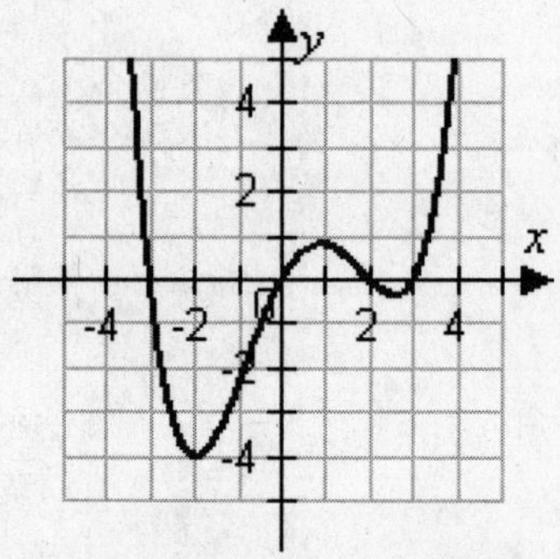

D)

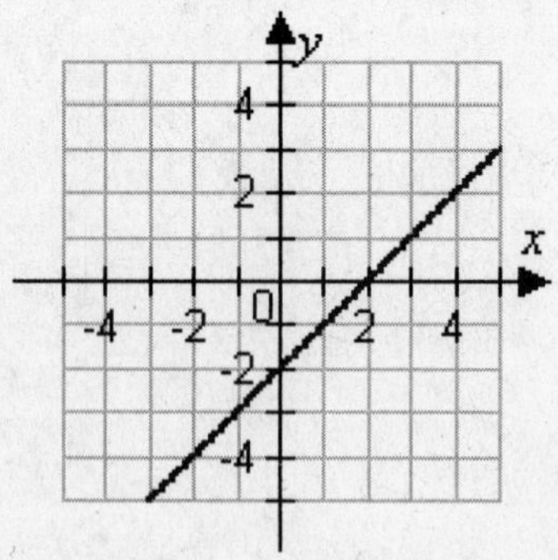

E) 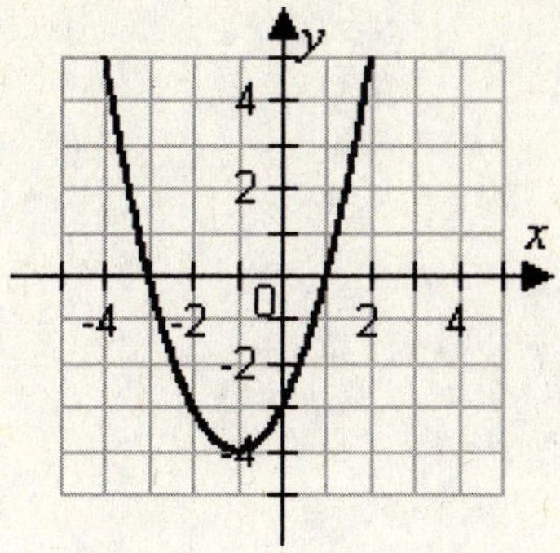

20. Write $f(x) = x^3 - 11x^2 + 18x + 25$ in the form $f(x) = (x-k)q(x) + r$ when $k = 6 + \sqrt{6}$.

A) $f(x) = \left[x + \left(6+\sqrt{6}\right)\right]\left[x^2 + \left(-5+\sqrt{6}\right)x - \left(6-\sqrt{6}\right)\right] - 5$

B) $f(x) = \left[x + \left(6+\sqrt{6}\right)\right]\left[x^2 + \left(-5+\sqrt{6}\right)x - \left(6-\sqrt{6}\right)\right] + 5$

C) $f(x) = \left[x - \left(6+\sqrt{6}\right)\right]\left[x^2 + \left(-5+\sqrt{6}\right)x - \left(6-\sqrt{6}\right)\right] + 5$

D) $f(x) = \left[x - \left(6+\sqrt{6}\right)\right]\left[x^2 + \left(-5+\sqrt{6}\right)x - \left(6-\sqrt{6}\right)\right] - 5$

E) $f(x) = \left[x + \left(6+\sqrt{6}\right)\right]\left[x^2 - \left(-5+\sqrt{6}\right)x - \left(6-\sqrt{6}\right)\right] - 5$

Answer Key

1. E
2. E
3. C
4. A
5. C
6. E
7. D
8. A
9. E
10. D
11. C
12. E
13. E
14. B
15. B
16. E
17. E
18. D
19. E
20. D

Name: ______________________________ Date: _______________

1. Find all the rational zeros of the function $f(x) = 2x^4 - 9x^3 - 41x^2 + 225x - 225$.

A) $x = -3, 5, -5, -\frac{2}{3}$

B) $x = 2, -15, 5$

C) $x = 3, 5, -5, \frac{3}{2}$

D) $x = -\frac{3}{5}, \frac{5}{2}, \frac{3}{2}, -5$

E) $x = 2, -15, \frac{5}{2}, \frac{3}{2}$

2. Using the factors $(x+4)$ and $(x+3)$, find the remaining factor(s) of $f(x) = x^3 + 6x^2 + 5x - 12$ and write the polynomial in fully factored form.

A) $f(x) = (x+4)(x+3)(x-1)$

B) $f(x) = (x+4)(x+3)^2$

C) $f(x) = (x+4)(x+3)(x+1)$

D) $f(x) = (x+4)^2(x+3)$

E) $f(x) = (x+4)(x+3)(x+3)$

3. Find the quadratic function *f* whose graph intersects the *x*-axis at $(1,0)$ and $(4,0)$ and the y-axis at $(0,4)$.

A) $f(x) = -x^2 - 3x - 4$

B) $f(x) = x^2 - 5x + 4$

C) $f(x) = x^2 + 3x + 4$

D) $f(x) = -x^2 + 3x + 4$

E) $f(x) = -x^2 + 5x + 4$

4. Sketch the graph of the rational function below.

$$f(x) = \frac{x^2 + 1}{x}$$

A)

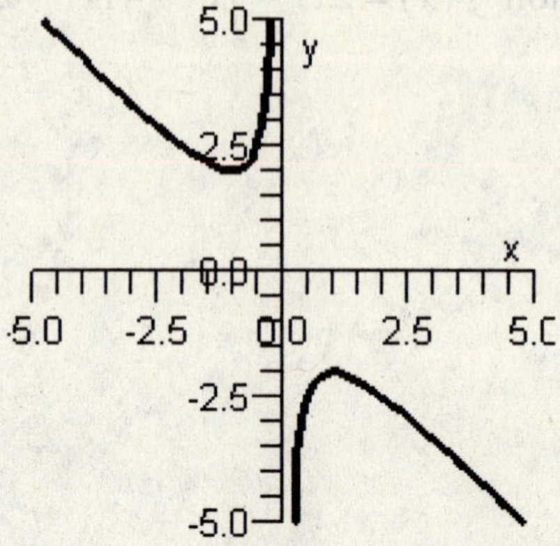

B)

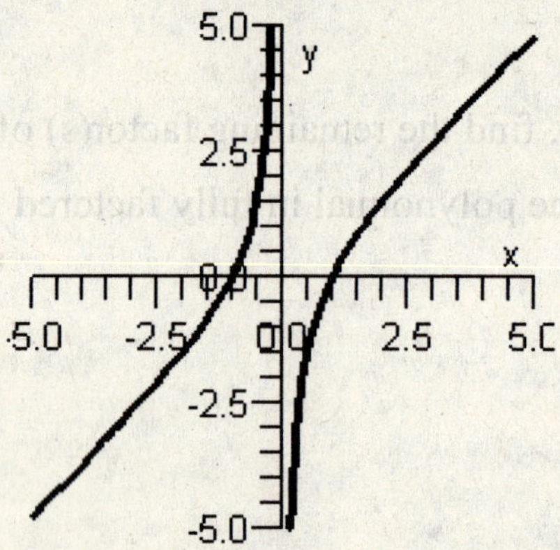

C)

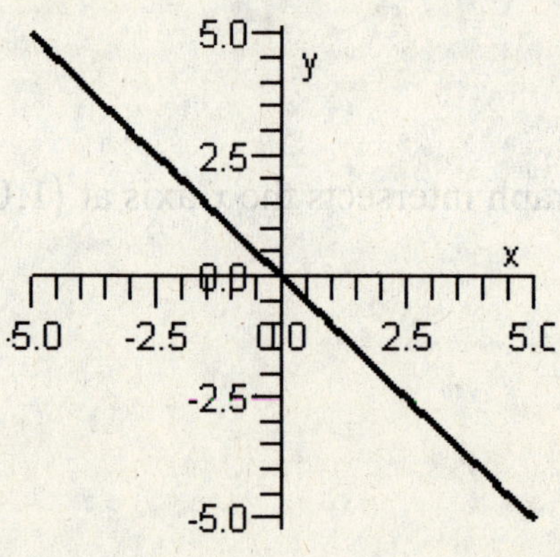

D) 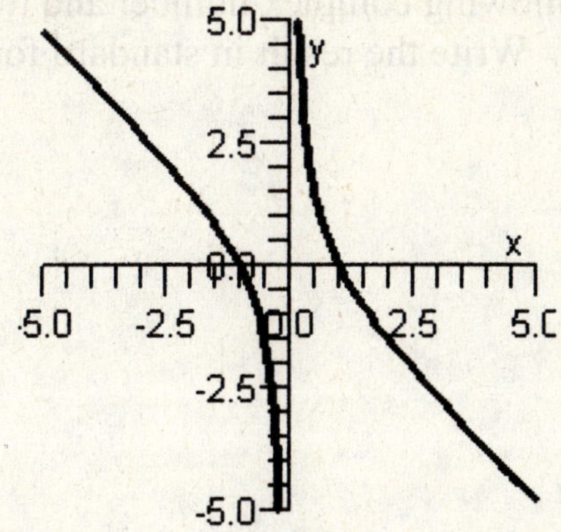

E) 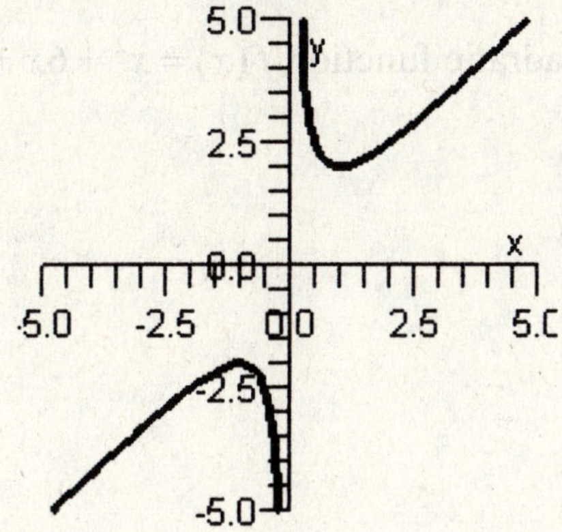

5. Find all zeros of the function $f(x)=x^2(x-2)(x^3-216)$.
   A) $x=2,216$
   B) $x=0,-2,-6$
   C) $x=0,2,6,-3-3\sqrt{3}i,-3+3\sqrt{3}i$
   D) $x=-2,-216$
   E) $x=0,2,6$

6. Simplify $\left(\sqrt{-3}\right)^9$ and write the answer in standard form.
   A) $-81\sqrt{3}\,i$
   B) $81\sqrt{3}\,i$
   C) $6561\sqrt{3}\,i$
   D) $81\sqrt{3}$
   E) The expression cannot be simplified.

7. Write the complex conjugate of the following complex number and then multiply the number by the complex conjugate. Write the result in standard form.

   $2+\sqrt{-27}$

   A) $2-27i;\ 25$
   B) $2-3\sqrt{3}i;\ 31$
   C) $-2-3\sqrt{3}i;\ 31$
   D) $-2-3\sqrt{3}i;\ 25$
   E) $2-3\sqrt{3}i;\ 31$

8. Determine the *x*-intercept(s) of the quadratic function $f(x)=x^2+6x+10$.

   A) $(-8,0),(1,0)$
   B) $(-3,0),(7,0)$
   C) $(-1,0),(-6,0)$
   D) $(-8,0),(-6,0)$
   E) no *x*-intercept(s)

9. Compare the graph of $p(x)=\left[-\frac{1}{3}(x+9)\right]^2-9$ with $p(x)=x^2$.

A) $p(x)=\left[-\frac{1}{3}(x+9)\right]^2-9$ shifts right 9 units, shifts downward 9 units, and shrinks by a factor of $-\frac{1}{9}$.

B) $p(x)=\left[-\frac{1}{3}(x+9)\right]^2-9$ shifts right 81 units, shifts upward 9 units, and shrinks by a factor of $\frac{1}{9}$.

C) $p(x)=\left[-\frac{1}{3}(x+9)\right]^2-9$ shifts left 9 units, shifts downward 9 units, and shrinks by a factor of $\frac{1}{9}$.

D) $p(x)=\left[-\frac{1}{3}(x+9)\right]^2-9$ shifts right 9 units, shifts upward 9 units, and shrinks by a factor of $\frac{1}{9}$.

E) $p(x)=\left[-\frac{1}{3}(x+9)\right]^2-9$ shifts left 81 units, shifts upward 9 units, and shrinks by a factor of $-\frac{1}{3}$.

10. Find a polynomial function of the lowest degree with real coefficients that has the zeros below and whose leading coefficient is one.

$$0,\ 2,\ 4-i$$

A) $f(x)=x^4-34x^3+32x^2-10x$
B) $f(x)=x^4-10x^3+33x^2-34x$
C) $f(x)=x^4+32x^3-10x^2-34x$
D) $f(x)=x^4-10x^3-34x^2+32x$
E) $f(x)=x^4+32x^3-34x^2-10x$

11. If $x = -1$ is a root of $x^3 + 2x^2 - x - 2 = 0$, use synthetic division to factor the polynomial completely and list all real solutions of the equation.
    A) $(x-2)(x+1)(x-1)$; $x = 2, -1, 1$
    B) $(x+2)(x+1)(x-1)$; $x = -2, -1, 1$
    C) $(x+2)(x+1)^2$; $x = -2, -1$
    D) $(x+2)^2(x+1)$; $x = -2, -1$
    E) $(x+2)(x-2)(x-1)$; $x = -2, 2, 1$

12. Determine the vertex of the graph of the quadratic function $f(x) = x^2 + 5$.
    A) $(0,-5)$
    B) $(5,0)$
    C) $(5,5)$
    D) $(0,5)$
    E) $(-5,0)$

13. Find all real zeros of the polynomial $f(x) = x^4 - 80x^2 + 1024$ and determine the mutiplicity of each.
    A) $x = 64$, multiplicity 2; $x = 16$, multiplicity 2
    B) $x = 8$, multiplicity 2; $x = 4$, multiplicity 2
    C) $x = 64$, multiplicity 2; $x = 4$, multiplicity 1
    D) $x = -8$, multiplicity 2; $x = -4$, multiplicity 2
    E) $x = 8$, multiplicity 1; $x = -8$, multiplicity 1; $x = 4$, multiplicity 1; $x = -4$, multiplicity 1

14. Determine the zeros (if any) of the rational function $g(x) = 5 + \dfrac{2}{x^2 + 5}$.
    A) $x = -\sqrt{5}, x = \sqrt{5}$
    B) $x = -2$
    C) $x = -\dfrac{2}{5}, x = \dfrac{2}{5}$
    D) $x = -5, x = 5$
    E) no zeros

15. Use the *regression* feature of a graphing utility to find a quadratic model for the data below.

x	y
–2	–7.3
–1	–2
0	–1
1	–4.8
2	–11.4
3	–22.8
4	–37.8

A) $y=-1.91x^2-1.21x-1.57$
B) $y=-1.81x^2-1.15x-1.42$
C) $y=-2.21x^2-1.04x-1$
D) $y=-2.11x^2-0.99x-1.07$
E) $y=-2.01x^2-1.1x-1.28$

16. Which of the following is the graph of the given equation?

$$f(x) = \frac{2 - x}{x + 3}$$

A)

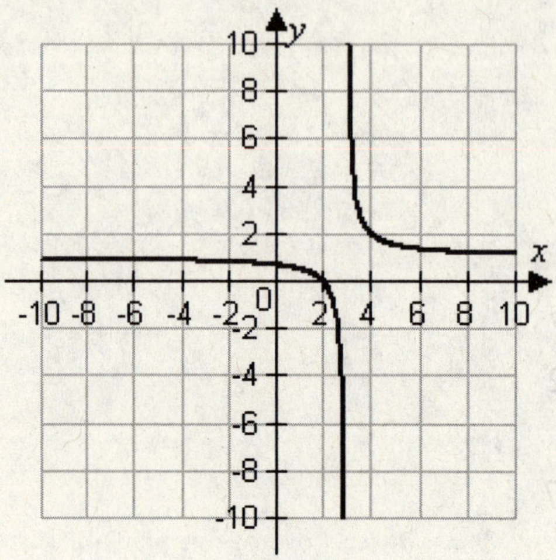

B)

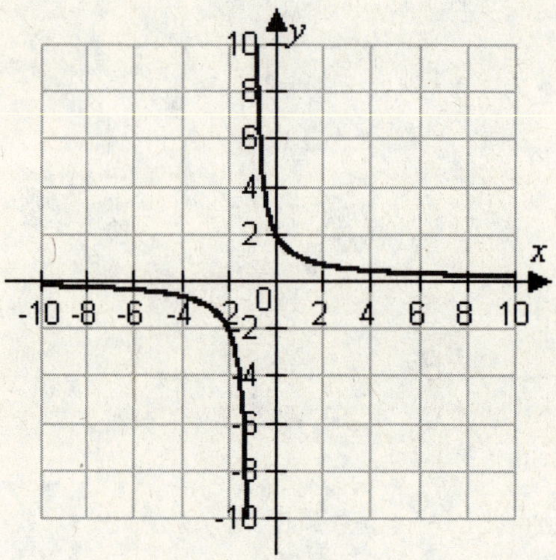

C)

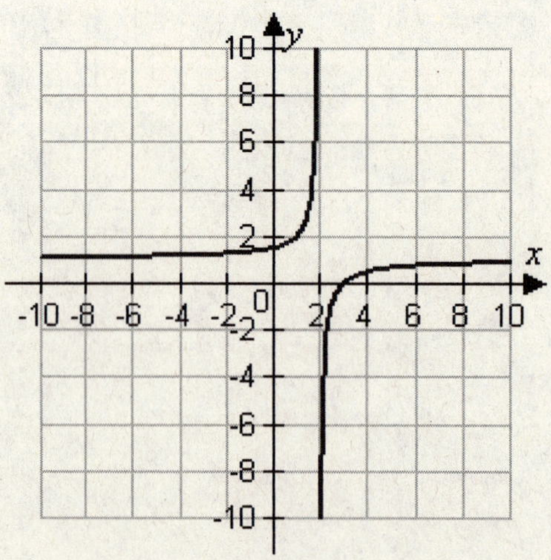

D)

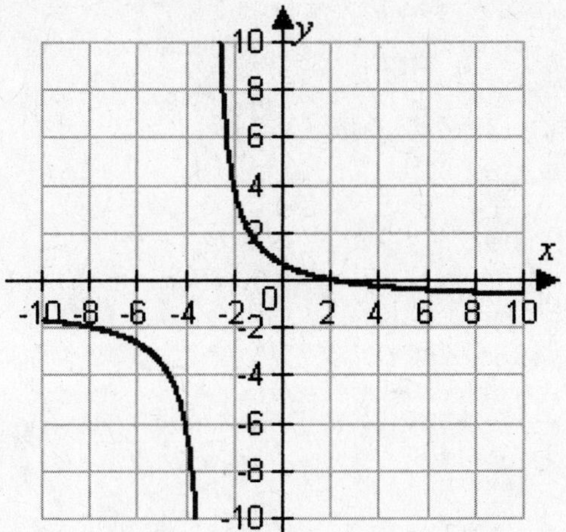

E)

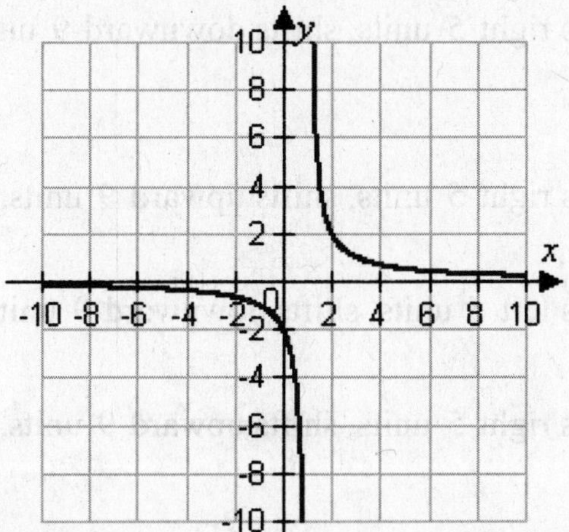

17. Simplify $\dfrac{1-2i}{5i}$ and write the answer in standard form.

A) $\dfrac{2}{5}-\dfrac{i}{5}$

B) $-\dfrac{2}{5}-\dfrac{i}{5}$

C) $-\dfrac{2}{5}+\dfrac{i}{5}$

D) $\dfrac{1}{5}-\dfrac{2i}{5}$

E) $-\dfrac{1}{5}-\dfrac{2i}{5}$

18. Use long division to divide.
$$\left(6x^2 - 20x + 6\right) \div (x - 3)$$
A) $6x - 38 + \dfrac{120}{x-3}$
B) $6x - 2$
C) $6x - 38 + \dfrac{40}{x-3}$
D) $6x - 40$
E) $-6x + 2$

19. Compare the graph of $s(x) = 5(x-5)^2 + 9$ with $s(x) = x^2$.
A) $s(x) = 5(x-5)^2 + 9$ shifts right 5 units, shifts downward 9 units, and shrinks by a factor of $\dfrac{1}{5}$.
B) $s(x) = 5(x-5)^2 + 9$ shifts right 5 units, shifts upward 9 units, and stretches by a factor of 5.
C) $s(x) = 5(x-5)^2 + 9$ shifts left 5 units, shifts downward 9 units, and stretches by a factor of 5.
D) $s(x) = 5(x-5)^2 + 9$ shifts right 5 units, shifts upward 9 units, and shrinks by a factor of $\dfrac{1}{5}$.
E) $s(x) = 5(x-5)^2 + 9$ shifts left 5 units, shifts upward 9 units, and stretches by a factor of 5.

20. Find the zeros of the function below algebraically, if any exist.
$$f(x) = x^6 + 28x^3 + 27$$
A) –3 and –1
B) –2 and –1
C) –3 and –2
D) –2, –1, 1, and 2
E) –3, –2, 2, and 3

Answer Key

1. C
2. A
3. B
4. E
5. C
6. B
7. E
8. E
9. C
10. B
11. B
12. D
13. E
14. E
15. E
16. D
17. B
18. B
19. B
20. A

Name: ______________________________ Date: _____________

1. Given $f(x)=\dfrac{5x+4}{5x^2+4x}$. Determine the domain of $f(x)$ and find any vertical asymptotes.

A) **domain:** all real numbers except $x=-\dfrac{4}{5}$
vertical asymptote: $x=0$

B) **domain:** all real numbers except $x=0$ and $x=-\dfrac{4}{5}$
vertical asymptote: $x=0$

C) **domain:** all real numbers
vertical asymptotes: $x=0$ and $x=-\dfrac{4}{5}$

D) **domain:** all real numbers except $x=0$ and $x=\dfrac{4}{5}$
vertical asymptote: $x=0$

E) **domain:** all real numbers except $x=\dfrac{4}{5}$
vertical asymptotes: $x=0$ and $x=-\dfrac{4}{5}$

2. Determine the value that $f(x)=\dfrac{3x-5}{x-6}$ approaches as x increases and decreases in magnitude without bound.

A) 6
B) 5
C) 4
D) 3
E) 2

3. Determine the x-intercept(s) of the quadratic function $f(x) = x^2 - 15x + 56$.
   A) $(8,0),(7,0)$
   B) $(-4,0),(-8,0)$
   C) $(-8,0),(-7,0)$
   D) $(4,0),(8,0)$
   E) no x-intercept(s)

4. Simplify $\dfrac{2+i}{5+2i}$ and write the answer in standard form.
   A) $-\dfrac{12}{29}+\dfrac{1}{29}i$
   B) $\dfrac{12}{29}-\dfrac{1}{29}i$
   C) $\dfrac{12}{29}+\dfrac{1}{29}i$
   D) $\dfrac{1}{29}+\dfrac{12}{29}i$
   E) $\dfrac{1}{29}-\dfrac{12}{29}i$

5. Write the polynomial in completely factored form. (Hint: One factor is x^2+1.)
   $f(x) = x^4 + 6x^3 + 14x^2 + 6x + 13$
   A) $f(x)=(x-2)(x+2)(x+3-i)(x+3+i)$
   B) $f(x)=(x-3)(x+3)(x-1-2i)(x-1+2i)$
   C) $f(x)=(x-1)(x+1)(x+3-2i)(x+3+2i)$
   D) $f(x)=(x-3i)(x+3i)(x-1-2i)(x-1+2i)$
   E) $f(x)=(x-i)(x+i)(x+3-2i)(x+3+2i)$

6. Use long division to divide.

   $\left(x^3+5x^2+36x+180\right)\div\left(x+5\right)$

   A) x^2+30

   B) $x^2+10x+44-\dfrac{244}{x+5}$

   C) $x^2+10x+86+\dfrac{110}{x+5}$

   D) $x^2+10x+44$

   E) x^2+36

7. Find the quadratic function *f* whose graph intersects the *x*-axis at $(-7,0)$ and $(1,0)$ and the y-axis at $(0,-14)$.

   A) $f(x)=-2x^2-16x-2$

   B) $f(x)=2x^2+12x-14$

   C) $f(x)=2x^2+16x-7$

   D) $f(x)=-2x^2+16x-14$

   E) $f(x)=-2x^2-12x-14$

8. Suppose the IQ scores (y, rounded to the nearest 10) for a group of people are summarized in the table below. Use the *regression* feature of a graphing utility to find a quadratic function of the form $y = ax^2 + bx + c$ for the data.

IQ Score y	Number of People x
70	53
80	72
90	93
100	90
110	78
120	47
130	16

A) $y = -0.04x^2 + 14.93x - 404.96$
B) $y = -0.06x^2 + 11.94x - 476.43$
C) $y = -0.08x^2 + 10.87x - 500.25$
D) $y = -0.07x^2 + 13.5x - 452.61$
E) $y = -0.09x^2 + 8.48x - 547.89$

9. Which of the following is the graph of the given equation?

$$f(x) = \frac{x-2}{x-3}$$

A)

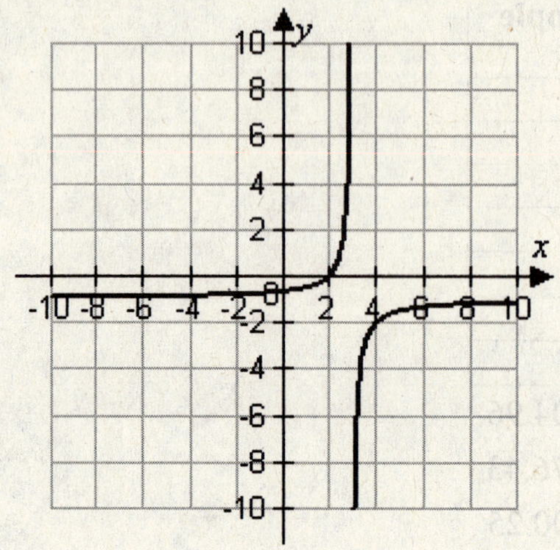

B)

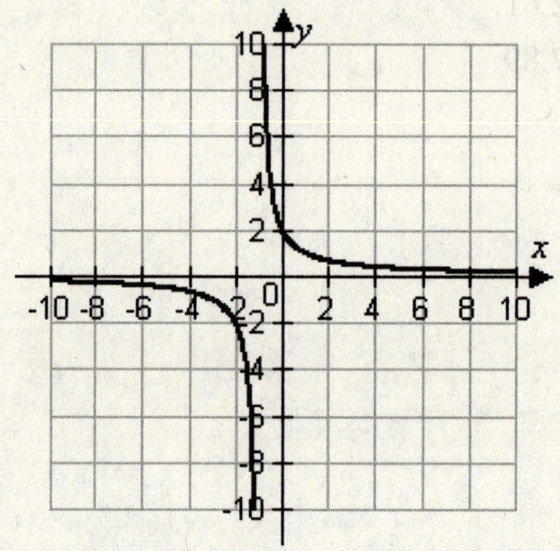

C)

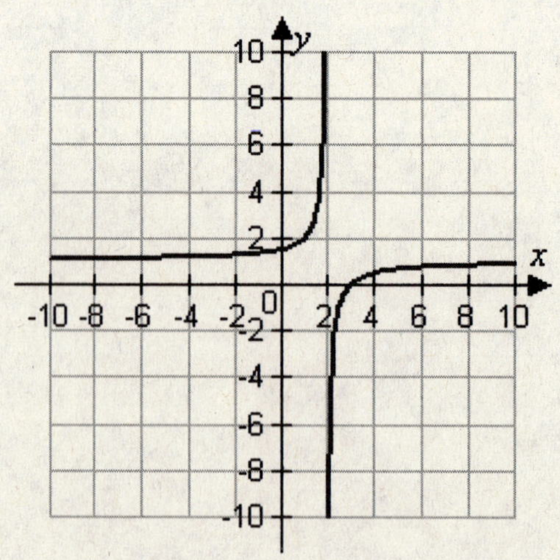

D)

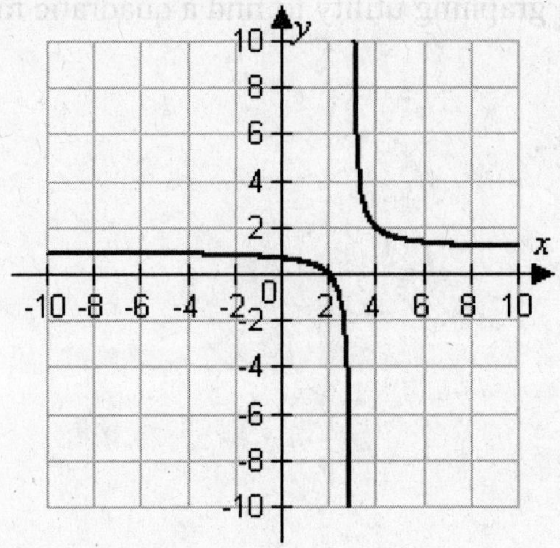
y
x
10
8
6
4
2
-2
-4
-6
-8
-10
-10 -8 -6 -4 -2 0 2 4 6 8 10

E)

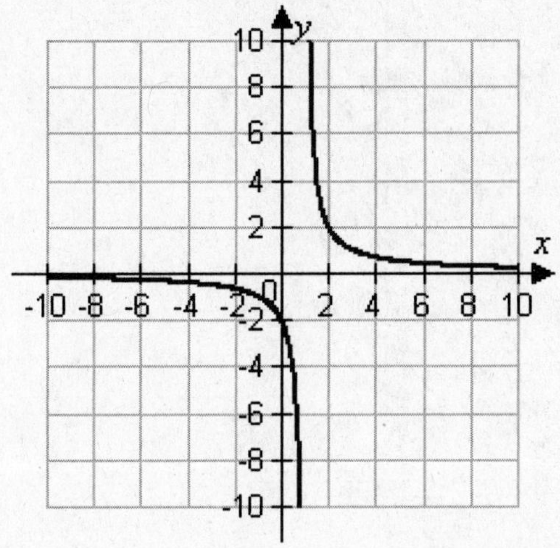
y
x
10
8
6
4
2
-2
-4
-6
-8
-10
-10 -8 -6 -4 -2 0 2 4 6 8 10

10. Use the *regression* feature of a graphing utility to find a quadratic model for the data below.

x	y
–2	15.7
–1	3.5
0	–2.7
1	–2.9
2	3.7
3	15.8
4	33.6

A) $y = 2.89x^2 - 3.35x - 2.9$

B) $y = 2.73x^2 - 3.2x - 2.75$

C) $y = 3.34x^2 - 2.9x - 2.7$

D) $y = 3.19x^2 - 2.74x - 2.4$

E) $y = 3.04x^2 - 3.05x - 2.61$

11. Sketch the graph of the rational function below.

$$f(x) = -\frac{x^2 - 1}{x}$$

A)

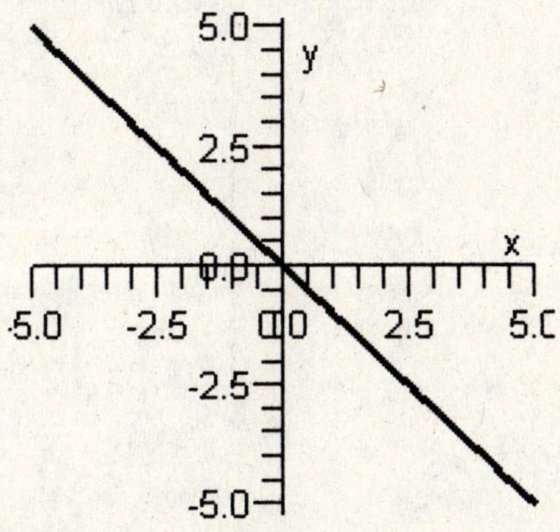

B)

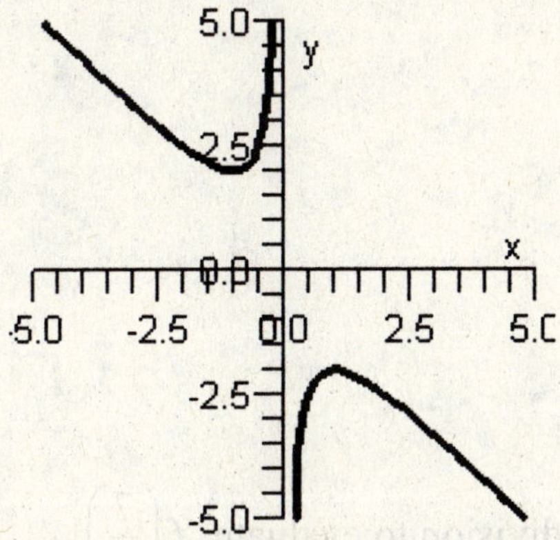

C)

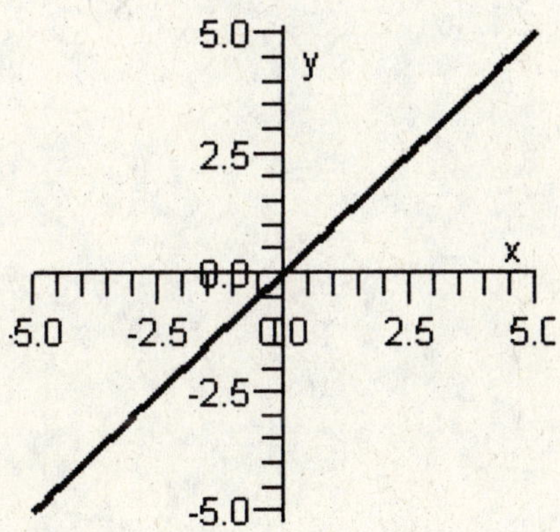

D)

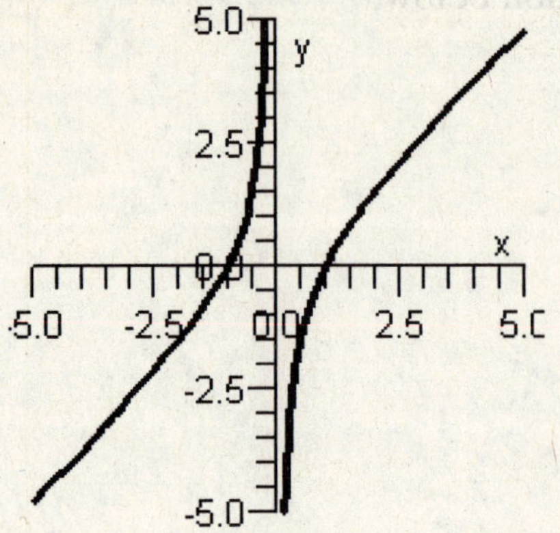

E)

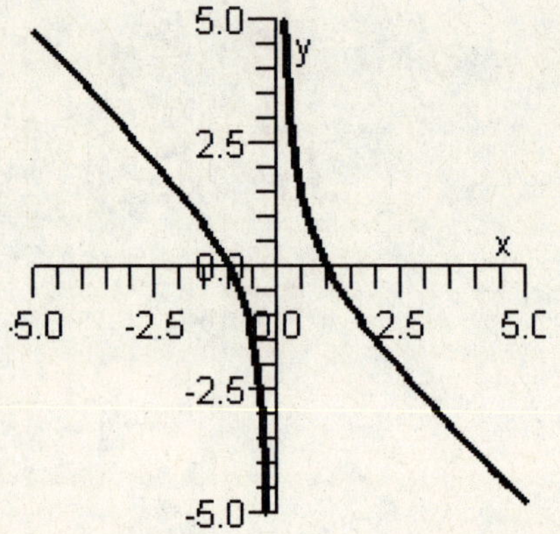

12. If $f(x)=4x^2-2x-7$, use synthetic division to evaluate $f\left(\dfrac{7}{8}\right)$.

A) $f\left(\dfrac{7}{8}\right)=\dfrac{21}{2}$

B) $f\left(\dfrac{7}{8}\right)=-\dfrac{21}{2}$

C) $f\left(\dfrac{7}{8}\right)=-\dfrac{91}{16}$

D) $f\left(\dfrac{7}{8}\right)=-\dfrac{35}{2}$

E) $f\left(\dfrac{7}{8}\right)=-\dfrac{77}{8}$

13. Find a polynomial function of the lowest degree with real coefficients that has the zeros below and whose leading coefficient is one.

$0,\ -3,\ 1+3i$

A) $f(x)=x^4+30x^3-5x^2+x$
B) $f(x)=x^4+x^3+4x^2+30x$
C) $f(x)=x^4-5x^3+x^2+30x$
D) $f(x)=x^4+x^3+30x^2-5x$
E) $f(x)=x^4-5x^3+30x^2+x$

14. Simplify $(-4+i)(4-5i)$ and write the answer in standard form.

A) $16+24i$
B) $-11-24i$
C) $-16+24i$
D) $-16-21i$
E) $-11+24i$

15. Write the standard form of the equation of the parabola that has a vertex at $\left(\frac{-2}{3},\frac{1}{9}\right)$ and passes through the point $(3,-4)$.

A) $f(x)=-\frac{37}{11}\left(x+\frac{2}{3}\right)^2+\frac{1}{9}$
B) $f(x)=-\frac{37}{121}\left(x-\frac{3}{2}\right)^2+\frac{1}{9}$
C) $f(x)=-\frac{37}{121}\left(x+\frac{2}{3}\right)^2+\frac{1}{9}$
D) $f(x)=-\frac{37}{11}\left(x-\frac{2}{3}\right)^2-\frac{1}{9}$
E) $f(x)=-\frac{37}{25}\left(x-\frac{3}{2}\right)^2-\frac{1}{9}$

16. Determine the vertex of the graph of the quadratic function $f(x)=x^2+5x+\dfrac{29}{4}$.

A) $\left(\dfrac{-5}{2},\dfrac{27}{2}\right)$
B) $\left(5,\dfrac{29}{4}\right)$
C) $\left(\dfrac{5}{2},\dfrac{29}{4}\right)$
D) $\left(\dfrac{5}{4},\dfrac{21}{4}\right)$
E) $\left(\dfrac{-5}{2},1\right)$

17. Use synthetic division to divide.
$\left(6+5x^3+23x+22x^2\right)\div(x+3)$

A) $5x^2+8x+3$
B) $5x^2+17x+10$
C) $5x^2+5x+6$
D) $5x^2+7x+2$
E) $5x^2+7x+5$

18. Find the zeros of the function below algebraically, if any exist.
$f(t)=\dfrac{1}{6}t^4-\dfrac{27}{2}$

A) –3, –1, 1, and 3
B) –3 and 3
C) –6 and 6
D) –6, –1, 1, and 6
E) No zeros exist.

19. Find real numbers a and b such that the equation $a+bi=4-9i$ is true.
   A) $a=-4,\ b=9$
   B) $a=4,\ b=9$
   C) $a=-4,\ b=-9$
   D) $a=4,\ b=-9$
   E) $a=13,\ b=-5$

20. Determine the vertex of the graph of the quadratic function $f(x)=x^2+2$.
   A) $(0,-2)$
   B) $(2,0)$
   C) $(2,2)$
   D) $(0,2)$
   E) $(-2,0)$

Answer Key

1. B
2. D
3. A
4. C
5. E
6. E
7. B
8. B
9. D
10. E
11. E
12. C
13. B
14. E
15. C
16. E
17. D
18. B
19. D
20. D

Name: ________________________ Date: ____________

1. Use the *regression* feature of a graphing utility to find a logarithmic model $y = a + b\ln x$ for the data.
   (1, 3), (2, 4), (3, 4.5), (4, 5), (5, 5.1), (6, 5.2), (7, 5.5)
   A) $y = 3.030 + 1.470\ln x$
   B) $y = 3.022 + 1.758\ln x$
   C) $y = 3.115 + 0.842\ln x$
   D) $y = 3.083 + 1.257\ln x$
   E) $y = 3.167 + 0.555\ln x$

2. Rewrite the logarithmic equation $\log_4 \frac{1}{16} = -2$ in exponential form.
   A) $4^{16} = -2$
   B) $4^{1/16} = -2$
   C) $4^{-2} = \frac{1}{16}$
   D) $\left(\frac{1}{16}\right)^{-2} = 4$
   E) $4^{-2} = -\frac{1}{16}$

3. Evaluate the logarithm $\log_{1/3} 0.113$ using the change of base formula. Round to 3 decimal places.
   A) –2.180
   B) 2.395
   C) 0.504
   D) 1.985
   E) –0.947

4. Solve the equation below algebraically. Round your result to three decimal places.

$$\frac{3+\ln x}{x^2}=0$$

A) –1.876
B) 1.061
C) –1.390
D) –2.512
E) 0.050

5. A sample contains 60 grams of carbon $\left({}^{14}C\right)$. ${}^{14}C$ has a half-life of 5715 years.

How much ${}^{14}C$ remains after 1900 years? Round your answer to three decimal places.

A) 40.052 grams
B) 47.651 grams
C) 33.913 grams
D) 19.948 grams
E) 32.913 grams

6. Match the function $y = 2e^{-x}$ with its graph.

Graph I:

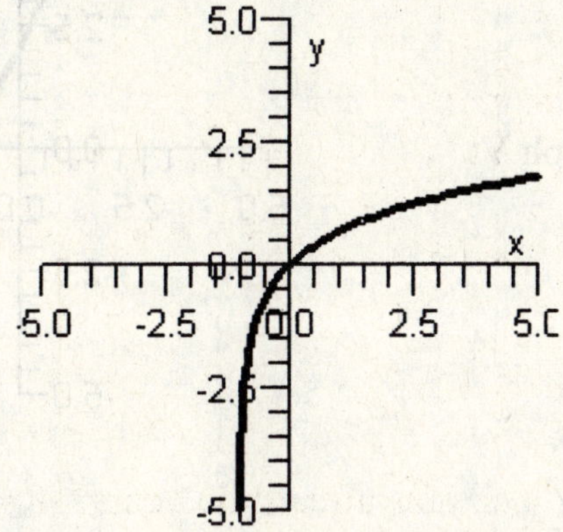

Graph II:

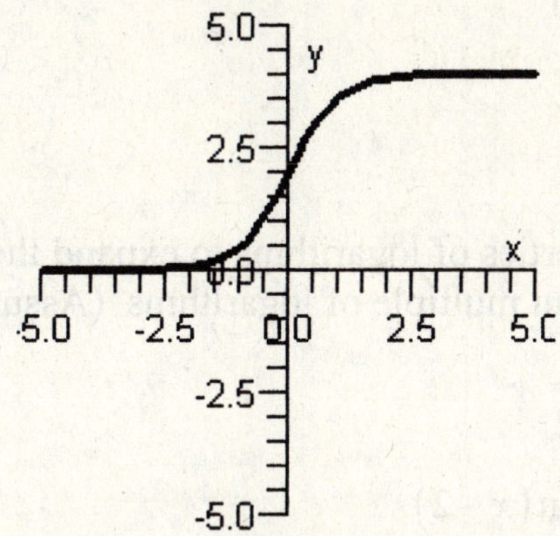

Graph III:

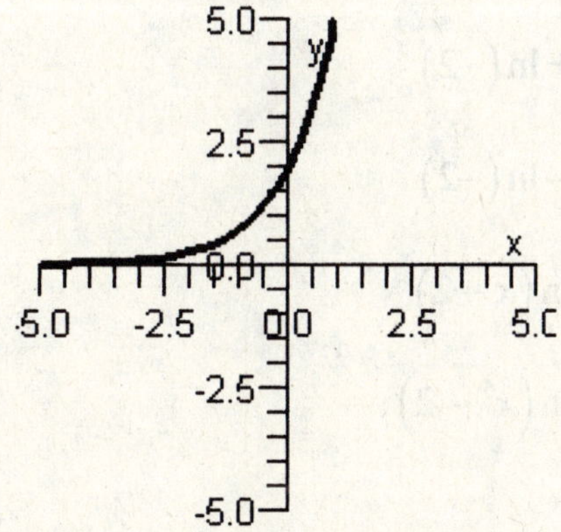

Graph IV:

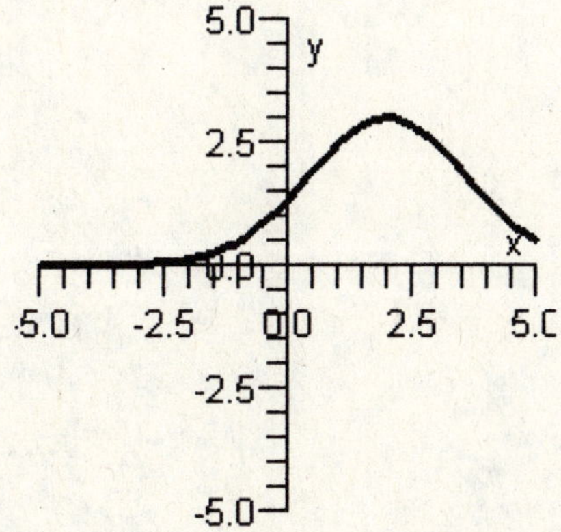

Graph V:

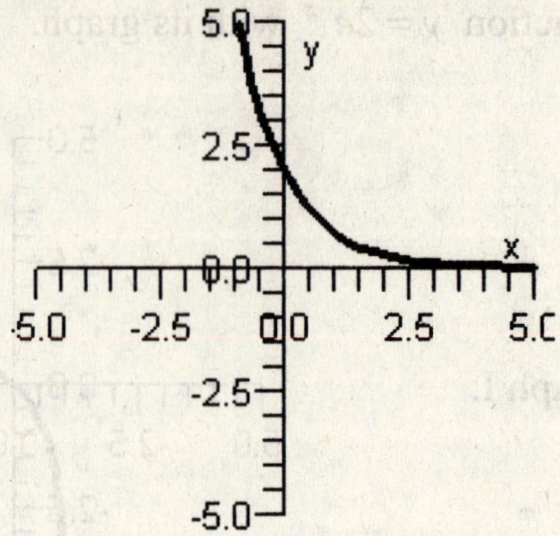

A) Graph II
B) Graph IV
C) Graph III
D) Graph V
E) Graph I

7. Use the properties of logarithms to expand the expression as a sum, difference, and/or constant multiple of logarithms. (Assume all variables are positive.)

$$\ln \frac{x}{\sqrt[3]{x^5 - 2}}$$

A) $\ln x + \frac{5}{3}\ln(x-2)$

B) $-\frac{2}{3}\ln x + \ln(-2)$

C) $-\frac{2}{3}\ln x - \ln(-2)$

D) $\ln x - \frac{5}{3}\ln(x-2)$

E) $\ln x - \frac{1}{3}\ln(x^5 - 2)$

8. What is the value of the function $f(x)=3.8e^{-1.5x}$ at $x = 2.5$? Round to 3 decimal places.
   A) 0.848
   B) 0.089
   C) 10.329
   D) –38.736
   E) 0.001

9. Solve for x: $e^x(8-e^x)=16$. Round to 3 decimal places.
   A) 3.079
   B) 2.079
   C) 1.386
   D) 2980.958
   E) no solution

10. Solve the equation below algebraically.
   $-2x^2e^{4x}-16xe^{4x}=0$
   A) –8 and 4
   B) 4 and 5
   C) –8 and 5
   D) 0 and 5
   E) –8 and 0

11. Determine which x-value below is a solution of the equation $\log_7\left(\frac{5}{3}x\right)=3$.
   $x=\frac{21}{125},\ x=\frac{1029}{5},\ x=\frac{189}{5},\ x=21,\ x=15$
   A) $x=\frac{21}{125}$
   B) $x=\frac{1029}{5}$
   C) $x=\frac{189}{5}$
   D) $x=21$
   E) $x=15$

12. Find the exact value of $\ln e^{2.50} - \ln \sqrt{e}$ without using a calculator.
    A) 1.25
    B) 5
    C) 2
    D) 3
    E) 2.5

13. Identify the graph that represents the function.

$y = 5e^{-2(x-1)^2}$

A)

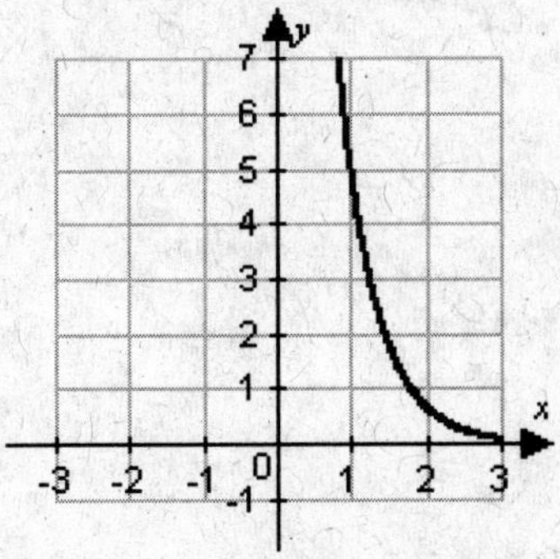

B)

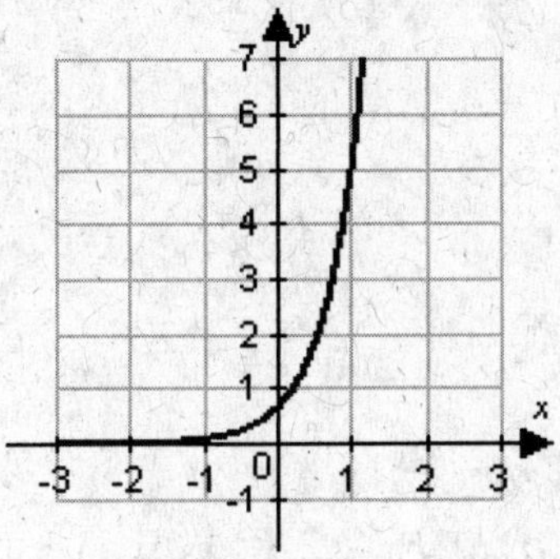

C)

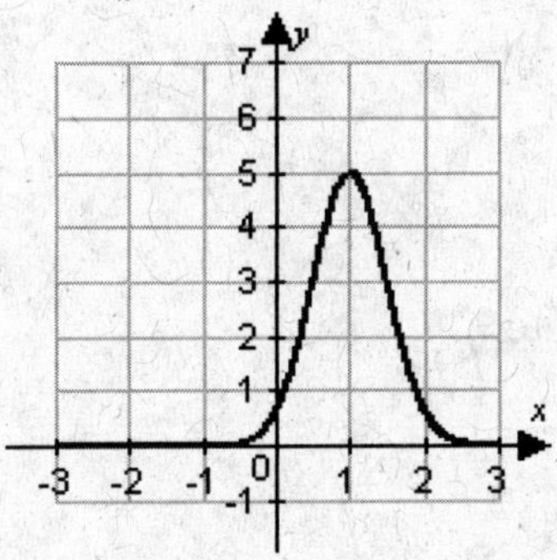

D)

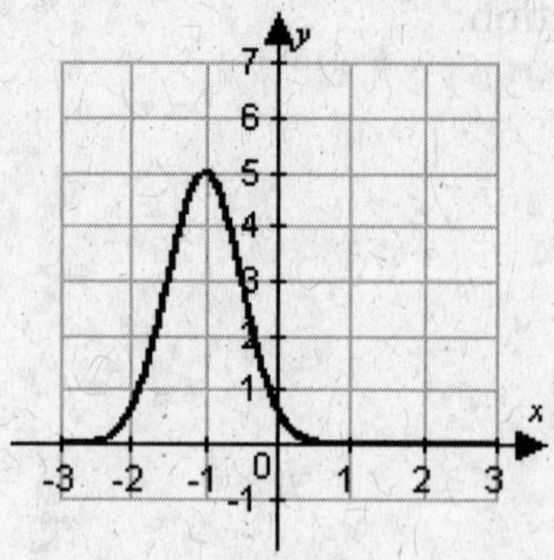

E)

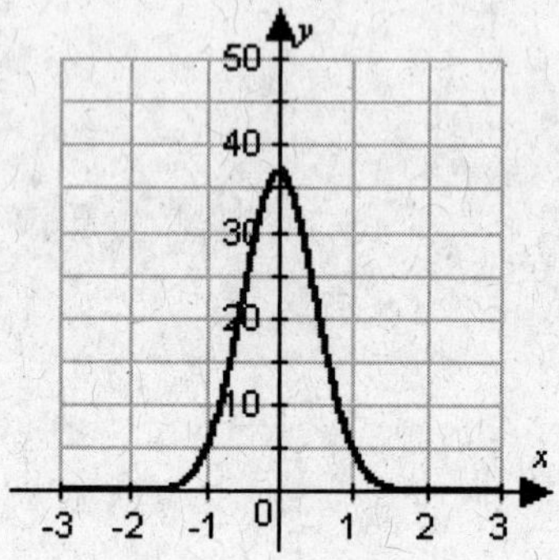

14. Identify the vertical asymptote of the function $f(x)=2+\log(x+3)$.
    A) $x=0$
    B) $x=-2$
    C) $x=-3$
    D) $x=3$
    E) The function has no vertical asymptote.

15. Find the exact value of $\log_8 \sqrt[3]{64}$ without using a calculator.

A) $\frac{64}{3}$
B) $\frac{3}{64}$
C) $\frac{16}{3}$
D) $\frac{2}{3}$
E) -1

16. Determine whether the scatter plot below could best be modeled by a linear model, a quadratic model, an exponential model, a logarithmic model, or a logistic model.

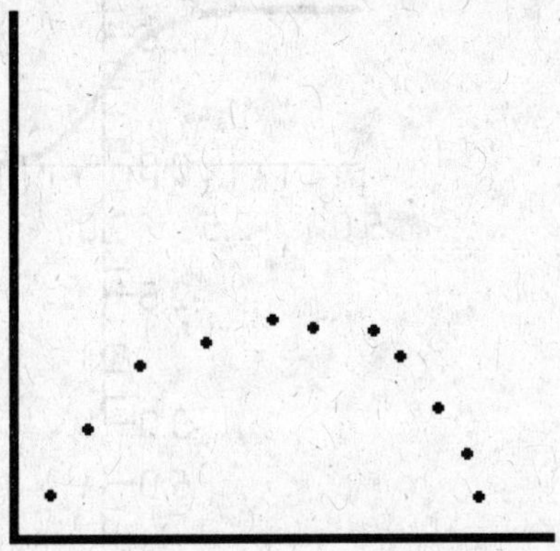

A) a linear model
B) a quadratic model
C) a logistic model
D) an exponential model
E) a logarithmic model

17. Identify the x-intercept of the function $y = 3 + \log_4 x$.

A) 64
B) $\frac{1}{64}$
C) -3
D) 12
E) The function has no x-intercept.

18. Match the function $y = \dfrac{3}{1+e^{-2x}}$ with its graph.

Graph I:

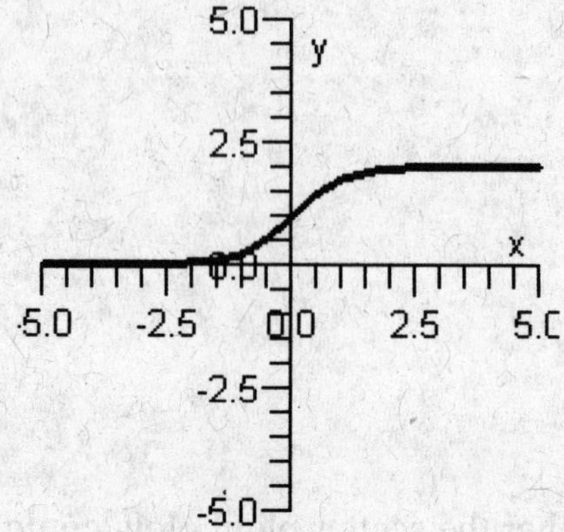

Graph II:

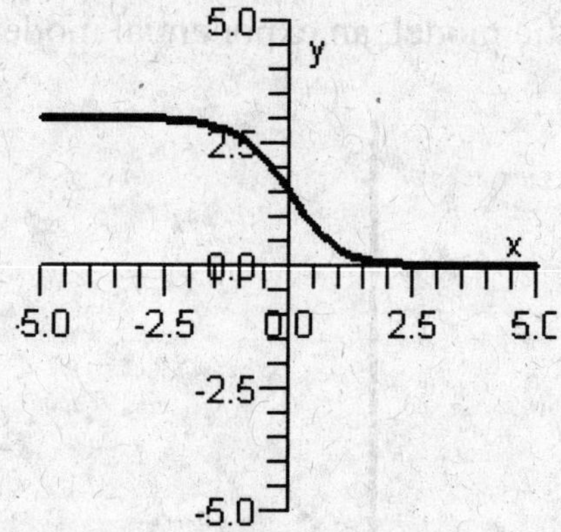

Graph III:

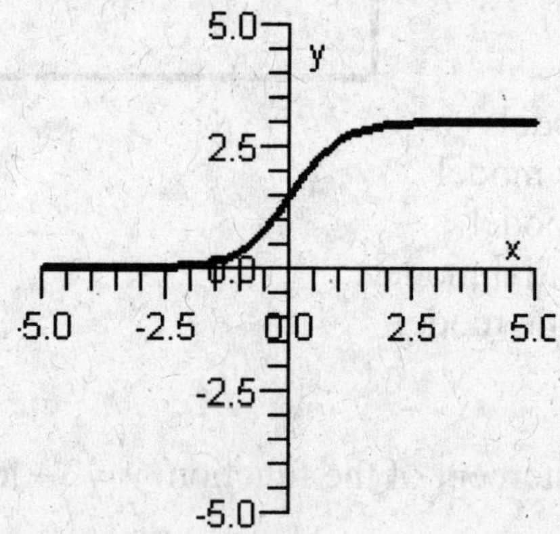

Graph IV:

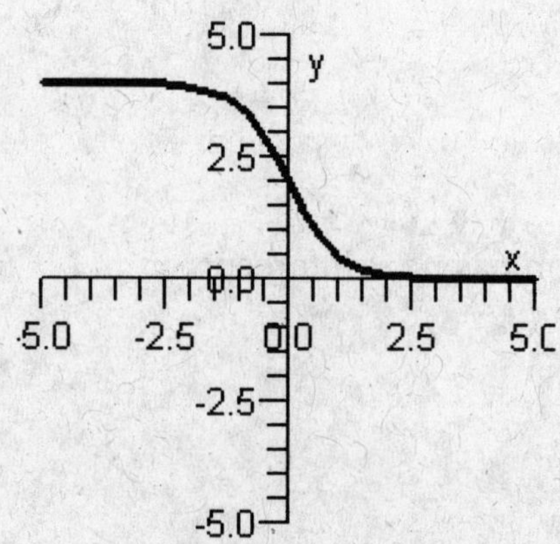

Graph V:

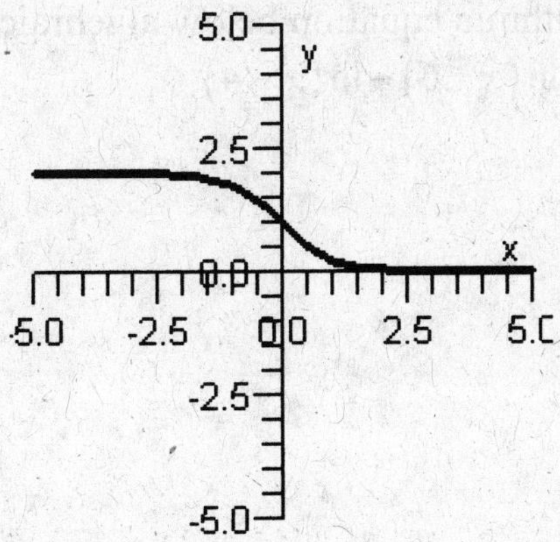

A) Graph V
B) Graph II
C) Graph I
D) Graph III
E) Graph IV

19. Solve the equation $f(x) = g(x)$ algebraically.

$f(x) = \ln e^{3x-9}$

$g(x) = x - 5$

A) 0
B) −1
C) 2
D) 5
E) 1

20. Solve the logarithmic equation below algebraically.

$$\ln(x-6) = \ln(x+6) - \ln(x-4)$$

A) –2 and 5
B) 2 and 9
C) 5 and 9
D) –2 and 9
E) 2 and 5

Answer Key

1. D
2. C
3. D
4. E
5. B
6. D
7. E
8. B
9. C
10. E
11. B
12. C
13. C
14. C
15. D
16. B
17. B
18. D
19. C
20. B

Name: ______________________________ Date: ______________

1. Find the exact value of the logarithm without using a calculator, if possible.

$$\ln \frac{1}{\sqrt[2]{e}}$$

A) -2
B) 2
C) $-\frac{1}{2}$
D) $\frac{1}{2}$
E) not possible without using a calculator

2. Find the exact value of $\ln e^{1.50} - \ln \sqrt{e}$ without using a calculator.
A) 0.75
B) 3
C) 1
D) 2
E) 1.5

3. What is the value of the function $f(x) = 3.3e^{-1.8x}$ at $x = 2.5$? Round to 3 decimal places.
A) 0.545
B) 0.037
C) 6.645
D) -40.366
E) 0.000

4. Solve the equation below algebraically. Round your result to three decimal places.

$$\frac{1+2\ln x}{x^5}=0$$

A) 0.862
B) –0.821
C) 1.855
D) –1.217
E) 0.607

5. Determine whether or not $x=\frac{1}{3}(e^{-3}+1)$ is a solution to $\ln(3x-1)=-3$.

A) yes
B) no

6. Solve $\ln x^2=3$ for x.

A) e^9
B) $10^{\sqrt{3}}$
C) $e^{\sqrt{3}}$
D) $-e^{3/2},e^{3/2}$
E) no solution

7. Solve the logarithmic equation below algebraically.

$$\log_6(2x+4)=\log_6(5x+1)$$

A) 3
B) –2
C) –4
D) 2
E) 1

8. Solve the logarithmic equation below algebraically. Round your result to three decimal places.

   $1+2\ln x=6$

   A) 8.318
   B) 10.682
   C) 12.182
   D) 15.377
   E) 11.366

9. Evaluate the logarithm $\log_{1/3} 0.211$ using the change of base formula. Round to 3 decimal places.

   A) –1.556
   B) 1.709
   C) 0.706
   D) 1.416
   E) –0.676

10. Use the properties of logarithms to expand the expression as a sum, difference, and/or constant multiple of logarithms. (Assume all variables are positive.)

    $\ln \sqrt[6]{t}$

    A) $\frac{1}{6}\ln t$
    B) $\ln t-\frac{1}{6}$
    C) $\ln t-6$
    D) $\frac{1}{3}\ln t$
    E) $\ln t-\frac{1}{3}$

11. Condense the expression below to the logarithm of a single quantity.

$$\frac{3}{2}\log_7(z+5)$$

A) $\log_7\left(\sqrt[3]{z}+5\right)$
B) $\log_7\left(\sqrt{z^3}+5\right)$
C) $\log_7\sqrt[3]{z+5}$
D) $\log_7\frac{3(z+5)}{2}$
E) $\log_7\sqrt{(z+5)^3}$

12. Find the exponential model $y=ae^{bx}$ that fits the points shown in the table below. Round parameters to the nearest thousandth.

x	0	1
y	5	10

A) $y=5e^{0.693x}$
B) $y=0.693e^{-5x}$
C) $y=-5e^{-0.693x}$
D) $y=5e^{-0.693x}$
E) $y=0.693e^{5x}$

13. Solve the exponential equation below algebraically. Round your result to three decimal places.

$$e^{2x}+3e^{x}-18=0$$

A) 1.099
B) 0.214
C) 2.077
D) –0.188
E) 3.271

14. Solve the exponential equation below algebraically. Round your result to three decimal places.

$$\frac{500}{1+e^{-x}}=150$$

A) 0.794
B) –1.483
C) –3.655
D) –3.796
E) –0.847

15. Simplify the expression below.

$$6+e^{\ln x^4}$$

A) $6+e^4$
B) $6+x^4$
C) $6+\ln 4$
D) $\ln 10$
E) $6+4\ln x$

16. The sales S (in thousands of units) of a cleaning solution after x hundred dollars is spent on advertising are given by $S=20\left(1-e^{kx}\right)$. When \$450 is spent on advertising, 2500 units are sold. Complete the model by solving for k and use the model to estimate the number of units that will be sold if advertising expenditures are raised to \$650. Round your answer to the nearest unit.

A) 3508 units
B) 6401 units
C) 5508 units
D) 2483 units
E) 8787 units

17. Identify the graph of the function.

$$f(x) = \left(\frac{1}{2}\right)^x$$

A) 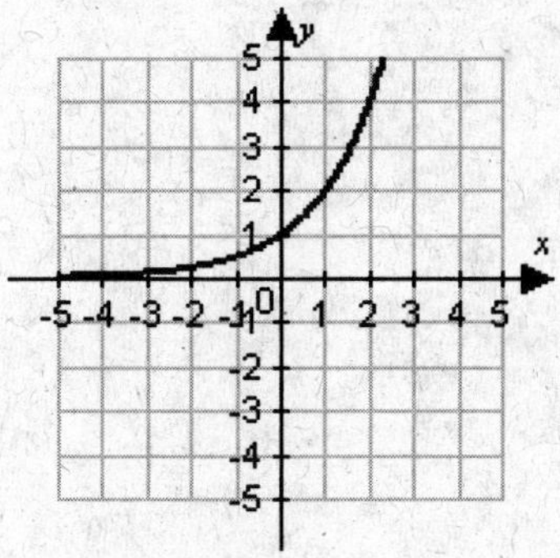

B) 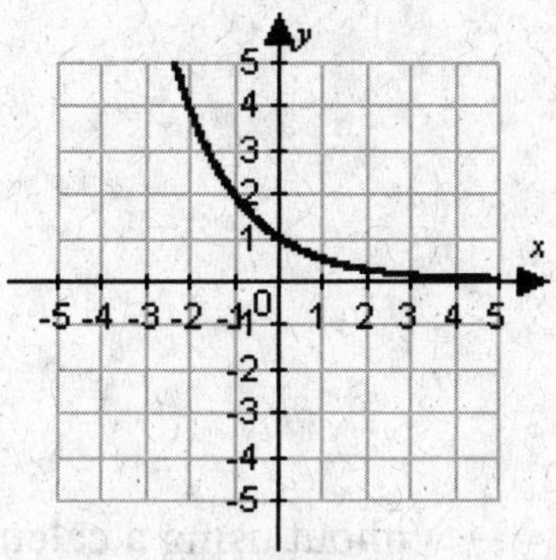

C) 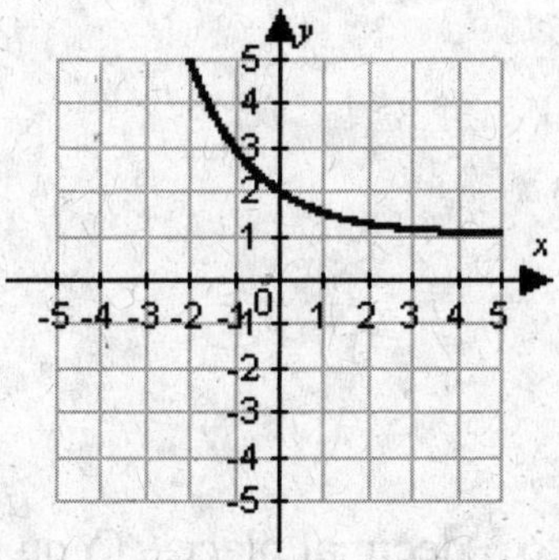

D)

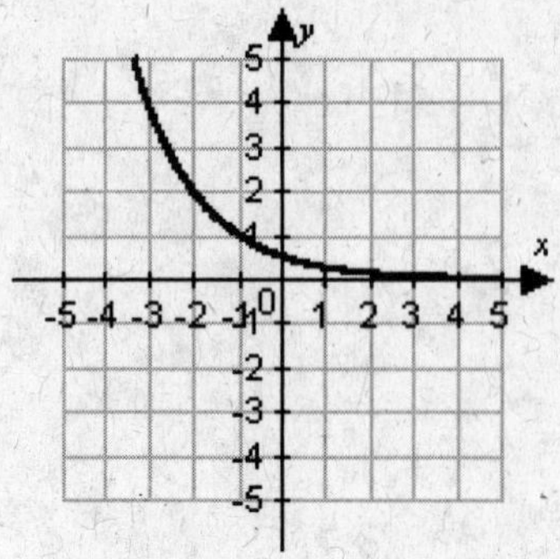

E)

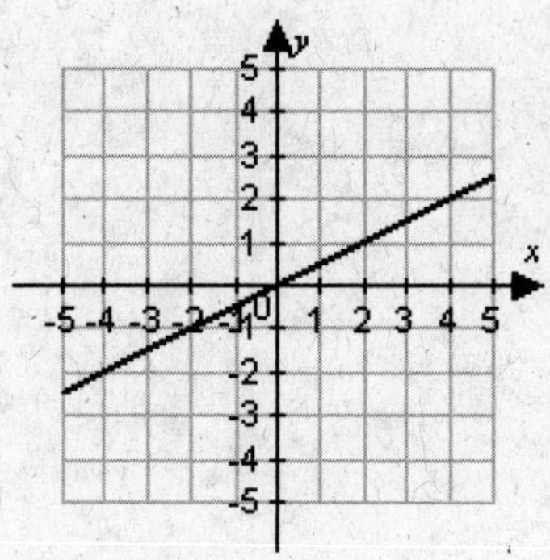

18. Evaluate the function $f(x)=\log_2 x$ at $x=\frac{1}{2}$ without using a calculator.

A) 0
B) –1
C) –2
D) 2
E) $\frac{1}{2}$

19. Approximate the solution of $9e^{x+3}=18$ to 3 decimal places. (You may use a graphing utility.)
A) –0.803
B) 4.389
C) –2.699
D) 3.693
E) –2.307

20. Find the domain of the function below.

$$f(x) = \ln\left(\frac{x+3}{x-5}\right)$$

A) $(-\infty, 5)$
B) $(-\infty, -3)$
C) $(-3, \infty)$
D) $(5, \infty)$
E) $(-\infty, -3) \cup (5, \infty)$

Answer Key

1. C
2. C
3. B
4. E
5. A
6. D
7. E
8. C
9. D
10. A
11. E
12. A
13. A
14. E
15. B
16. A
17. B
18. B
19. E
20. E

Name: ______________________________ Date: ____________

1. Find the exact value of $\log_4 12 - \log_4 3$ without using a calculator.

   A) $\frac{1}{2}$
   B) 1
   C) 3
   D) $\frac{3}{2}$
   E) 4

2. Use the properties of logarithms to expand the expression as a sum, difference, and/or constant multiple of logarithms. (Assume all variables are positive.)
   $\ln \sqrt[4]{t}$

   A) $\frac{1}{4}\ln t$
   B) $\ln t - \frac{1}{4}$
   C) $\ln t - 4$
   D) $\frac{1}{2}\ln t$
   E) $\ln t - \frac{1}{2}$

3. Determine whether or not $x = \frac{6}{7}$ is a solution to $3^{6x-3} = 81$.

   A) yes
   B) no

4. Identify the graph of the function.

$$f(x) = \left(\frac{1}{2}\right)^{1-x}$$

A)

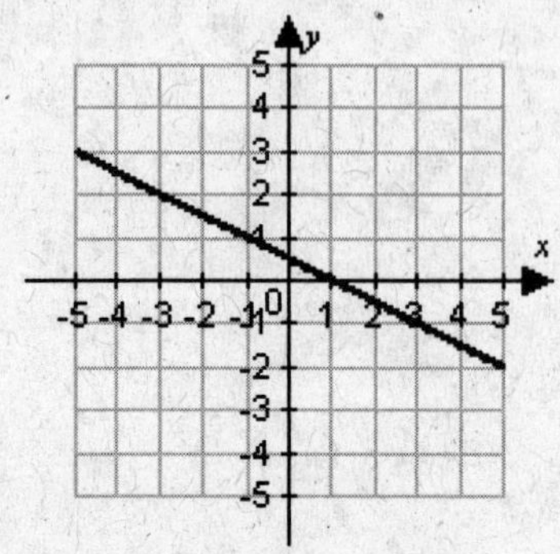

B)

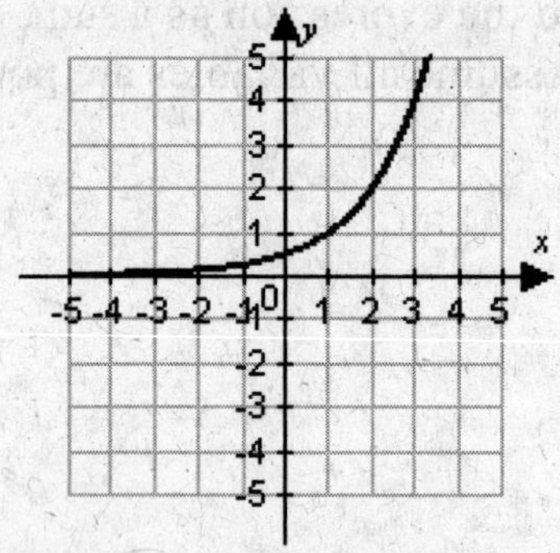

C)

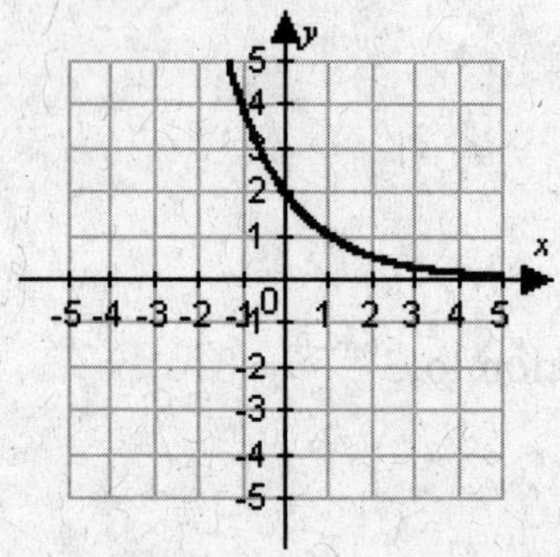

D) 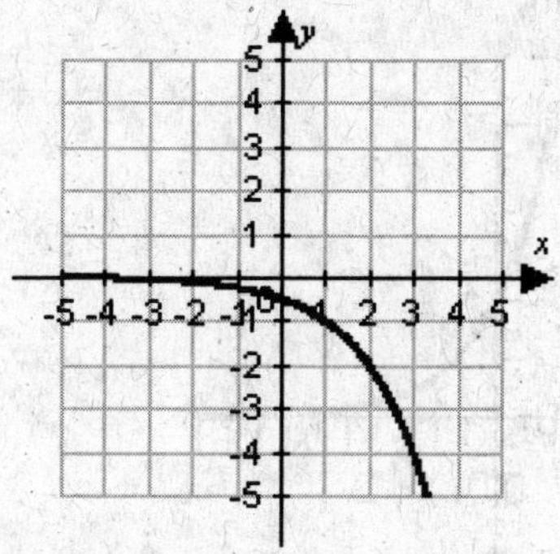

E) 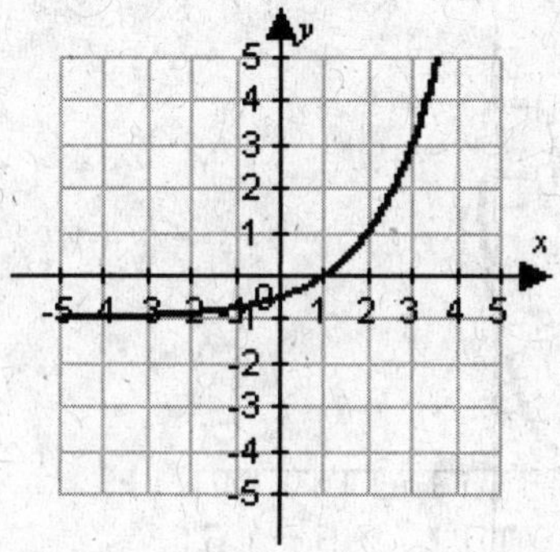

5. Use the properties of logarithms to expand the expression as a sum, difference, and/or constant multiple of logarithms. (Assume all variables are positive.)

$\log_4 \frac{y}{2}$

A) $2\log_4 y$
B) $y-\log_4 2$
C) $\log_4 y-2$
D) $\frac{\log_4 y}{2}$
E) $\log_4 y-\log_4 2$

6. Match the function $y = 2e^{-x}$ with its graph.

Graph I: 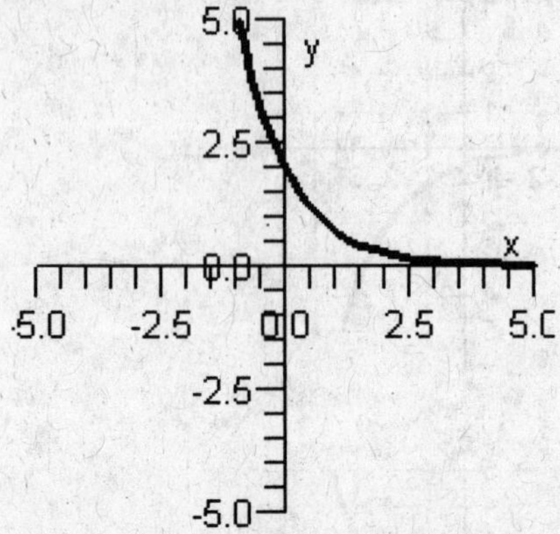

Graph II: 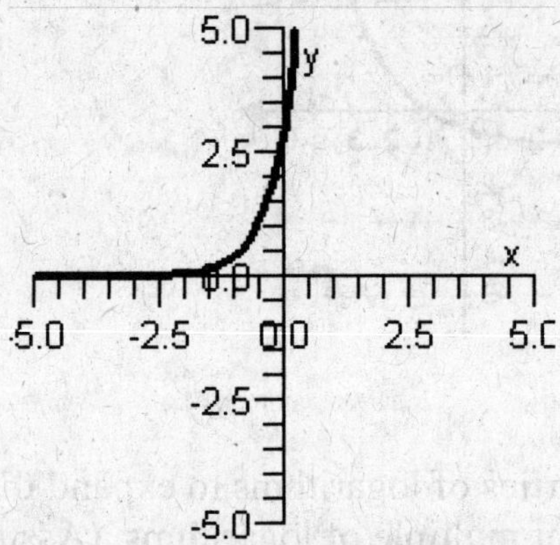

Graph III: 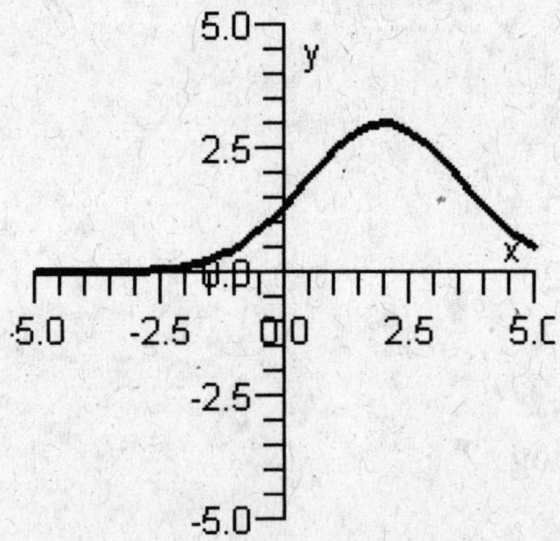

Graph IV: 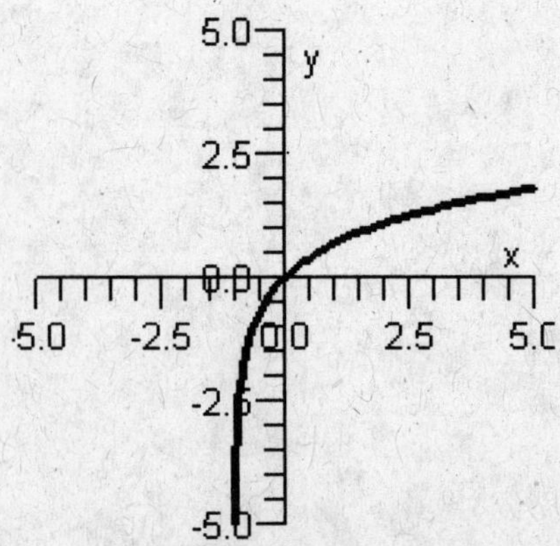

Graph V: 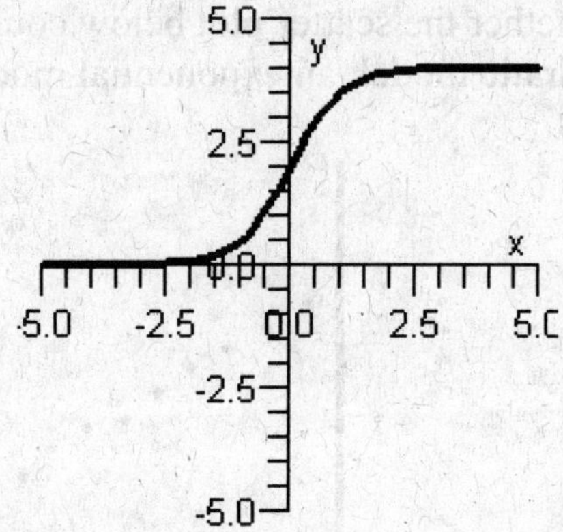

A) Graph III
B) Graph V
C) Graph IV
D) Graph I
E) Graph II

7. Identify the x-intercept of the function $f(x) = 3\ln(x-4)$.
A) $x = 4$
B) $x = 0$
C) $x = 3$
D) $x = 5$
E) The function has no x-intercept.

8. Determine whether the scatter plot below could best be modeled by a linear model, a quadratic model, an exponential model, a logarithmic model, or a logistic model.

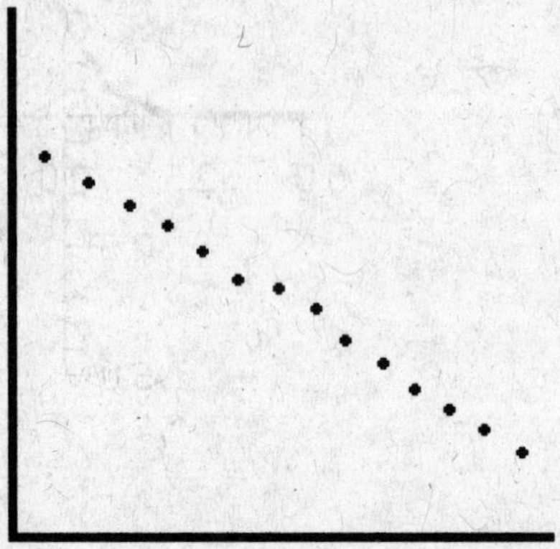

A) a logistic model
B) an exponential model
C) a linear model
D) a logarithmic model
E) a quadratic model

9. Evaluate the function $f(x) = \log_2 x$ at $x = \frac{1}{4}$ without using a calculator.

A) -1
B) -2
C) -3
D) 4
E) $\frac{1}{4}$

10. Solve the logarithmic equation below algebraically. Round your result to three decimal places.

$$\ln\left(x^2 + 6\right) = 5$$

A) ± 11.604
B) ± 14.688
C) ± 15.639
D) ± 12.762
E) ± 11.934

11. Evaluate the logarithm $\log_7 126$ using the change of base formula. Round to 3 decimal places.
A) 4.836
B) 0.402
C) 2.485
D) 9.411
E) 2.100

12. Find the domain of the function below.
$f(x)=\sqrt{\ln(x+2)}$
A) $(-\infty,-2]$
B) $(-\infty,-1]$
C) $[-2,\infty)$
D) $[-1,\infty)$
E) $(0,\infty)$

13. Solve $\ln x^2 = 13$ for x.
A) e^{169}
B) $10^{\sqrt{3}}$
C) $e^{\sqrt{13}}$
D) $-e^{13/2}, e^{13/2}$
E) no solution

14. Solve the equation below for x.
$\log_5 5^7 = x$
A) 8
B) 2
C) 5
D) 9
E) 7

15. The sales S (in thousands of units) of a cleaning solution after x hundred dollars is spent on advertising are given by $S = 20\left(1 - e^{kx}\right)$. When $450 is spent on advertising, 2500 units are sold. Complete the model by solving for k and use the model to estimate the number of units that will be sold if advertising expenditures are raised to $650. Round your answer to the nearest unit.
   A) 3508 units
   B) 6401 units
   C) 2160 units
   D) 5663 units
   E) 8787 units

16. Match the function $y=1+\ln(x+4)$ with its graph.

Graph I:

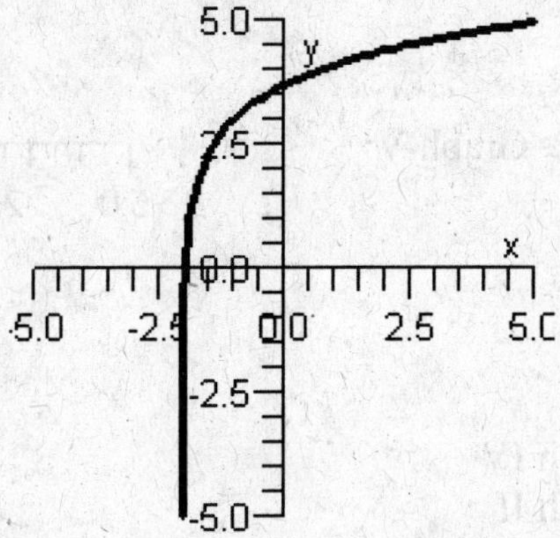

Graph II:

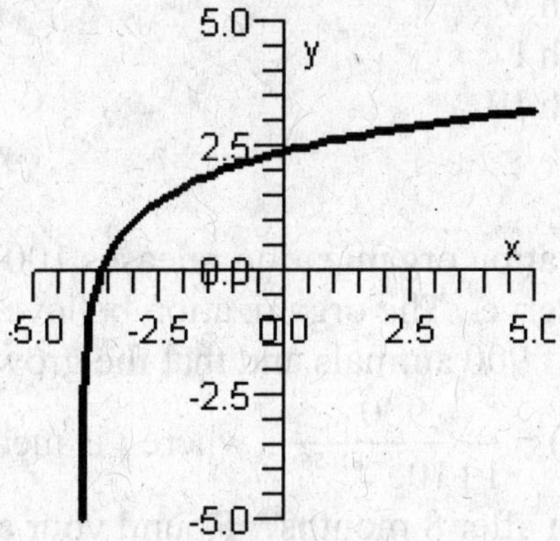

Graph III:

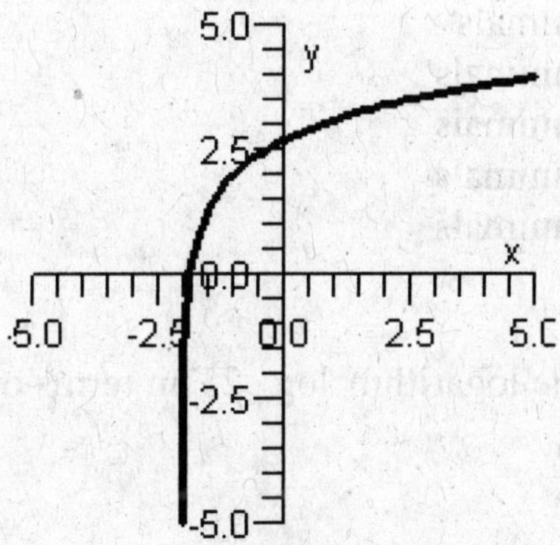

Graph IV:

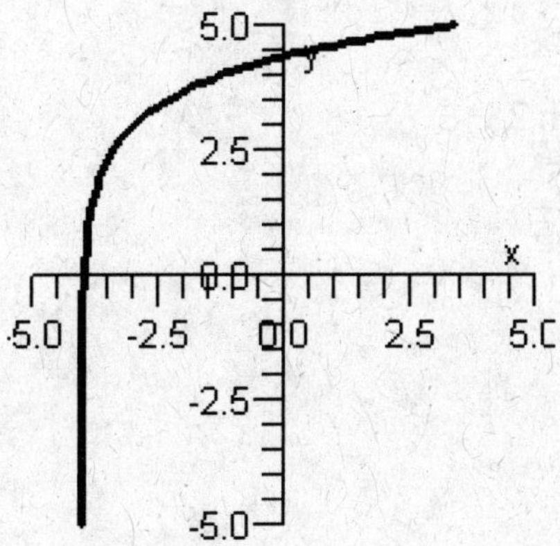

Graph V:

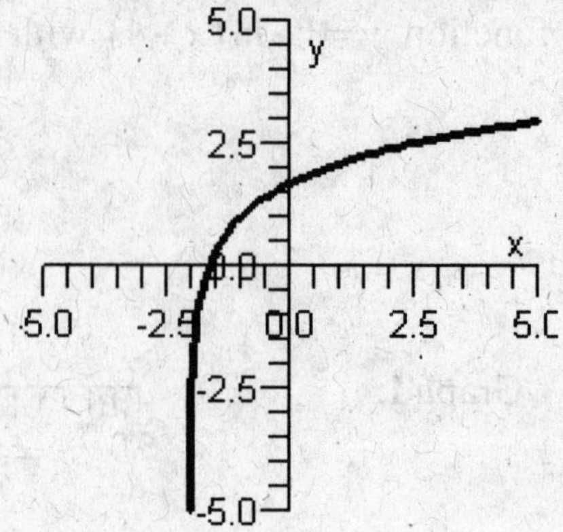

A) Graph IV
B) Graph II
C) Graph V
D) Graph I
E) Graph III

17. A conservation organization releases 100 animals of an endangered species into a game preserve. The organization believes that the preserve has a carrying capacity of 900 animals and that the growth of the herd will follow the logistic curve $p(t) = \dfrac{900}{1+10e^{-0.1656t}}$, where t is measured in months. What is the population after 8 months? Round your answer to the nearest animal.
A) 23 animals
B) 243 animals
C) 159 animals
D) 308 animals
E) 246 animals

18. Rewrite the logarithm $\log_4 25$ in terms of the natural logarithm.

A) $\dfrac{\ln 25}{\ln 4}$

B) $\dfrac{\ln 4}{\ln 25}$

C) $\ln 4 \ln 25$

D) $\dfrac{\ln 25}{\log_4 e}$

E) $\ln 25$

19. Solve the equation below algebraically. Round your result to three decimal places.

$$2x\ln\left(\frac{1}{x}\right)-x=0$$

A) 0.607
B) 0.000
C) –2.009
D) 2.086
E) –1.058

20. Solve the logarithmic equation below algebraically.

$$\ln(x+3)=\ln(x+1)-\ln(x+7)$$

A) 1 and 6
B) –5 and –4
C) –4 and 1
D) –4 and 6
E) –5 and 1

Answer Key

1. B
2. A
3. B
4. B
5. E
6. D
7. D
8. C
9. B
10. E
11. C
12. D
13. D
14. E
15. A
16. B
17. E
18. A
19. A
20. B

Name: ______________________________ Date: ____________

1. A sample contains 75 grams of carbon $\left({}^{14}C\right)$. ${}^{14}C$ has a half-life of 5715 years. How much ${}^{14}C$ remains after 100 years? Round your answer to three decimal places.
   A) 73.688 grams
   B) 74.096 grams
   C) 49.524 grams
   D) 1.312 grams
   E) 37.524 grams

2. Determine whether or not $x=\frac{4}{7}$ is a solution to $3^{4x-3}=81$.
   A) yes
   B) no

3. Solve the exponential equation below algebraically. Round your result to three decimal places.
   $$\frac{375}{1+e^{x}}=125$$
   A) 2.697
   B) 2.883
   C) 3.303
   D) −1.086
   E) 0.693

4. Solve the logarithmic equation below algebraically. Round your result to three decimal places.
   $$\ln\left(x^{2}+3\right)=4$$
   A) ±9.301
   B) ±6.772
   C) ±9.442
   D) ±4.318
   E) ±7.183

5. Solve the logarithmic equation below algebraically. Round your result to three decimal places.

   $\ln\sqrt{x-4}=2$

   A) 38.489
   B) 28.421
   C) 3.668
   D) 74.354
   E) 58.598

6. Evaluate the function $f(x)=\log_4 x$ at $x=\frac{1}{16}$ without using a calculator.

   A) –1
   B) –2
   C) –3
   D) 16
   E) $\frac{1}{16}$

7. Match the function $y = 1 + \ln(x+2)$ with its graph.

Graph I:

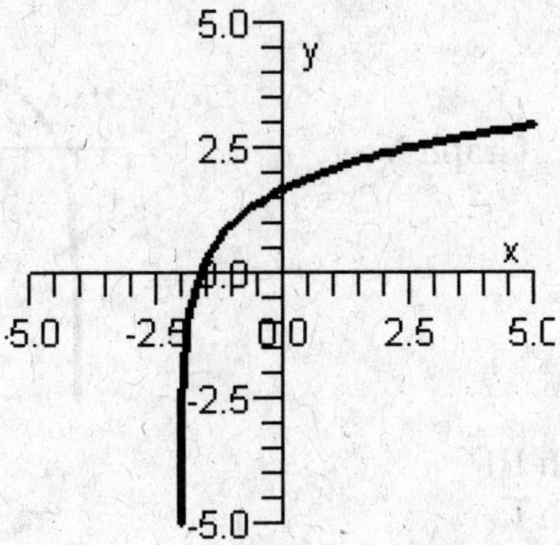

Graph II:

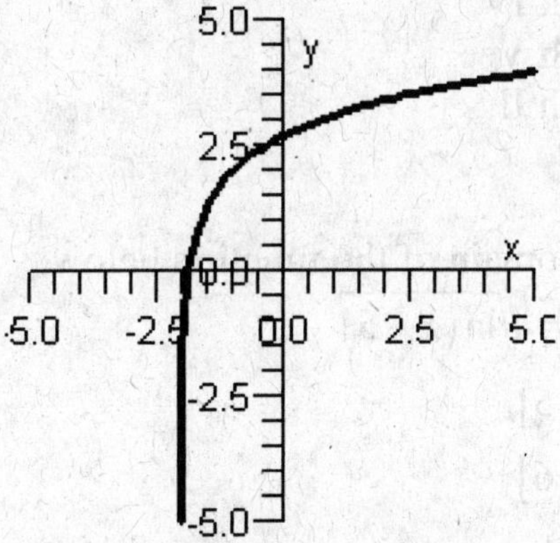

Graph III:

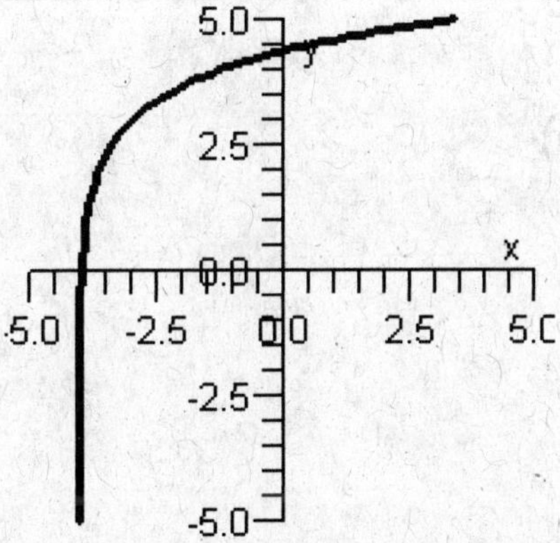

Graph IV:

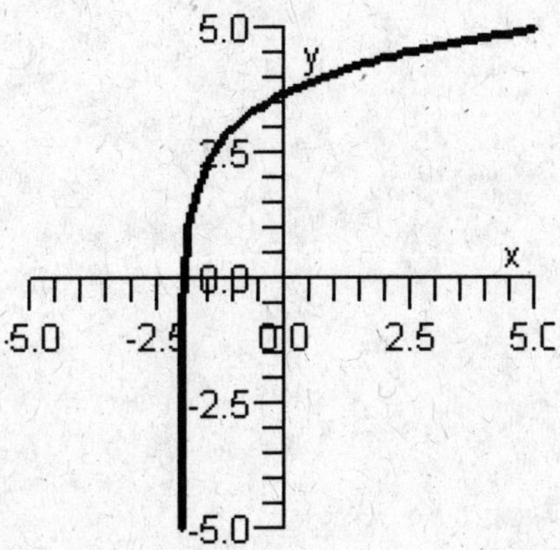

Graph V: 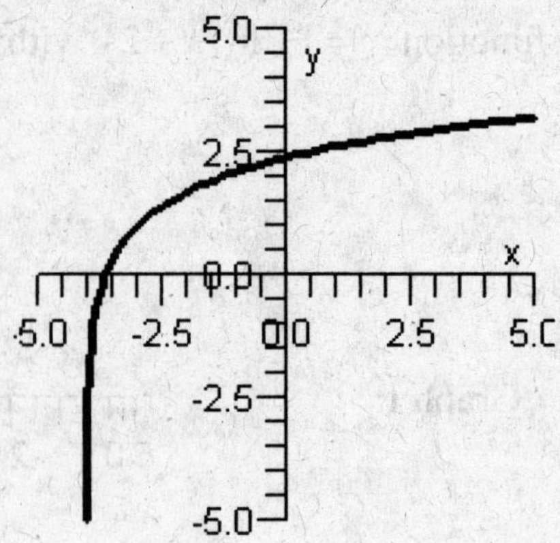

A) Graph III
B) Graph I
C) Graph IV
D) Graph V
E) Graph II

8. Find the domain of the function below.

$$f(x)=\sqrt{\ln(x-5)}$$

A) $(-\infty,5]$
B) $(-\infty,6]$
C) $[5,\infty)$
D) $[6,\infty)$
E) $(0,\infty)$

9. Match the function $y = \dfrac{3}{1+e^{-2x}}$ with its graph.

Graph I:

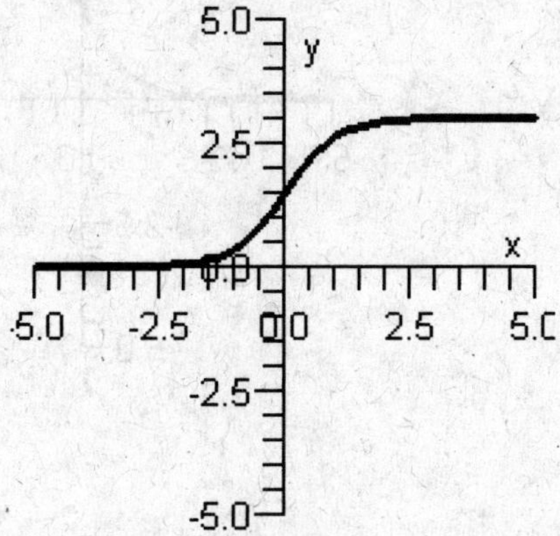

Graph II:

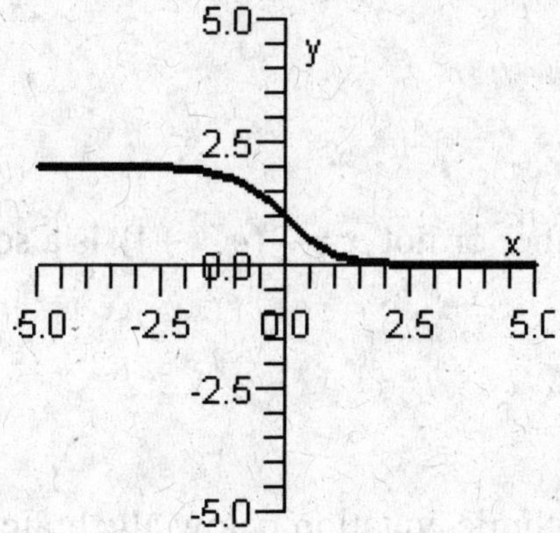

Graph III:

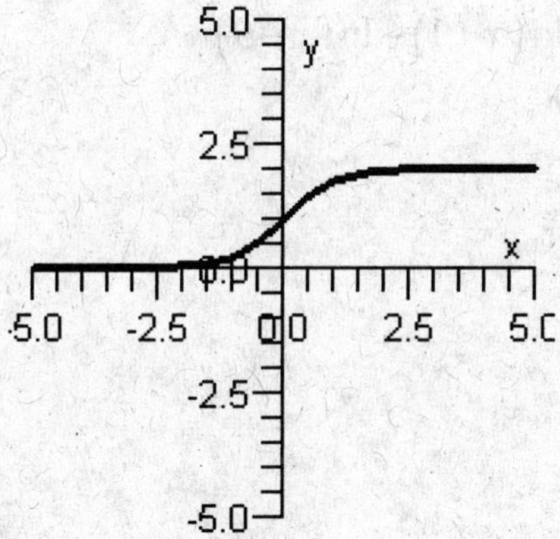

Graph IV:

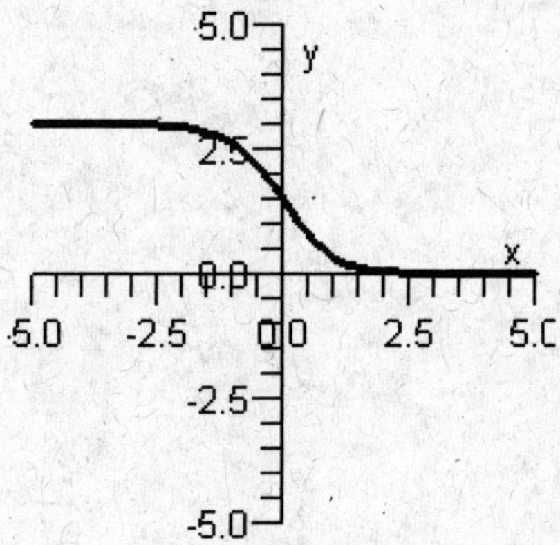

Graph V:

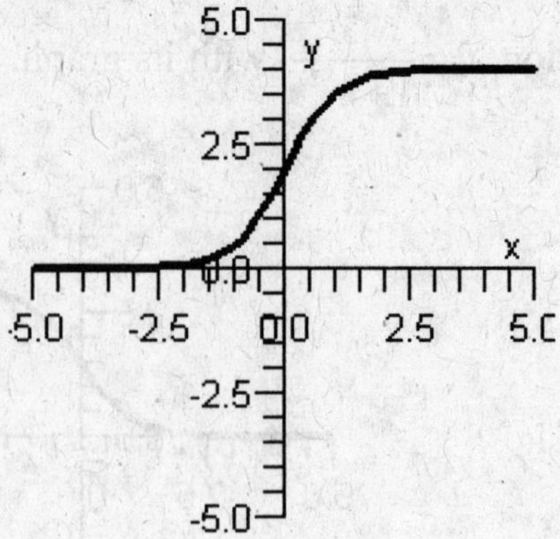

A) Graph III
B) Graph V
C) Graph IV
D) Graph I
E) Graph II

10. Determine whether or not $x = \frac{1}{3}(e^{-2} + 1)$ is a solution to $\ln(3x - 1) = -2$.

A) yes
B) no

11. Solve the logarithmic equation below algebraically.

$$\ln(x-5) = \ln(x+1) - \ln(x-9)$$

A) –12 and –9
B) 4 and 11
C) –12 and 11
D) –9 and 11
E) –12 and 4

12. Solve the equation $f(x) = g(x)$ algebraically.

$$f(x) = \ln e^{2x+2}$$
$$g(x) = 3x + 6$$

A) 5
B) 4
C) –4
D) –5
E) 2

13. Solve $\ln x^2 - \ln 3 = 0$ for x.

A) 9
B) $-\sqrt{3}, \sqrt{3}$
C) e^9
D) $e^{3/2}$
E) no solution

14. Determine whether the scatter plot below could best be modeled by a linear model, a quadratic model, an exponential model, a logarithmic model, or a logistic model.

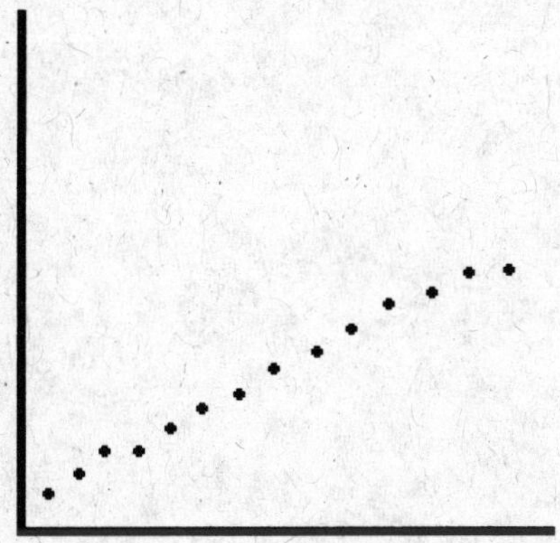

A) a logistic model
B) an exponential model
C) a linear model
D) a logarithmic model
E) a quadratic model

15. Solve the exponential equation below algebraically. Round your result to three decimal places.

$$150e^{-0.01x} = 140,000$$

A) –704.498
B) –695.069
C) –682.289
D) –683.876
E) –707.153

16. Solve the exponential equation below algebraically. Round your result to three decimal places.

$$250\left[\frac{(1+0.03)^x}{0.03}\right] = 160,000$$

A) 86.873
B) 99.967
C) 111.803
D) 97.718
E) 87.344

17. Solve the exponential equation below algebraically.

$$e^{-6x} = e^{x^2+5}$$

A) –6 and –1
B) –5 and –1
C) –6 and –5
D) –5 and 6
E) –1 and 6

18. The atmospheric pressure decreases with increasing altitude. At sea level, the average air pressure is approximately 1.03323 kilograms per square centimeter, and this pressure is called one atmosphere. Variations in weather conditions cause changes in the atmospheric pressure of up to ±5 percent. The table below shows the pressures p (in atmospheres) for various altitudes h (in kilometers). Use the *regression* feature of a graphing utility to find the logarithmic model $h = a + b \ln p$ for the data. Use the model to estimate the altitude at which the pressure is 0.28 atmosphere. Round your answer to two decimal places.

Altitude, h	Pressure, p
0	1
5	0.55
10	0.25
15	0.12
20	0.06
25	0.02

A) 4.54 kilometers
B) 10.88 kilometers
C) 7.44 kilometers
D) 9.07 kilometers
E) 8.34 kilometers

19. Determine whether the scatter plot below could best be modeled by a linear model, a quadratic model, an exponential model, a logarithmic model, or a logistic model.

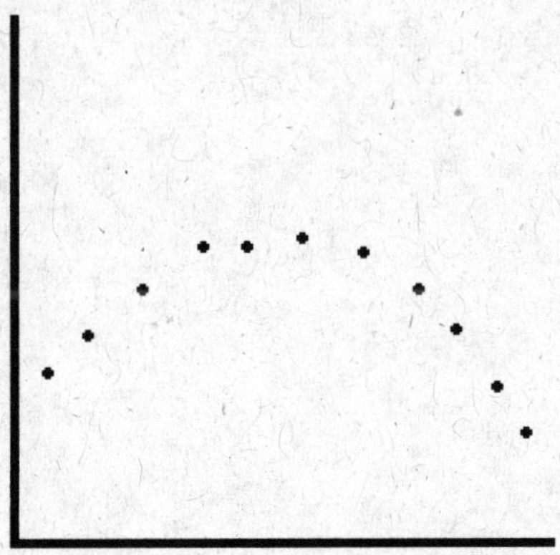

A) a linear model
B) a quadratic model
C) a logistic model
D) an exponential model
E) a logarithmic model

20. Solve $\left(\frac{1}{4}\right)^x = 64$ for x.

A) 1
B) −1
C) −3
D) −4
E) no solution

Answer Key

1. B
2. B
3. E
4. E
5. E
6. B
7. B
8. D
9. D
10. A
11. B
12. C
13. B
14. C
15. D
16. B
17. B
18. D
19. B
20. C

Name: ____________________________ Date: ______________

1. Identify the x-intercept of the function $f(x) = 3\ln(x-4)$.
   A) $x = 4$
   B) $x = 0$
   C) $x = 3$
   D) $x = 5$
   E) The function has no x-intercept.

2. Find the exact value of the logarithm without using a calculator, if possible.
   $\log_4 128 + \log_4 8$
   A) 1
   B) 2
   C) 3
   D) 4
   E) not possible without using a calculator

3. Solve the logarithmic equation below algebraically. Round your result to three decimal places.
   $4\log_3(x-4) = 13$
   A) 41.280
   B) 43.869
   C) 43.623
   D) 35.658
   E) 39.534

4. Simplify the expression $\log_3\left(\frac{1}{27}\right)^3$.
   A) 3
   B) –9
   C) 0
   D) –81
   E) The expression cannot be simplified.

5. Determine whether the scatter plot below could best be modeled by a linear model, a quadratic model, an exponential model, a logarithmic model, or a logistic model.

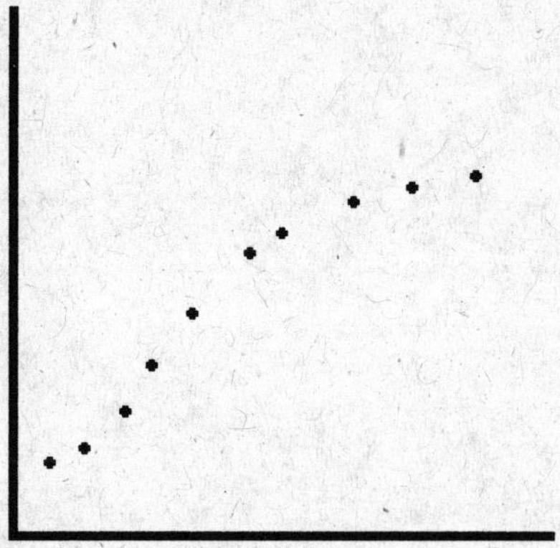

A) an exponential model
B) a logarithmic model
C) a linear model
D) a quadratic model
E) a logistic model

6. Condense the expression below to the logarithm of a single quantity.

$$\frac{5}{2}\log_8(z-1)$$

A) $\log_8\left(\sqrt[5]{z}-1\right)$

B) $\log_8\left(\sqrt{z^5}-1\right)$

C) $\log_8\sqrt[5]{z-1}$

D) $\log_8\dfrac{5(z-1)}{2}$

E) $\log_8\sqrt{(z-1)^5}$

7. Find the exponential model $y = ae^{bx}$ that fits the points shown in the table below. Round parameters to the nearest thousandth.

x	–2	0
y	48	3

A) $y = 3e^{-1.386x}$
B) $y = -1.386e^{-3x}$
C) $y = -3e^{1.386x}$
D) $y = 3e^{1.386x}$
E) $y = -1.386e^{3x}$

8. Condense the expression $\log_5 x + \log_5 7$ to the logarithm of a single term.

A) $\log(7x)^5$
B) $\log_5 7x$
C) $\log_5 7^x$
D) $\log_5 x^7$
E) $\log_5 (x+7)$

9. Rewrite the logarithmic equation $\log_2 \frac{1}{4} = -2$ in exponential form.

A) $2^4 = -2$
B) $2^{1/4} = -2$
C) $2^{-2} = \frac{1}{4}$
D) $\left(\frac{1}{4}\right)^{-2} = 2$
E) $2^{-2} = -\frac{1}{4}$

10. Condense the expression $\frac{1}{5}[\log_4 x + \log_4 6] - [\log_4 y]$ to the logarithm of a single term.

A) $\log_4 \frac{(6x)^5}{y}$
B) $\log_4 \frac{6x}{5y}$
C) $\log_4 \sqrt[5]{\frac{6x}{y}}$
D) $\log_4 \frac{\sqrt[5]{6x}}{y}$
E) $\log_4 \sqrt[5]{6x} - \log_4 y$

11. Evaluate the function $f(x) = \frac{1}{4}\ln x$ at $x = 14.67$. Round to 3 decimal places. (You may use your calculator.)

A) 1.546
B) 0.671
C) 1.280
D) –0.347
E) undefined

12. Match the function $y = 3 + \ln(x+2)$ with its graph.

Graph I:

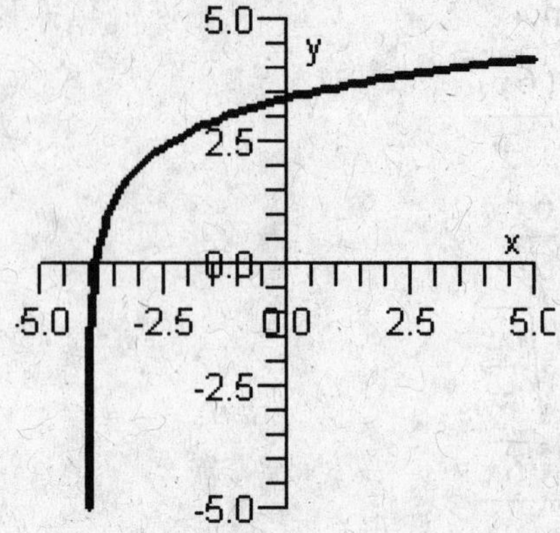

Graph II:

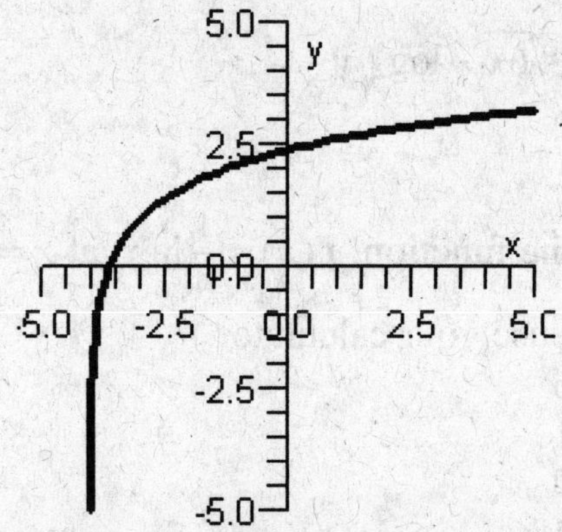

Graph III:

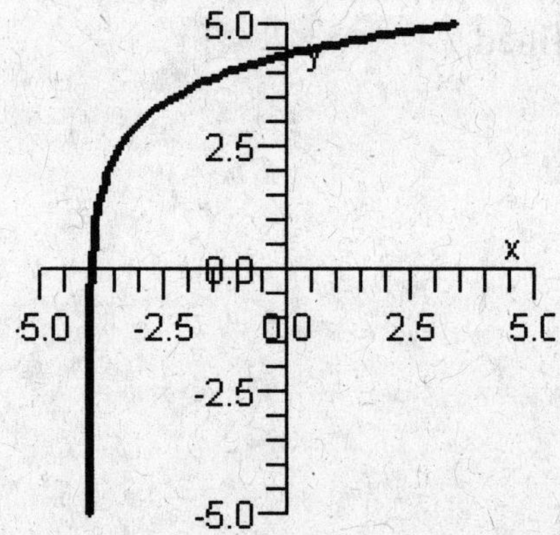

Graph IV:

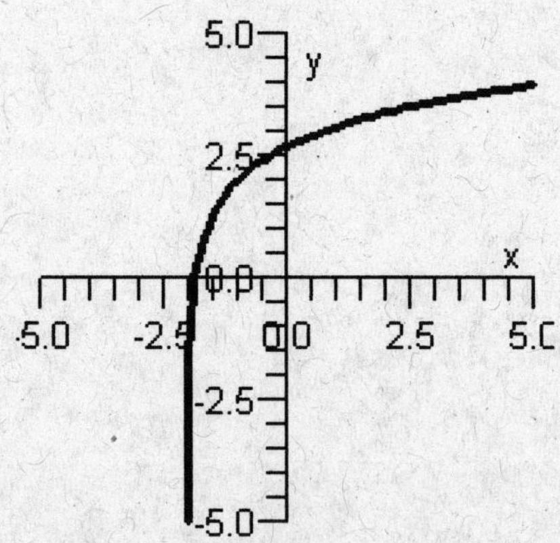

Graph V:

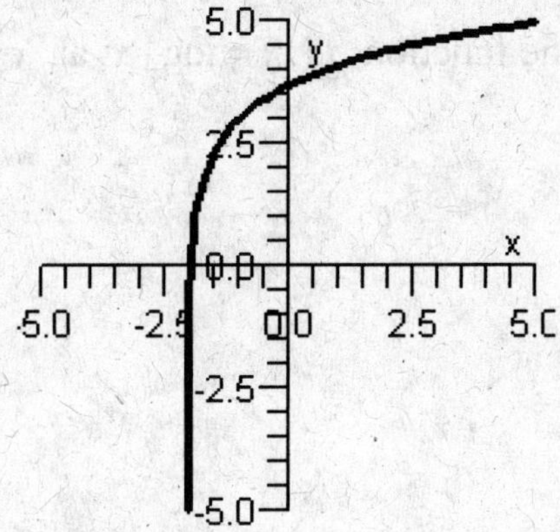

A) Graph II
B) Graph V
C) Graph III
D) Graph IV
E) Graph I

13. Find the exact value of $\log_5 \sqrt[3]{25}$ without using a calculator.

A) $\frac{25}{3}$

B) $\frac{3}{25}$

C) $\frac{10}{3}$

D) $\frac{2}{3}$

E) -1

14. Simplify the expression $\log_5 150$.

A) $2+\log_5 6$
B) $30\log_5 2$
C) 6
D) $2\log_5 6$
E) The expression cannot be simplified.

15. Evaluate the function $f(x) = \log_4 x$ at $x = \dfrac{1}{64}$ without using a calculator.

A) –2
B) –3
C) –4
D) 64
E) $\dfrac{1}{64}$

16. Rewrite the logarithm $\log_4 25$ in terms of the natural logarithm.

A) $\dfrac{\ln 25}{\ln 4}$
B) $\dfrac{\ln 4}{\ln 25}$
C) $\ln 4 \ln 25$
D) $\dfrac{\ln 25}{\log_4 e}$
E) $\ln 25$

17. Identify the value of the function $f(x) = \log_{10} x$ at $x = 715$. Round to 3 decimal places.

A) 2.854
B) 3.354
C) 3.854
D) 4.354
E) 6.572

18. Determine whether the scatter plot below could best be modeled by a linear model, a quadratic model, an exponential model, a logarithmic model, or a logistic model.

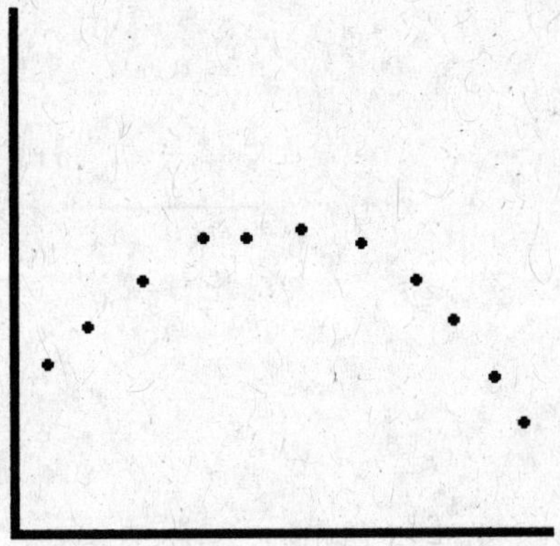

A) a linear model
B) a quadratic model
C) a logistic model
D) an exponential model
E) a logarithmic model

19. Condense the expression below to the logarithm of a single quantity.

$$5\left[\ln x - \ln(x+2) - \ln(x-2)\right]$$

A) $\ln\left(x^5 - \left(x^2-4\right)^5\right)$

B) $\ln\left(x^5 - (x+2)^5 - (x-2)^5\right)$

C) $\ln\dfrac{x^5}{x^2-4}$

D) $\ln\left(\left(\dfrac{x}{x^2-4}\right)^5\right)$

E) $\ln\dfrac{x}{\left(x^2-4\right)^5}$

20. Solve for x: $5^{-x/2} = 0.0052$. Round to 3 decimal places.
A) 6.535
B) 10.518
C) 13.737
D) –13.737
E) –3.268

Answer Key

1. D
2. E
3. E
4. B
5. E
6. E
7. A
8. B
9. C
10. D
11. B
12. B
13. D
14. A
15. B
16. A
17. A
18. B
19. D
20. A

Name: ______________________ Date: ___________

1. Determine whether the scatter plot below could best be modeled by a linear model, a quadratic model, an exponential model, a logarithmic model, or a logistic model.

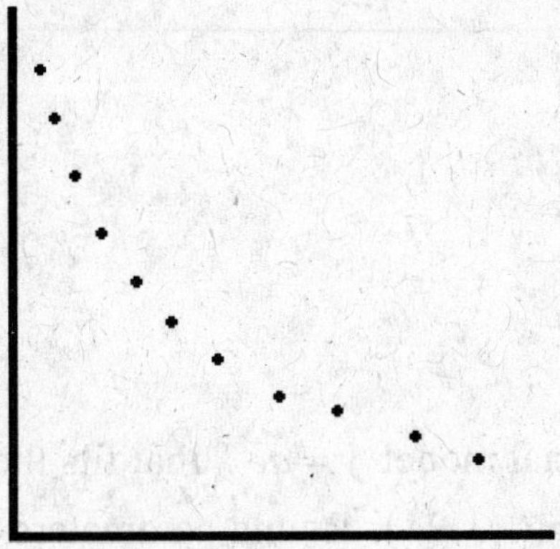

A) a logarithmic model
B) a quadratic model
C) a logistic model
D) a linear model
E) an exponential model

2. Simplify the expression $\log_5 150$.

A) $2+\log_5 6$
B) $30\log_5 2$
C) 6
D) $2\log_5 6$
E) The expression cannot be simplified.

3. Write the logarithmic equation below in exponential form.

$$\ln \sqrt[6]{e} = \frac{1}{6}$$

A) $\frac{1}{e^6} = \sqrt[6]{e}$
B) $e^{\frac{1}{6}} = \sqrt[6]{e}$
C) $\frac{e}{6} = \sqrt[6]{e}$
D) $e^6 = \sqrt[6]{e}$
E) $e^{-6} = \sqrt[6]{e}$

4. Find the exponential model $y = ae^{bx}$ that fits the points shown in the graph when $A = (0,3)$ and $B = (-3,24)$. Round parameters to the nearest thousandth.

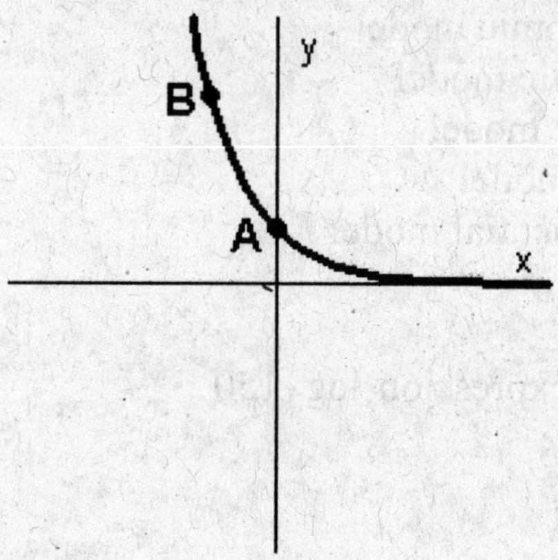

A) $y = -3e^{0.693x}$
B) $y = -0.693e^{-3x}$
C) $y = 3e^{-0.693x}$
D) $y = 3e^{0.693x}$
E) $y = -0.693e^{3x}$

5. Solve the logarithmic equation below algebraically. Round your result to three decimal places.

   $\ln\sqrt{x+6}=2$

   A) 60.262
   B) 97.681
   C) 65.968
   D) 67.927
   E) 48.598

6. Find the domain of the function below.

   $f(x)=\ln\left(\dfrac{x-4}{x+5}\right)$

   A) $(-\infty,4)$
   B) $(-\infty,-5)$
   C) $(-5,\infty)$
   D) $(4,\infty)$
   E) $(-\infty,-5)\cup(4,\infty)$

7. Identify the graph of the function.

$$f(x) = \left(\frac{1}{2}\right)^{x^2}$$

A)

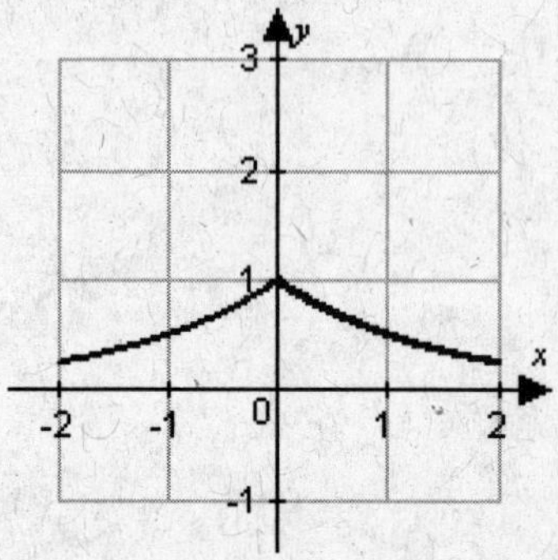

B)

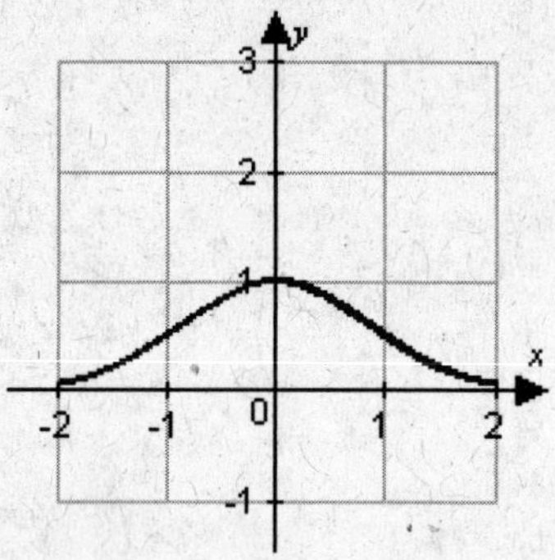

C)

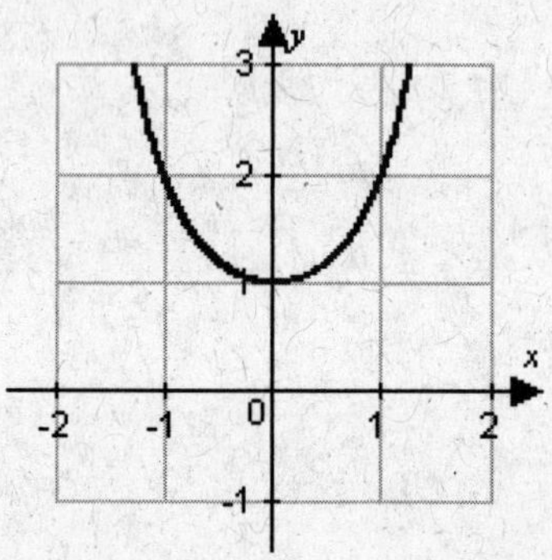

D)

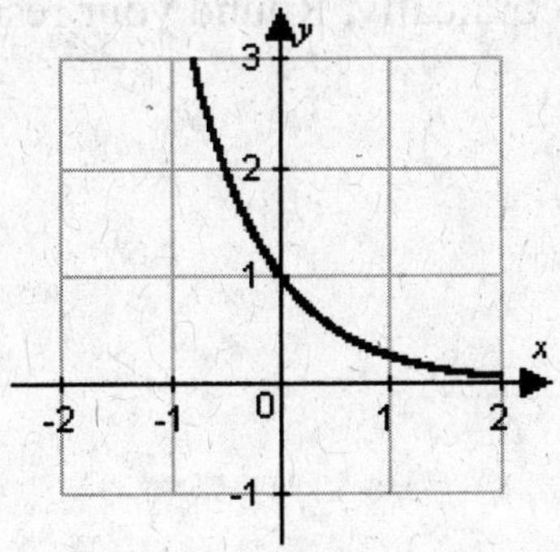

E)

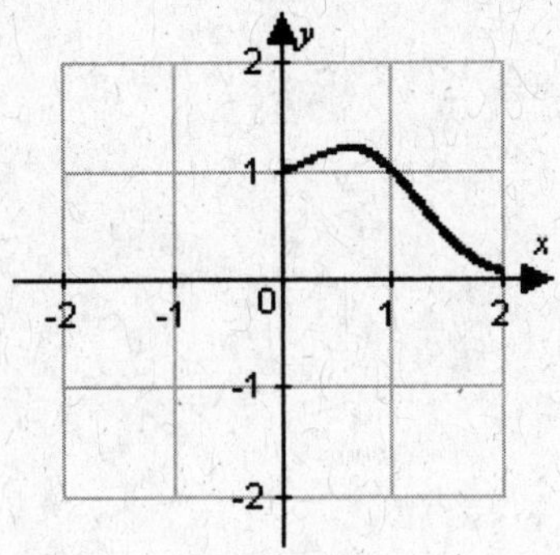

8. Identify the x-intercept of the function $y = 3 + \log_4 x$.
   A) 64
   B) $\frac{1}{64}$
   C) –3
   D) 12
   E) The function has no x-intercept.

9. Solve the equation below algebraically. Round your result to three decimal places.
   $e^x + 4xe^x = 0$
   A) –0.367
   B) –0.001
   C) –1.036
   D) –0.136
   E) –0.250

10. Solve the logarithmic equation below algebraically. Round your result to three decimal places.

$$\log_9 4x - \log_9 \left(1+\sqrt{x}\right) = 2$$

A) 449.651
B) 618.792
C) 849.024
D) 5.103
E) 161.208

11. Find the domain of the function below.

$$f(x) = \ln\left(\frac{x}{x^2+16}\right)$$

A) $(4,\infty)$
B) $(-\infty,-4)\cup(4,\infty)$
C) $(-\infty,0)\cup(4,\infty)$
D) $(0,\infty)$
E) $(-\infty,-4)\cup(0,\infty)$

12. Solve the equation below algebraically.

$$-x^2e^{2x} - 7xe^{2x} = 0$$

A) –7 and –6
B) –6 and 6
C) –7 and 6
D) 0 and 6
E) –7 and 0

13. Evaluate the function $f(x) = \log_2 x$ at $x = \frac{1}{4}$ without using a calculator.

A) –1
B) –2
C) –3
D) 4
E) $\frac{1}{4}$

14. Condense the expression below to the logarithm of a single quantity.

$$4\left[\ln x - \ln(x+4) - \ln(x-4)\right]$$

A) $\ln\left(x^4 - \left(x^2-16\right)^4\right)$
B) $\ln\left(x^4 - (x+4)^4 - (x-4)^4\right)$
C) $\ln\dfrac{x^4}{x^2-16}$
D) $\ln\left(\left(\dfrac{x}{x^2-16}\right)^4\right)$
E) $\ln\dfrac{x}{\left(x^2-16\right)^4}$

15. Solve the exponential equation below algebraically.

$$e^{-x^2} = e^{2x^2-9x}$$

A) 0 and 3
B) –2 and 0
C) –2 and 3
D) –6 and 3
E) –6 and 0

16. Simplify the expression below.

$$1+e^{\ln x^6}$$

A) $1+e^6$
B) $1+x^6$
C) $1+\ln 6$
D) $\ln 7$
E) $1+6\ln x$

17. Use the properties of logarithms to expand the expression as a sum, difference, and/or constant multiple of logarithms. (Assume all variables are positive.)

$$\ln \frac{x}{\sqrt[3]{x^2-3}}$$

A) $\ln x + \frac{2}{3}\ln(x-3)$

B) $\frac{1}{3}\ln x + \ln(-3)$

C) $\frac{1}{3}\ln x - \ln(-3)$

D) $\ln x - \frac{2}{3}\ln(x-3)$

E) $\ln x - \frac{1}{3}\ln(x^2-3)$

18. Identify the graph that represents the function.

$$y = \ln \frac{1}{x+1}$$

A)

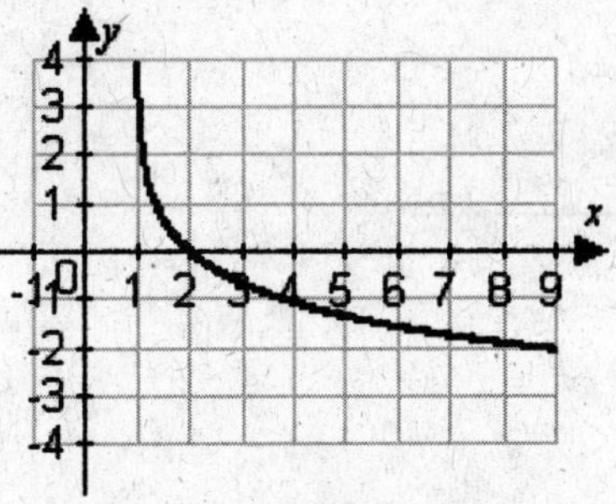

B)

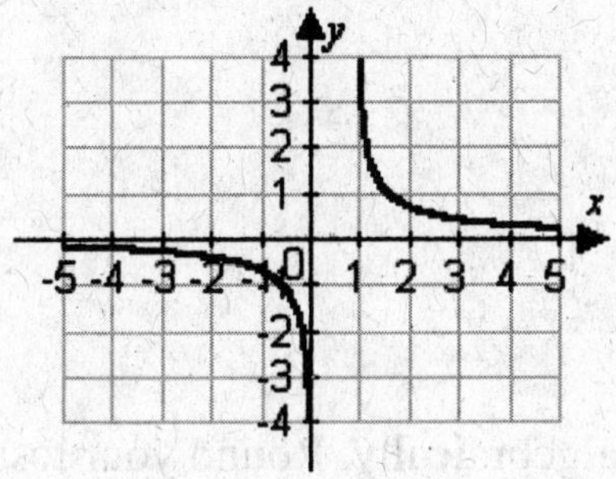

C)

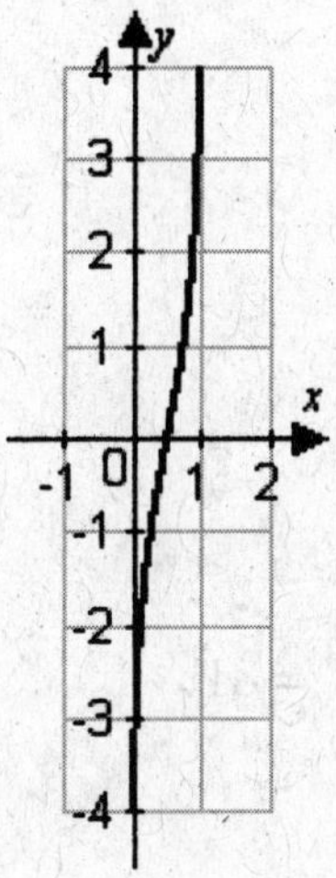

D)

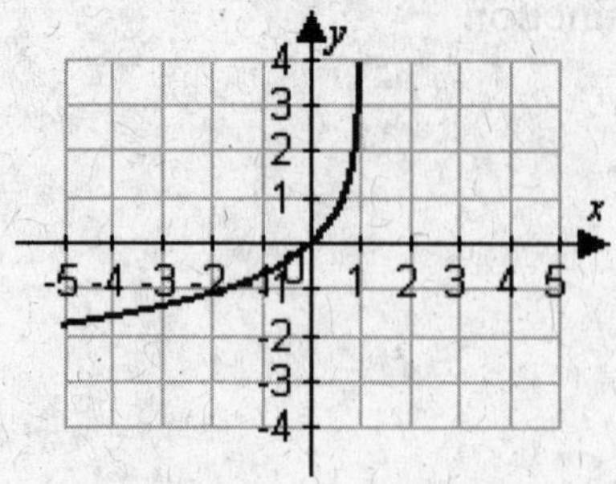

E)

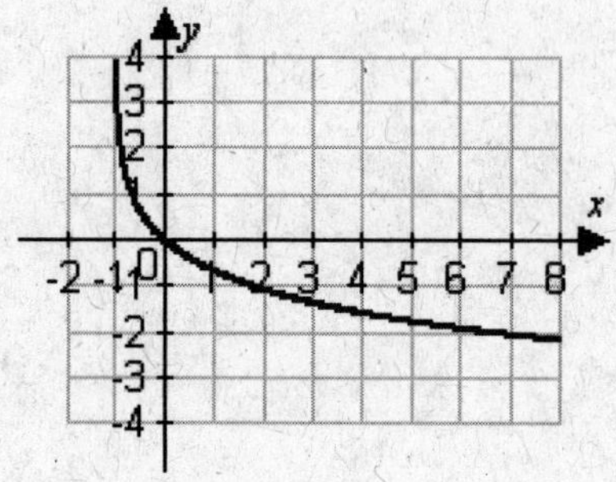

19. Solve the logarithmic equation below algebraically. Round your result to three decimal places.

$$5\log_4(x+1)=12$$

A) 21.326
B) 22.244
C) 23.558
D) 23.291
E) 26.858

20. Determine whether or not $x=\frac{4}{7}$ is a solution to $3^{4x-3}=81$.

A) yes
B) no

Answer Key

1. E
2. A
3. B
4. C
5. E
6. E
7. B
8. B
9. E
10. A
11. D
12. E
13. B
14. D
15. A
16. B
17. E
18. E
19. E
20. B

Name: ___________________________ Date: ______________

1. Find the radian measure of the central angle of a circle of radius r that intercepts an arc of length s.

   radius: $r = 12$ kilometers   arc length: $s = 64$ kilometers

   A) $\frac{3}{16}$

   B) $\frac{3\pi}{16}$

   C) $\frac{16\pi}{3}$

   D) $\frac{8}{3}$

   E) $\frac{16}{3}$

2. Determine the period and amplitude of the following function.

$$y = 4\cos\left(\frac{3x}{4} + \frac{\pi}{4}\right)$$

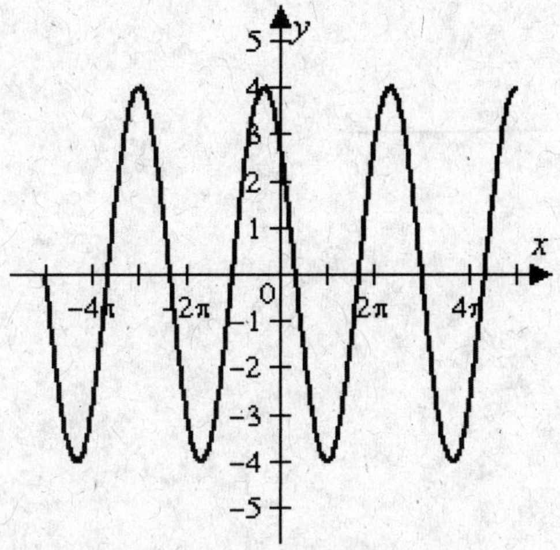

A) period: 2π, amplitude: 2
B) period: 10π, amplitude: 2
C) period: 5π, amplitude: 4
D) period: $\frac{8\pi}{3}$; amplitude: 4
E) period: 3π, amplitude: 4

3. A certain satellite orbits 13,000 miles above Earth's surface (see figure). Find the angle of depression α from the satellite to the horizon. Assume the radius of the Earth is 4000 miles. Round your answer to the nearest hundredth of a degree.

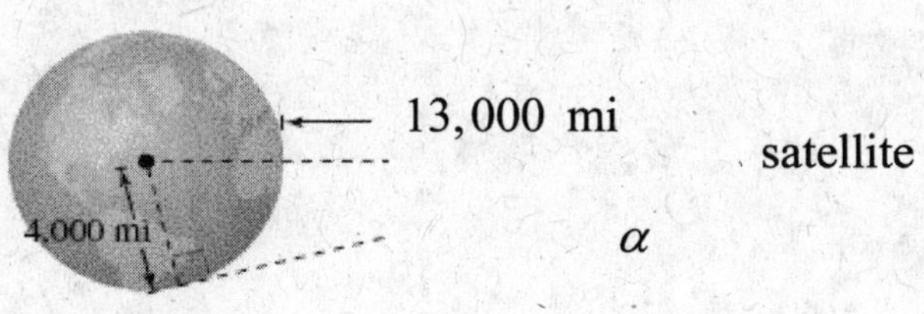

A) 13.61°
B) 17.92°
C) 72.08°
D) 76.39°
E) 13.24°

4. Convert the angle measure to decimal degree form.
   427°50'55"
   A) 426.151°
   B) 427.050°
   C) 427.849°
   D) 7.454°
   E) 24,471.314°

5. Use a graphing utility to graph the function below, making sure to show at least two periods.

$\sec\frac{x}{2}$

A)

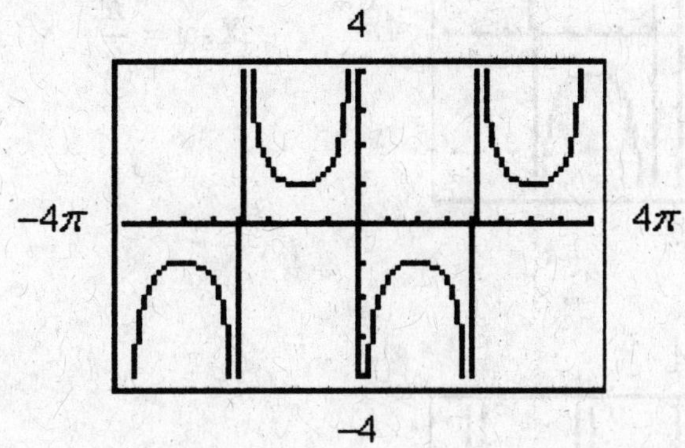

$\text{Xscl} = \frac{\pi}{2}$

B)

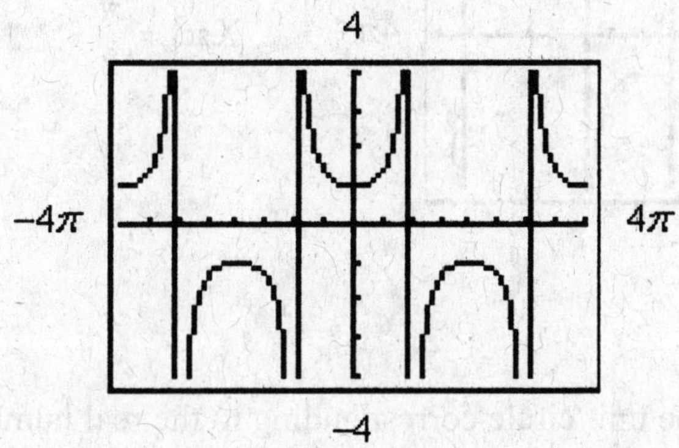

$\text{Xscl} = \frac{\pi}{2}$

C)

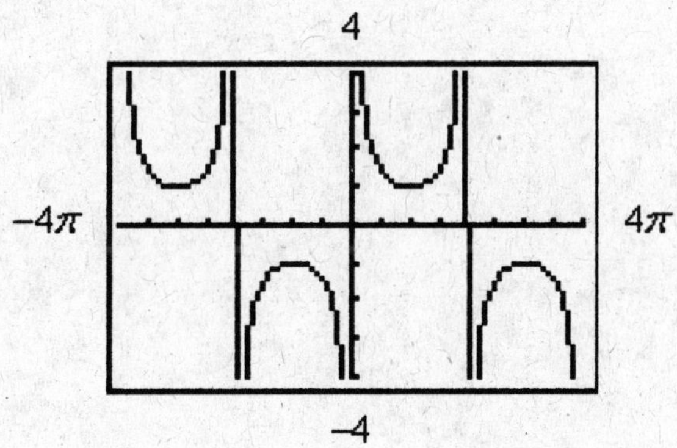

$\text{Xscl} = \frac{\pi}{2}$

D)

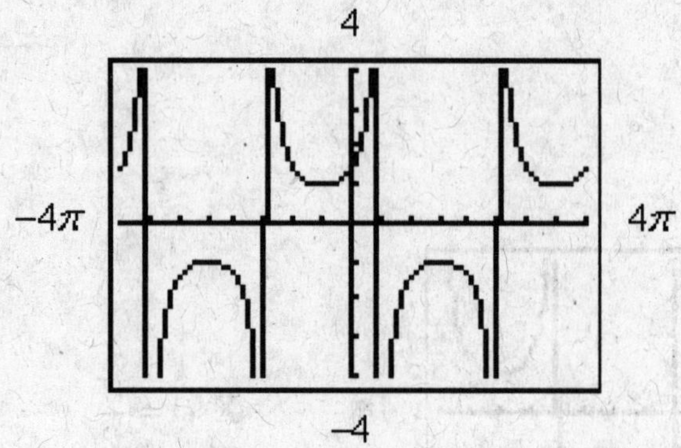

$\text{Xscl} = \dfrac{\pi}{2}$

E)

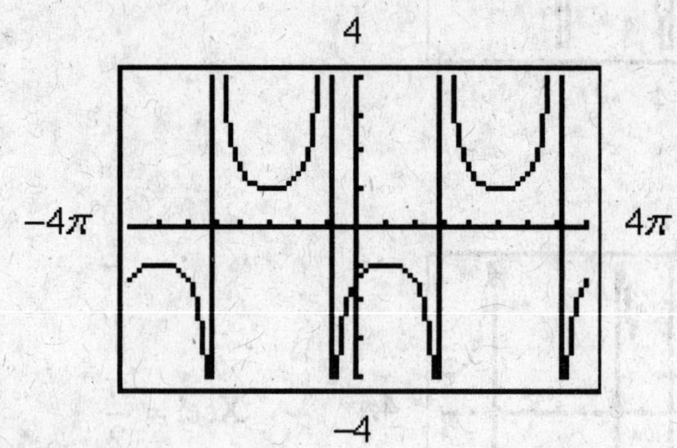

$\text{Xscl} = \dfrac{\pi}{2}$

6. Find the point (x, y) on the unit circle corresponding to the real number $t = \dfrac{11\pi}{6}$.

A) $\left(-\dfrac{1}{2}, \dfrac{\sqrt{3}}{2}\right)$

B) $\left(\dfrac{\sqrt{3}}{2}, -\dfrac{1}{2}\right)$

C) $\left(-\dfrac{\sqrt{3}}{2}, \dfrac{1}{2}\right)$

D) $\left(\dfrac{1}{2}, -\dfrac{\sqrt{3}}{2}\right)$

E) $\left(\dfrac{\sqrt{2}}{2}, \dfrac{\sqrt{2}}{2}\right)$

7. Find (if possible) the complement of $\frac{\pi}{7}$.

A) $\frac{3\pi}{7}$
B) $\frac{2\pi}{7}$
C) $\frac{6\pi}{7}$
D) $\frac{5\pi}{14}$
E) not possible

8. Determine the value of x.

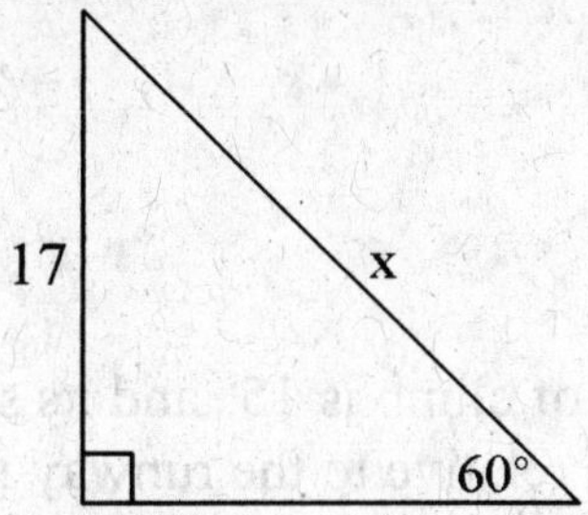

A) $\frac{34\sqrt{3}}{3}$
B) $\frac{1}{34}$
C) 34
D) $\frac{\sqrt{3}}{34}$
E) $\frac{17}{60}$

9. Determine two coterminal angles (one positive and one negative) for the given angle. Give your answer in degrees.

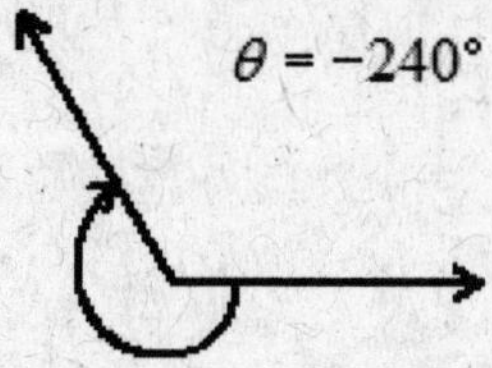

10. Rewrite the angle $\frac{\pi}{6}$ radians in degree measure.

A) 60°
B) 20°
C) 30°
D) 15°
E) 330°

11. When an airplane leaves the runway, its angle of climb is 15° and its speed is 275 feet per second. Find the plane's altitude relative to the runway in feet after 1 minute. Round your answer to the nearest foot.
A) 3417 feet
B) 2563 feet
C) 5979 feet
D) 4271 feet
E) 5125 feet

12. Determine a pair of angles (one positive and one negative) in degree measure coterminal to the angle 117°.
A) 297°, −63°
B) 477°, −243°
C) 417°, −183°
D) 234°, −117°
E) 297°, −243°

13. Determine two coterminal angles (one positive and one negative) for $\theta = \dfrac{5\pi}{6}$.

A) $\dfrac{11\pi}{6}, -\dfrac{13\pi}{6}$

B) $\dfrac{19\pi}{6}, -\dfrac{17\pi}{6}$

C) $\dfrac{17\pi}{6}, -\dfrac{7\pi}{6}$

D) $\dfrac{13\pi}{6}, -\dfrac{7\pi}{12}$

E) $\dfrac{5\pi}{3}, -\dfrac{7\pi}{18}$

14. Determine the graph of $y = \frac{1}{3}\cos(x)$.

A)

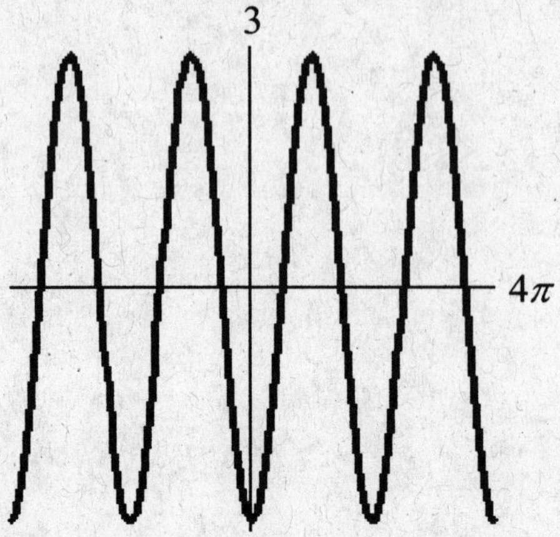

B)

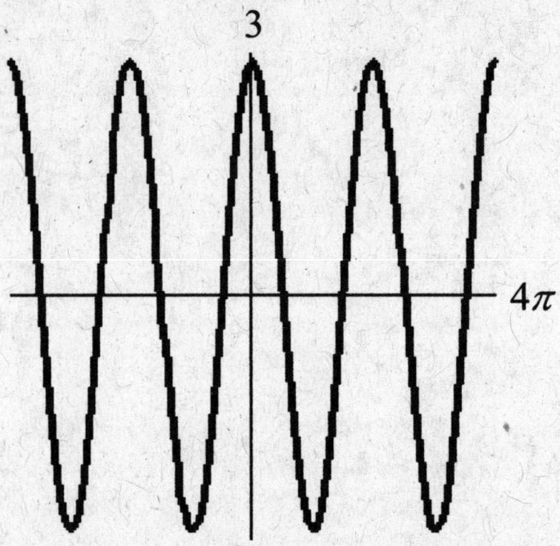

C)

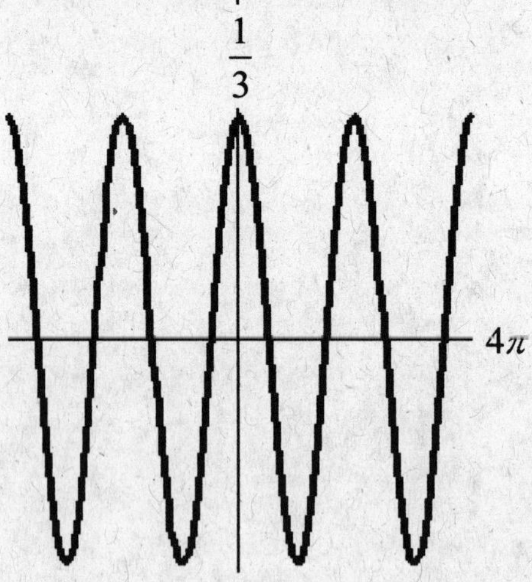

D) 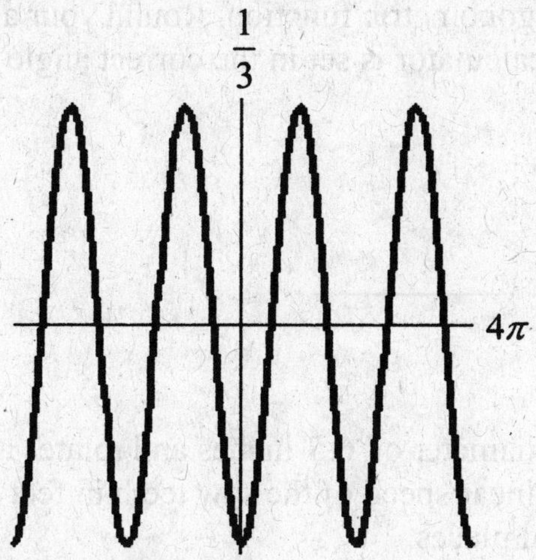

E) 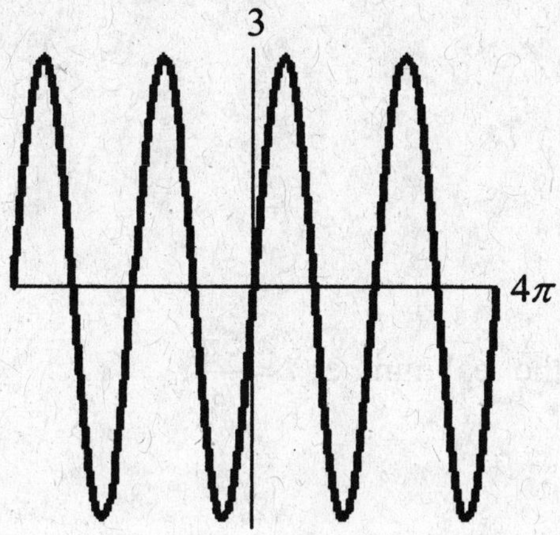

15. Find the exact value of $\sec\left(\arctan\frac{8}{15}\right)$.

A) $\frac{8}{15}$
B) $\frac{32}{15}$
C) $\frac{17}{15}$
D) $\frac{17}{32}$
E) $\frac{15}{8}$

16. Use a calculator to evaluate the trigonometric function. Round your answer to four decimal places. (Be sure the calculator is set in the correct angle mode.)
    $\tan 1.7$
    A) 0.0297
    B) 0.9917
    C) –0.1288
    D) –7.6966
    E) 1.0084

17. The circular blade of a saw has a diameter of 6.5 inches and rotates at 2000 revolutions per minute. Find the linear speed of the saw teeth in feet per second. Round your answer to two decimal places.
    A) 113.45 feet per second
    B) 680.68 feet per second
    C) 9.03 feet per second
    D) 56.72 feet per second
    E) 209.44 feet per second

18. Find the cotangent (if it exists) of the real number $t = \frac{5\pi}{4}$.
    A) $-\frac{\sqrt{2}}{2}$
    B) 1
    C) –1
    D) 0
    E) $\cot\left(\frac{5\pi}{4}\right)$ does not exist

19. Find the amplitude of $y = -\frac{\pi}{7}\cos\left(\frac{3x}{2}\right)$.

A) $-\frac{\pi}{7}$
B) $\frac{4\pi}{3}$
C) $\frac{3}{2}$
D) $\frac{1}{7}$
E) $\frac{\pi}{7}$

20. Evaluate the tangent of the angle without using a calculator.
$-120°$
A) $\sqrt{3}$
B) $-\frac{\sqrt{3}}{3}$
C) $\frac{\sqrt{3}}{3}$
D) $\frac{1}{2}$
E) 0

Answer Key

1. E
2. D
3. D
4. C
5. B
6. B
7. D
8. A
9. Answers may vary. One possible response is given below.
   $-600°, 120°$
10. C
11. D
12. B
13. C
14. C
15. C
16. D
17. D
18. B
19. E
20. A

Name: ____________________________ Date: ______________

1. Determine the exact value of $\csc\theta$ when $\cos\theta = \frac{5}{13}$ and $\cot\theta < 0$.

A) $\csc\theta = -\frac{1}{12}$

B) $\csc\theta = -\frac{13}{12}$

C) $\csc\theta = -\frac{12}{13}$

D) $\csc\theta = -\frac{13}{6}$

E) $\csc\theta = -\frac{14}{11}$

2. Find the frequency of the simple harmonic motion described by the function below.

$d = -3\cos(18\pi t)$

A) 9

B) $-\frac{2\pi}{3}$

C) 18π

D) $-\frac{2}{3}$

E) –3

3. Rewrite the given angle in radian measure as a multiple of π. (Do not use a calculator.)

72°

A) $\frac{7\pi}{5}$

B) π

C) $\frac{\pi}{5}$

D) $\frac{2\pi}{5}$

E) $\frac{3\pi}{5}$

4. Use a calculator to evaluate the function. Round your answers to four decimal places. (Be sure the calculator is in the correct angle mode.)
   $\csc 57°27'$
   A) 0.8429
   B) 1.2753
   C) 1.1863
   D) 1.8586
   E) –1.4725

5. Evaluate the sine of the angle without using a calculator.
   150°
   A) $\frac{\sqrt{2}}{2}$
   B) $-\frac{\sqrt{2}}{2}$
   C) $\frac{\sqrt{3}}{2}$
   D) $\frac{1}{2}$
   E) 0

6. Determine the period and amplitude of $y = 5\cos\left(\frac{x}{13} + \frac{\pi}{5}\right)$.
   A) period: $\frac{2\pi}{13}$; amplitude: 5
   B) period: 26π ; amplitude: 5
   C) period: 13π ; amplitude: 10
   D) period: $\frac{\pi}{13}$; amplitude: 5
   E) period: $-\frac{2\pi}{13}$; amplitude: –5

7. Use a calculator to evaluate the function. Round your answers to four decimal places. (Be sure the calculator is in the correct angle mode.)
   $\sin 78.9°$
   A) –0.3524
   B) 0.1146
   C) 0.9813
   D) 0.1925
   E) 5.0970

8. Use trigonometric identities to transform the left side of the equation into the right side. Assume all angles are positive acute angles, and show all of your work.
   $(\csc x - \cot x)(\csc x + \cot x) = 1$

9. Find the cotangent (if it exists) of the real number $t = -\frac{5\pi}{4}$.

   A) $\frac{\sqrt{2}}{2}$
   B) –1
   C) 1
   D) 0
   E) $\cot\left(-\frac{5\pi}{4}\right)$ does not exist

10. Find the exact value of the given trigonometric function of the angle θ shown in the figure. (Use the Pythagorean Theorem to find the third side of the triangle.) Find: $\sin\theta$

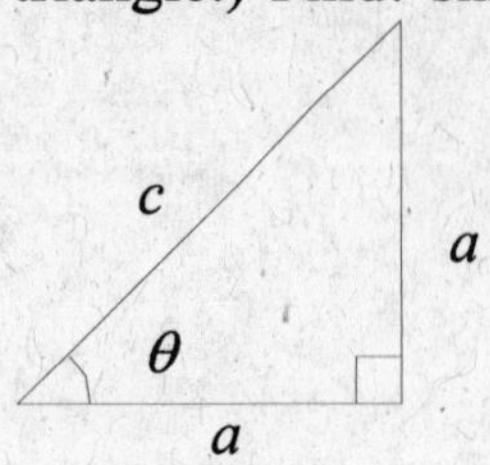

$a = 8$

A) $8\sqrt{2}$
B) $\sqrt{2}$
C) $\dfrac{\sqrt{2}}{2}$
D) 1
E) $\dfrac{\sqrt{2}}{8}$

11. Find the reference angle for the angle 6.1 radians. Round your answer to one decimal place.
A) –3.0
B) –0.2
C) 3.0
D) 0.2
E) –12.4

12. Write an algebraic expression that is equivalent to $\sin\left(\arctan\dfrac{x}{9}\right)$.

A) $\dfrac{9}{\sqrt{x^2+81}}$
B) $\dfrac{9}{x}$
C) $\dfrac{\sqrt{x^2+81}}{x}$
D) $\dfrac{\sqrt{x^2+81}}{9}$
E) $\dfrac{x}{\sqrt{x^2+81}}$

13. Find (if possible) the complement and supplement of the given angle.
    51°
    A) complement: 129°; supplement: 39°
    B) complement: 51°; supplement: 129°
    C) complement : 129°; supplement: 309°
    D) complement: 39°; supplement: 309°
    E) complement: 39°; supplement: 129°

14. A submarine, cruising at a depth $d = 65$ meters, is on a trajectory that passes directly below a ship (see figure). If θ is the angle of depression from the ship to the submarine, find the distance L from the ship to the sub when $\theta = 70°$. Round to the nearest meter.

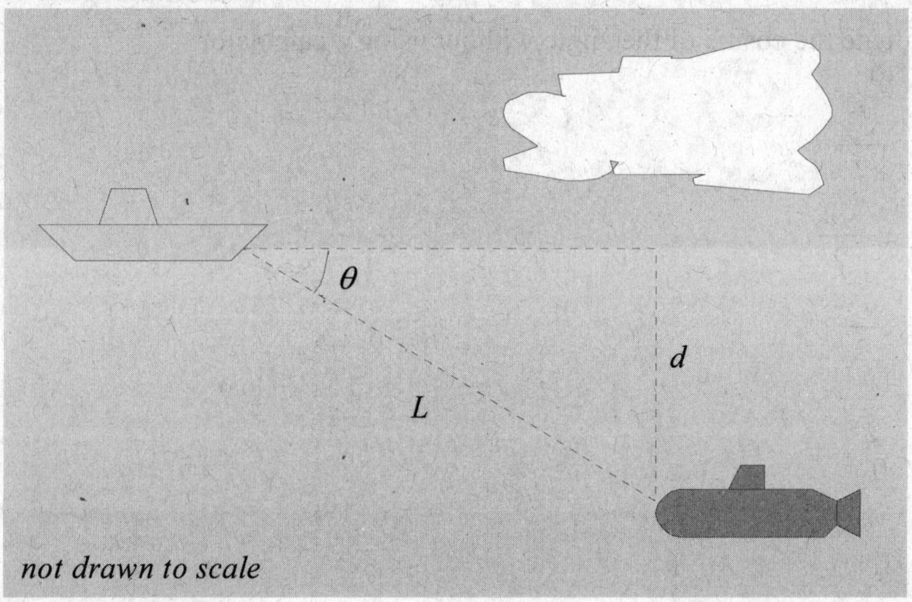

    A) $L = 0$ meters
    B) $L = 84$ meters
    C) $L = 69$ meters
    D) $L = 190$ meters
    E) $L = 24$ meters

15. Approximate $\tan^{-1}(3)$. Round your answer to four decimal places.
    A) 1.2490
    B) 0.3463
    C) –0.1425
    D) 0.8006
    E) –7.0153

16. Determine the exact value of $\arctan(-1)$.

A) $\frac{\pi}{4}$

B) $-\frac{\pi}{2}$

C) $-\frac{\pi}{4}$

D) 0

E) $\frac{\pi}{2}$

17. Evaluate the cosine of the angle without using a calculator.

30°

A) $\frac{\sqrt{2}}{2}$

B) $-\frac{\sqrt{2}}{2}$

C) $\frac{\sqrt{3}}{2}$

D) $\frac{1}{2}$

E) 0

18. Use a graphing utility to graph the function below, making sure to show at least two periods.

$\sec \frac{x}{2}$

A)

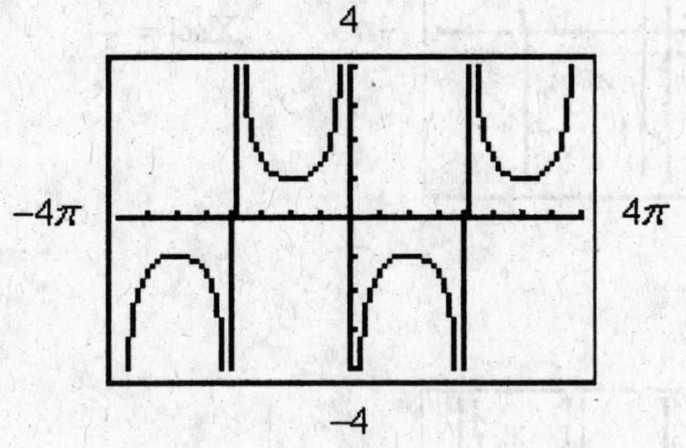

$\text{Xscl} = \frac{\pi}{2}$

B)

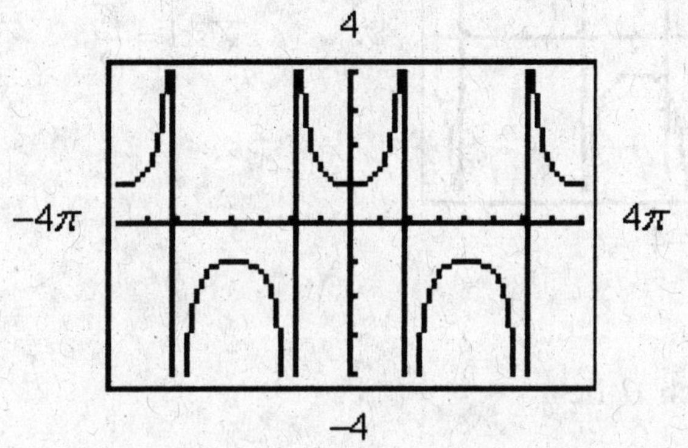

$\text{Xscl} = \frac{\pi}{2}$

C)

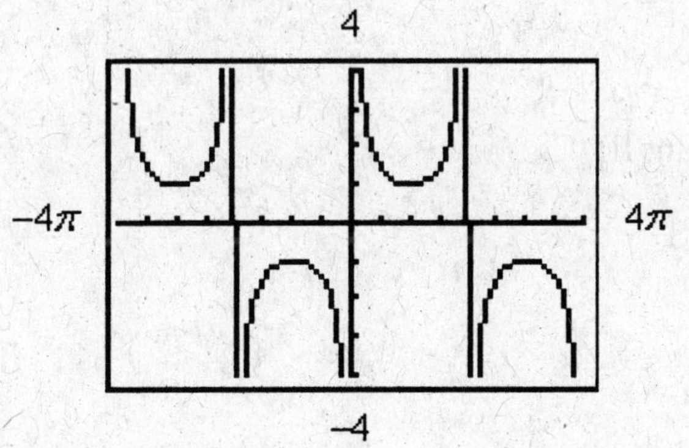

$\text{Xscl} = \frac{\pi}{2}$

D)

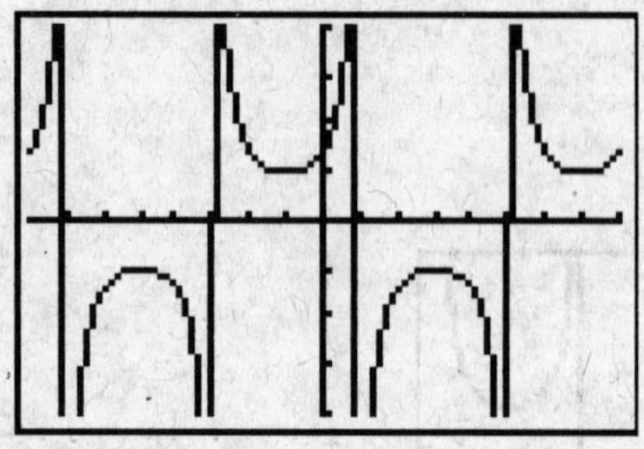

$\text{Xscl} = \dfrac{\pi}{2}$

E)

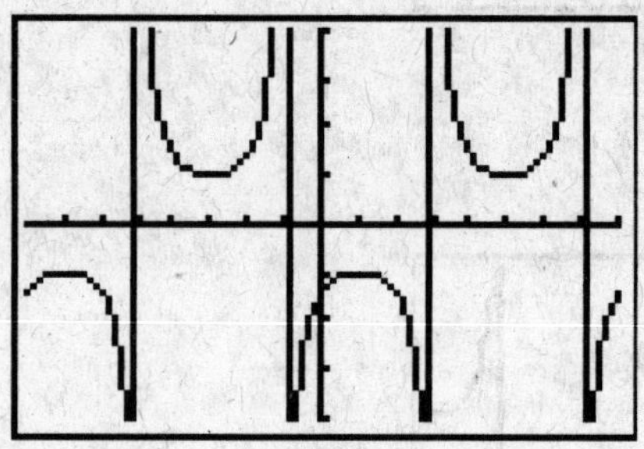

$\text{Xscl} = \dfrac{\pi}{2}$

19. State the quadrant in which θ lies.
    $\cot(\theta) < 0$ and $\sec(\theta) > 0$
    A) Quadrant III
    B) Quadrant I
    C) Quadrant IV
    D) Quadrant II
    E) Quadrant I or Quadrant III

20. The terminal side of θ lies on the the line $12x+5y=0$ in the second quadrant. Find the exact value of $\tan\theta$.

A) $\tan\theta=-\frac{5}{12}$

B) $\tan\theta=-\frac{7}{5}$

C) $\tan\theta=\frac{12}{7}$

D) $\tan\theta=-\frac{12}{5}$

E) $\tan\theta=-\frac{37}{14}$

Answer Key

1. B
2. A
3. D
4. C
5. D
6. B
7. C
8. $$\begin{aligned}(\csc x - \cot x)(\csc x + \cot x) &= \csc^2 x + \csc x \cot x - \csc x \cot x - \cot^2 x\\ &= \csc^2 x - \cot^2 x\\ &= \left(1 + \cot^2 x\right) - \cot^2 x\\ &= 1\end{aligned}$$
9. B
10. C
11. D
12. E
13. E
14. C
15. A
16. C
17. C
18. B
19. C
20. D

Name: ___________________________ Date: _____________

1. The terminal side of θ lies on the the line $15x + 8y = 0$ in the second quadrant. Find the exact value of $\tan\theta$.
   A) $\tan\theta = -\dfrac{8}{15}$
   B) $\tan\theta = -\dfrac{7}{8}$
   C) $\tan\theta = \dfrac{15}{7}$
   D) $\tan\theta = -\dfrac{15}{8}$
   E) $\tan\theta = -\dfrac{61}{31}$

2. Determine the exact value of $\cos^{-1}\left(\dfrac{\sqrt{2}}{2}\right)$.
   A) $\dfrac{\pi}{2}$
   B) $\dfrac{\pi}{4}$
   C) $\dfrac{\pi}{3}$
   D) $\dfrac{\pi}{6}$
   E) $\dfrac{3\pi}{4}$

3. Find the reference angle θ' for the given angle θ.
   $\theta = 312°$
   A) 138°
   B) –222°
   C) 58°
   D) 48°
   E) 38°

4. Find the point (x, y) on the unit circle corresponding to the real number $t = -\frac{\pi}{6}$.

A) $\left(-\frac{1}{2}, \frac{\sqrt{3}}{2}\right)$

B) $\left(\frac{\sqrt{3}}{2}, -\frac{1}{2}\right)$

C) $\left(-\frac{\sqrt{3}}{2}, \frac{1}{2}\right)$

D) $\left(\frac{1}{2}, -\frac{\sqrt{3}}{2}\right)$

E) $\left(\frac{\sqrt{2}}{2}, \frac{\sqrt{2}}{2}\right)$

5. Write an algebraic expression that is equivalent to $\tan(\arccos 3x)$.

A) $\frac{1}{3x}$

B) $\frac{\sqrt{1-9x^2}}{3x}$

C) $\sqrt{1-9x^2}$

D) $\frac{1}{\sqrt{1-9x^2}}$

E) $3x$

6. The terminal side of θ lies on the given line in the specified quadrant. Find the value of the given trigonometric function of θ by finding a point on the line.

Line	*Quadrant*	*Evaluate*:
$2x+y=0$	IV	$\csc\theta$

A) $\frac{\sqrt{5}}{2}$
B) $-\frac{2}{\sqrt{5}}$
C) $-\frac{\sqrt{5}}{2}$
D) $\frac{2}{\sqrt{5}}$
E) $-\frac{2}{1}$

7. Approximate $\sin^{-1}(-0.82)$. Round your answer to four decimal places.
A) –1.0401
B) –1.3677
C) –0.7311
D) –0.9614
E) –0.9389

8. Rewrite the angle $\frac{2\pi}{3}$ radians in degree measure.

A) 240°
B) 80°
C) 120°
D) 60°
E) 300°

9. Determine two angles (one positive and one negative, in radian measure) coterminal to the angle $\frac{\pi}{4}$.

A) $\frac{5\pi}{4}, -\frac{7\pi}{4}$

B) $\frac{17\pi}{4}, -\frac{11\pi}{4}$

C) $\frac{5\pi}{4}, -\frac{19\pi}{4}$

D) $\frac{13\pi}{4}, -\frac{11\pi}{4}$

E) $\frac{17\pi}{4}, -\frac{7\pi}{4}$

10. Determine a pair of coterminal angles (in radian measure) to the angle $\frac{\pi}{3}$.

A) $\frac{4\pi}{3}, -\frac{2\pi}{3}$

B) $\frac{7\pi}{3}, \frac{4\pi}{3}$

C) $\frac{10\pi}{3}, -\frac{2\pi}{3}$

D) $\frac{7\pi}{3}, -\frac{5\pi}{3}$

E) $\frac{7\pi}{3}, -\frac{2\pi}{3}$

11. Use the function value and constraint below to evaluate the given trigonometric function.

Function Value	*Constraint*	*Evaluate*:
$\sec\theta = -5$	$\tan\theta < 0$	$\cot\theta$

A) $-2\sqrt{6}$
B) $2\sqrt{6}$
C) $-\frac{1}{2\sqrt{6}}$
D) $-\frac{1}{5}$
E) Undefined

12. Use the figure and a straightedge to approximate the value of $\cos 1.75$.

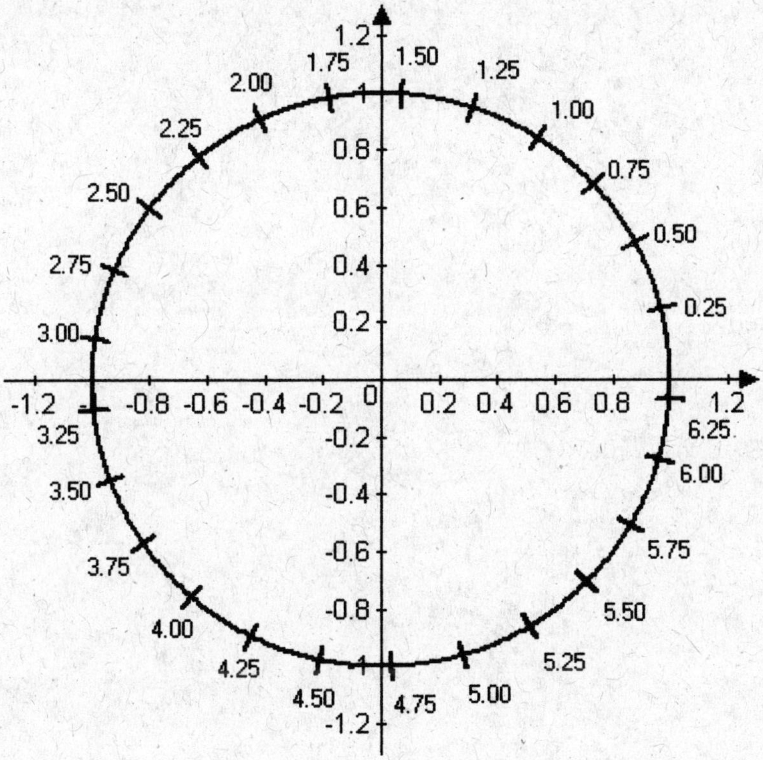

A) 1.00
B) 0.98
C) –0.18
D) –5.52
E) –5.61

13. Use a calculator to evaluate the function. Round your answers to four decimal places. (Be sure the calculator is in the correct angle mode.)

$\sec 41.3°$

A) –1.1156
B) 0.7513
C) 1.3311
D) –0.7671
E) 0.8785

14. Find the indicated trigonometric value in the specified quadrant.

Function	*Quadrant*	*Trigonometric Value*
$\csc\theta = -\dfrac{10}{9}$	III	$\tan\theta$

A) $\dfrac{9}{\sqrt{19}}$
B) $\dfrac{\sqrt{19}}{10}$
C) $\dfrac{10}{\sqrt{19}}$
D) $\dfrac{9}{10}$
E) Undefined

15. Find the exact value of the given trigonometric function of the angle θ shown in the figure. (Use the Pythagorean Theorem to find the third side of the triangle.)
Find: $\cos\theta$

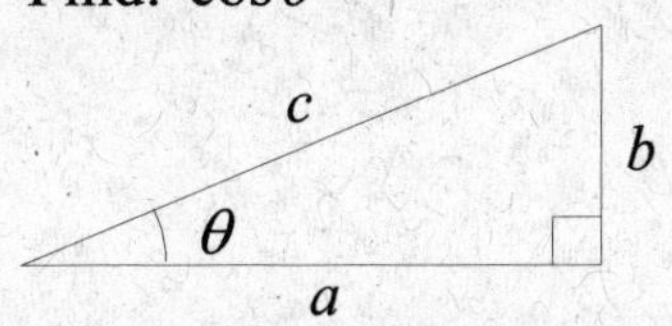

$b = 24$, $c = 51$

A) $\frac{8}{17}$
B) $\frac{15}{17}$
C) $\frac{8}{15}$
D) $\frac{15}{8}$
E) $\frac{17}{15}$

16. Evaluate the cosine of the angle without using a calculator.
$330°$

A) $\frac{\sqrt{2}}{2}$
B) $-\frac{\sqrt{2}}{2}$
C) $\frac{\sqrt{3}}{2}$
D) $\frac{1}{2}$
E) 0

17. Use the properties of inverse trigonometric functions to evaluate $\arcsin\left[\sin\left(\frac{4\pi}{9}\right)\right]$.

A) $-\frac{5\pi}{9}$
B) $\frac{4\pi}{5}$
C) $\frac{9\pi}{4}$
D) $\frac{4\pi}{9}$
E) $\frac{\pi}{9}$

18. Use a calculator to evaluate the trigonometric function. Round your answer to four decimal places. (Be sure the calculator is set to the correct angle mode.)

$\sec\left(\frac{-5\pi}{8}\right)$

A) 1.0006
B) –1.0824
C) 1.2331
D) –2.6131
E) 0.4142

19. Given the equation below, determine two solutions such that $0° \le \theta < 360°$.

$\sin\theta = -\frac{\sqrt{3}}{2}$

A) $\theta = 120°, 240°$
B) $\theta = 240°, 300°$
C) $\theta = 150°, 210°$
D) $\theta = 135°, 225°$
E) $\theta = 225°, 315°$

20. Evaluate $\sec(4.9)$. Round your answer to four decimal places.

A) –1.0179
B) 0.1865
C) –0.9825
D) 5.3616
E) 0.9792

Answer Key

1. D
2. B
3. D
4. B
5. B
6. C
7. D
8. C
9. E
10. D
11. C
12. C
13. C
14. A
15. B
16. C
17. D
18. D
19. B
20. D

Name: ____________________________ Date: ____________

1. The terminal side of θ lies on the given line in the specified quadrant. Find the value of the given trigonometric function of θ by finding a point on the line.

Line	*Quadrant*	*Evaluate*:
$y=10x$	I	$\cos\theta$

A) $\sqrt{101}$
B) $\dfrac{1}{\sqrt{101}}$
C) $\dfrac{1}{10}$
D) $\dfrac{10}{\sqrt{101}}$
E) $-\dfrac{10}{\sqrt{101}}$

2. Use a calculator to evaluate the function. Round your answers to four decimal places. (Be sure the calculator is in the correct angle mode.)
$\cot 53°30'$
A) 1.3514
B) 10.7304
C) 0.7400
D) –1.1859
E) –0.8433

3. Find (if possible) the complement and supplement of the given angle.
1.2
A) complement: 1.2; supplement: $\pi-1.2 \approx 1.9$
B) complement: $\dfrac{\pi}{2}-2.5 \approx 0.4$; no supplement
C) complement: $\dfrac{\pi}{2}-1.2 \approx 0.4$; supplement: $\pi-1.2 \approx 1.9$
D) no complement; supplement: $\pi-1.2 \approx 1.9$
E) no complement; no supplement

4. Determine the exact value, if it exists, of $\sec(-\pi)$.

A) $-\frac{\sqrt{3}}{2}$
B) -1
C) $\frac{\sqrt{3}}{2}$
D) 1
E) The value does not exist.

5. Use a calculator to evaluate $\tan 5.1$. Round your answer to four decimal places.
A) –0.5903
B) –3.1994
C) –2.4494
D) 0.0892
E) 0.5892

6. Evaluate, if possible, the given trigonometric function at the indicated value.

$\sin t, t = \frac{4\pi}{3}$

A) $-\frac{\sqrt{3}}{2}$
B) $-\frac{1}{2}$
C) $-\frac{2\sqrt{3}}{3}$
D) $\frac{\sqrt{3}}{2}$
E) not possible

7. Determine the exact value of $\csc\theta$ when $\cos\theta = \frac{11}{61}$ and $\sin\theta < 0$.

A) $\csc\theta = -\frac{1}{60}$

B) $\csc\theta = -\frac{61}{60}$

C) $\csc\theta = -\frac{60}{61}$

D) $\csc\theta = -\frac{61}{30}$

E) $\csc\theta = -\frac{62}{59}$

8. Use a calculator to evaluate $\tan 78°30'$. Round your answer to four decimal places.

A) 4.9152
B) 4.8288
C) –0.2445
D) –0.0398
E) 2.3649

9. Evaluate the trigonometric function using its period as an aid.

$\sin\left(\frac{13\pi}{3}\right)$

A) $\frac{\sqrt{3}}{2}$

B) $-\frac{\sqrt{3}}{2}$

C) $\frac{1}{2}$

D) $-\frac{1}{2}$

E) $\frac{2\sqrt{3}}{3}$

10. If the sides of a rectangular solid are as shown, and $s = 6$, determine the angle, θ, between the diagonal of the base of the solid and the diagonal of the solid. Round answer to two decimal places.

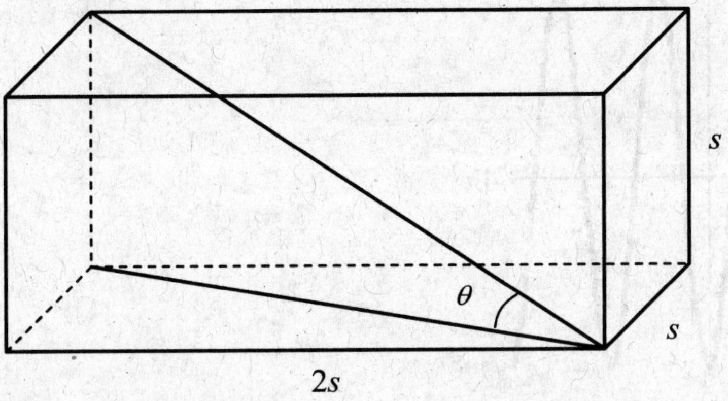

A) 17.21°
B) 19.86°
C) 21.91°
D) 24.09°
E) 26.28°

11. The terminal side of θ lies on the given line in the specified quadrant. Find the value of the given trigonometric function of θ by finding a point on the line.

Line	*Quadrant*	*Evaluate*:
$11x+3y=0$	II	$\csc\theta$

A) $-\dfrac{\sqrt{130}}{11}$

B) $\dfrac{11}{\sqrt{130}}$

C) $\dfrac{\sqrt{130}}{11}$

D) $-\dfrac{11}{\sqrt{130}}$

E) $\dfrac{11}{3}$

12. Determine the graph of $y = \frac{1}{5}\cos(x)$.

A)

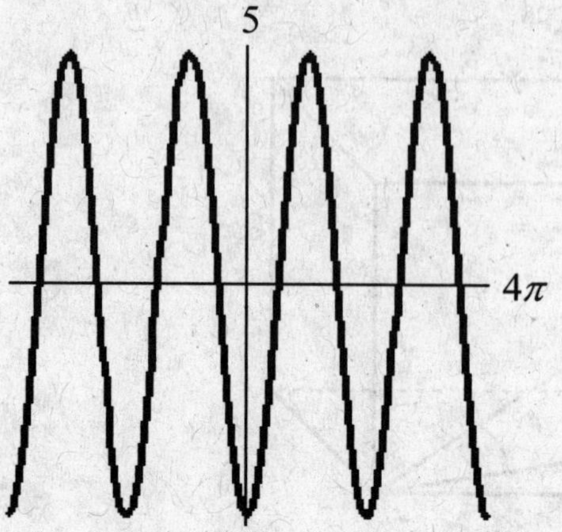

B)

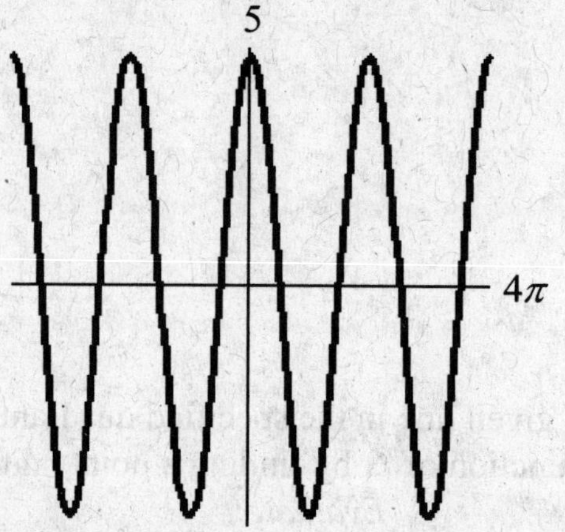

C)

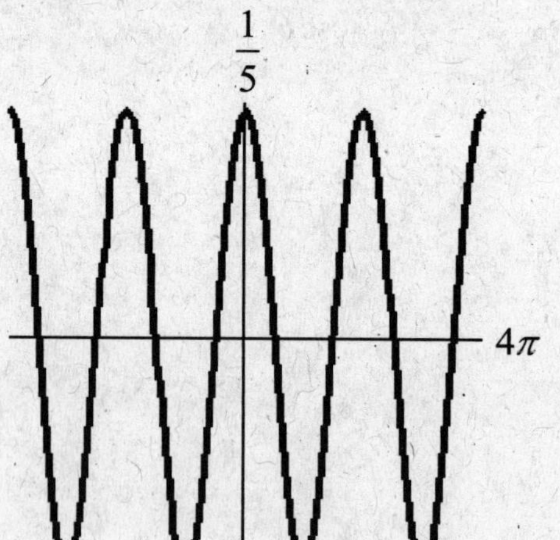

D)

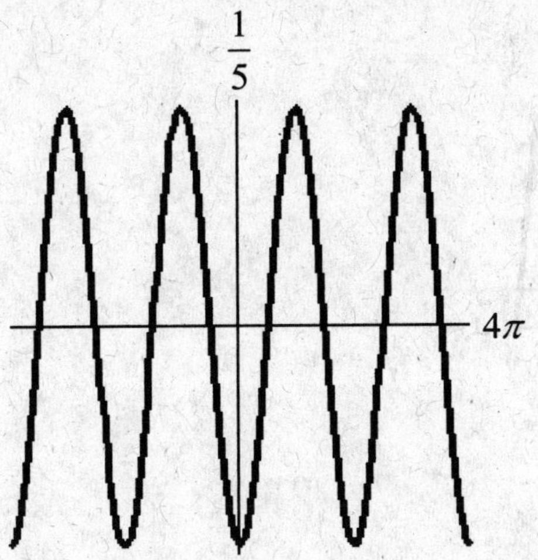

E)

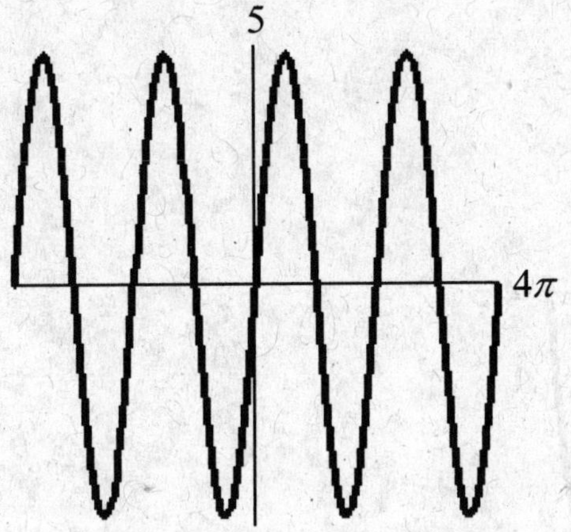

13. The circular blade of a saw has a diameter of 8 inches and rotates at 2000 revolutions per minute. Find the linear speed of the saw teeth in feet per second. Round your answer to two decimal places.
A) 139.63 feet per second
B) 837.76 feet per second
C) 11.11 feet per second
D) 69.81 feet per second
E) 209.44 feet per second

14. Determine the graph of $y = 5\sin(x)$.

A)

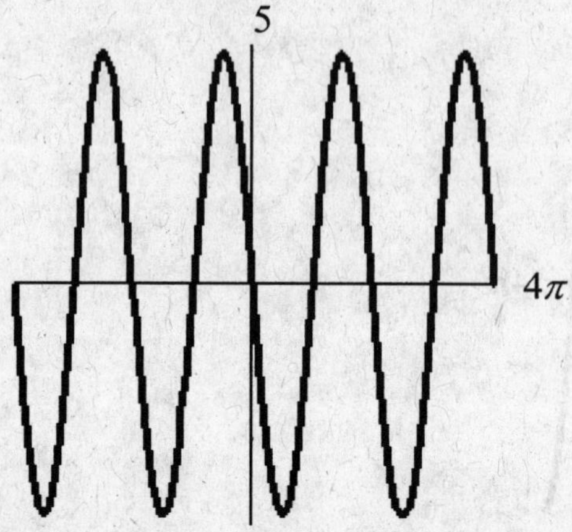

B)

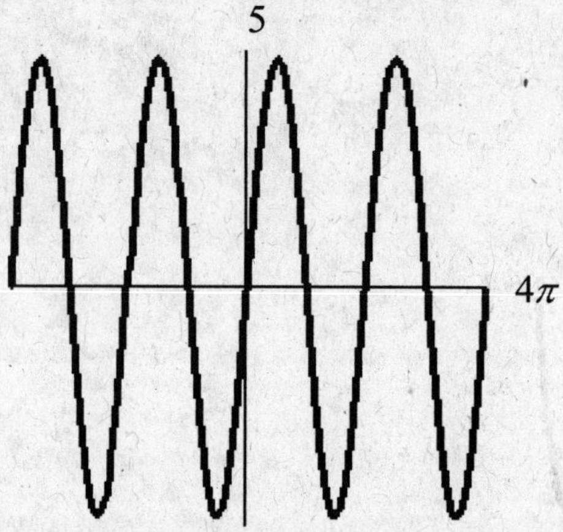

C)

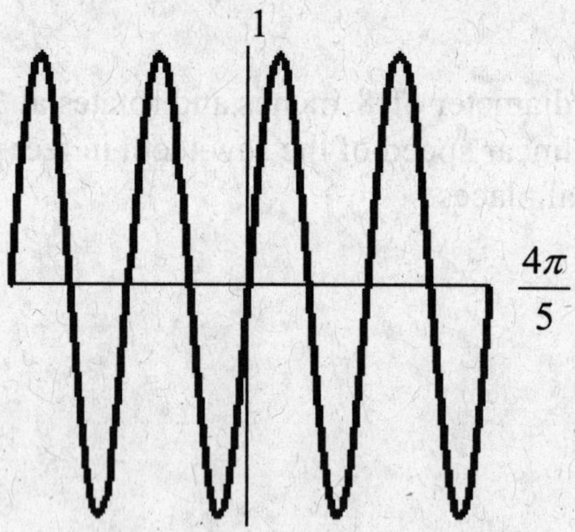

D) 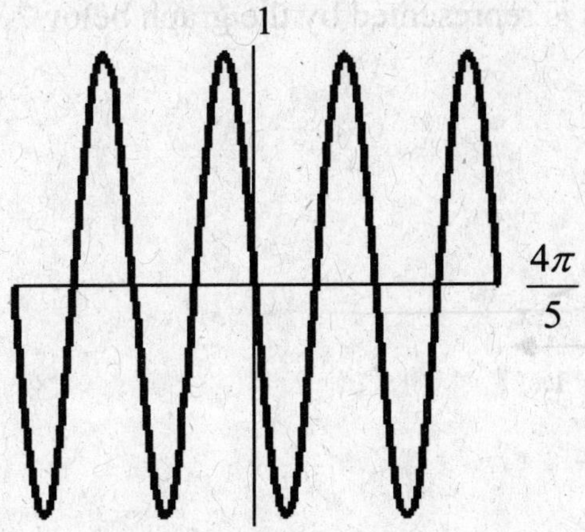

E) 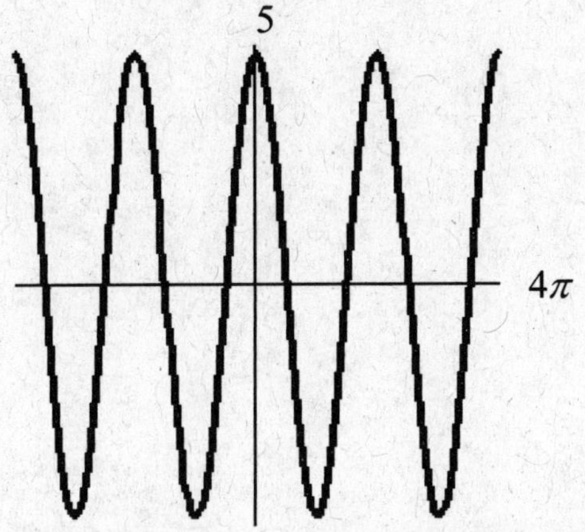

15. Find the radian measure of the central angle of a circle of radius r that intercepts an arc of length s.

    radius: $r = 10$ inches arc length: $s = 47$ inches

    A) $\frac{10}{47}$

    B) $\frac{10\pi}{47}$

    C) $\frac{47\pi}{10}$

    D) $\frac{47}{20}$

    E) $\frac{47}{10}$

16. Which of the following functions is represented by the graph below?

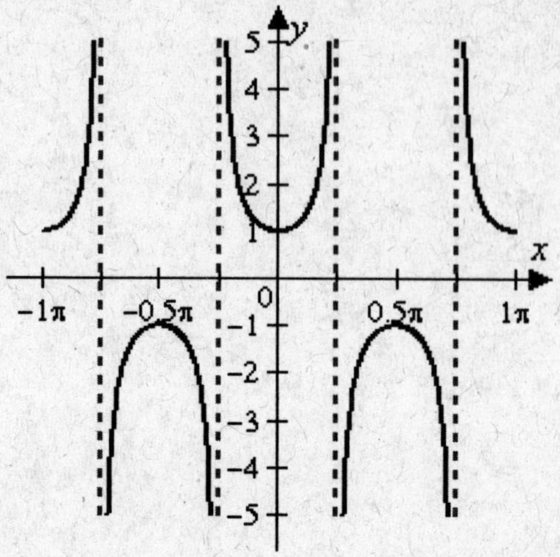

A) $y = \tan\frac{x}{3}$

B) $y = -3\csc\frac{\pi x}{2}$

C) $y = \frac{1}{3}\cot\frac{\pi x}{2}$

D) $y = \sec 2x$

E) $y = \tan(x + \pi)$

17. Find the exact value of the expression below.

$$\sin^{-1}\left[\sin\left(\frac{7\pi}{2}\right)\right]$$

A) $\frac{7\pi}{2}$

B) $-\pi$

C) $-\frac{\pi}{2}$

D) $\frac{\pi}{2}$

E) $-\frac{7\pi}{2}$

18. Evaluate $\arccos\frac{\sqrt{3}}{2}$ without using a calculator.

A) $-\frac{\pi}{6}$

B) $\frac{\pi}{4}$

C) $-\frac{3\pi}{4}$

D) $\frac{\pi}{6}$

E) $\frac{\pi}{3}$

19. Determine the quadrant in which the angle $-191°$ lies.
A) Quadrant III
B) Quadrant II
C) Quadrant IV
D) Quadrant I

20. Find the radius of a circular sector with an arc length 26 feet and a central angle $\frac{3\pi}{2}$ radians. Round your answer to two decimal places.
A) 5.52 feet
B) 12.41 feet
C) 3.40 feet
D) 0.18 foot
E) 0.08 foot

Answer Key

1. B
2. C
3. C
4. B
5. C
6. A
7. B
8. A
9. A
10. D
11. C
12. C
13. D
14. B
15. E
16. D
17. C
18. D
19. B
20. A

Name: ______________________ Date: ___________

1. Find the point (x, y) on the unit circle corresponding to the real number $t = -\frac{7\pi}{6}$.

A) $\left(\frac{1}{2}, -\frac{\sqrt{3}}{2}\right)$
B) $\left(-\frac{\sqrt{3}}{2}, \frac{1}{2}\right)$
C) $\left(\frac{\sqrt{3}}{2}, -\frac{1}{2}\right)$
D) $\left(-\frac{1}{2}, \frac{\sqrt{3}}{2}\right)$
E) $\left(-\frac{\sqrt{2}}{2}, -\frac{\sqrt{2}}{2}\right)$

2. Evaluate, if possible, the given trigonometric function at the indicated value.

$\sec t, t = \frac{5\pi}{6}$

A) $-\frac{2\sqrt{3}}{3}$
B) 2
C) -2
D) $-\frac{\sqrt{3}}{2}$
E) not possible

3. Determine the exact value, if it exists, of $\sec(-\pi)$.

A) $-\frac{\sqrt{3}}{2}$
B) -1
C) $\frac{\sqrt{3}}{2}$
D) 1
E) The value does not exist.

4. Approximate $\tan^{-1}(3)$. Round your answer to four decimal places.
   A) 1.2490
   B) 0.3463
   C) –0.1425
   D) 0.8006
   E) –7.0153

5. Given that $\cos\theta = \frac{14}{15}$, find $\csc\theta$.
   (*Hint:* Sketch a right triangle corresponding to the trigonometric function of the acute angle θ; then use the Pythagorean Theorem to determine the third side.)
   A) $\frac{\sqrt{29}}{15}$
   B) $\frac{15}{14}$
   C) $\frac{14}{\sqrt{29}}$
   D) $15\sqrt{29}$
   E) $\frac{15}{\sqrt{29}}$

6. A communications company erects a 91-foot tall cellular telephone tower on level ground. Determine the angle of depression, θ (in degrees), from the top of the tower to a point 55 feet from the base of the tower. Round answer to two decimal places.
   A) 41.85°
   B) 50.35°
   C) 58.85°
   D) 30.73°
   E) 55.73°

7. Determine the quadrant in which the angle –170.8° lies.
   A) Quadrant IV
   B) Quadrant III
   C) Quadrant I
   D) Quadrant II

8. State the quadrant in which θ lies if $\sec\theta < 0$ and $\tan\theta < 0$.
   A) Quadrant III
   B) Quadrant IV
   C) Quadrant I
   D) Quadrant II

9. Determine the quadrant in which the angle 3, given in radians, lies.
   A) 1
   B) 3
   C) 4
   D) 2

10. A plane is 65 miles south and 60 miles west of an airport. The pilot wants to fly directly to the airport. What bearing should be taken? Round your answer to the nearest degree.
    A) 317°
    B) 47°
    C) 223°
    D) 137°
    E) 43°

11. Use a calculator to evaluate $\cot\left(\frac{\pi}{6}\right)$. Round your answer to four decimal places.
    A) 1.7321
    B) 0.5773
    C) –0.1561
    D) –6.4061
    E) 5.9443

12. Determine the exact value of $\arcsin(-0.5)$.

A) $\frac{\pi}{6}$
B) 0
C) $-\frac{\pi}{6}$
D) $\frac{\pi}{3}$
E) $-\frac{\pi}{3}$

13. Determine the graph of $y = \cos\left(\dfrac{5x}{3}\right)$.

A)

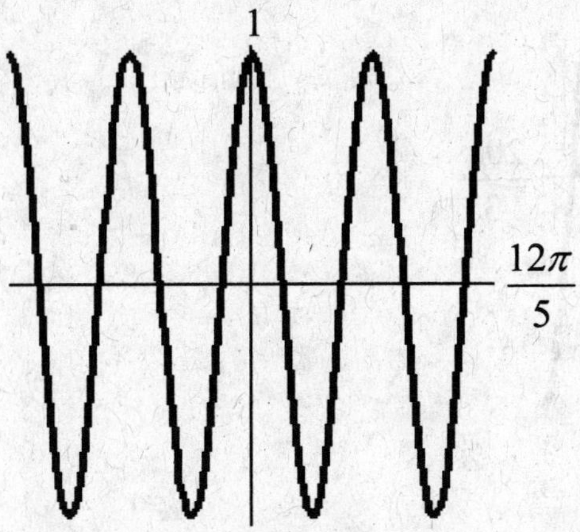

B)

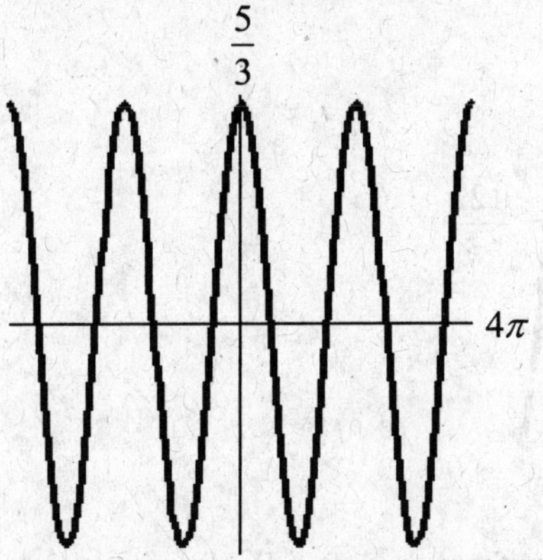

C)

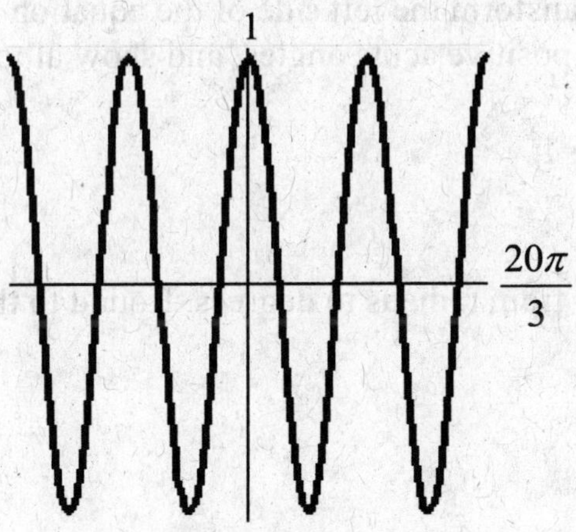

D)

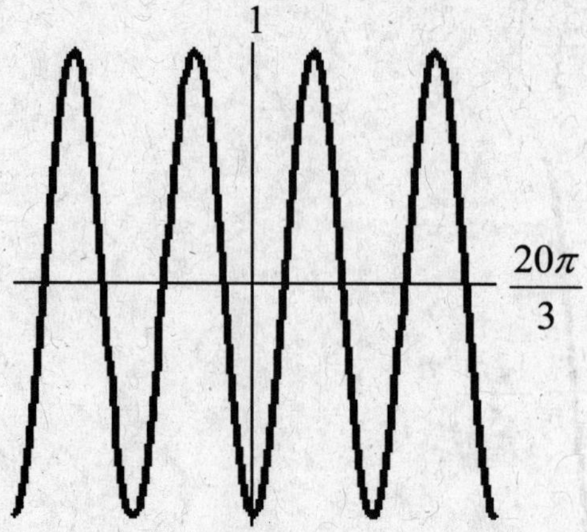

E)

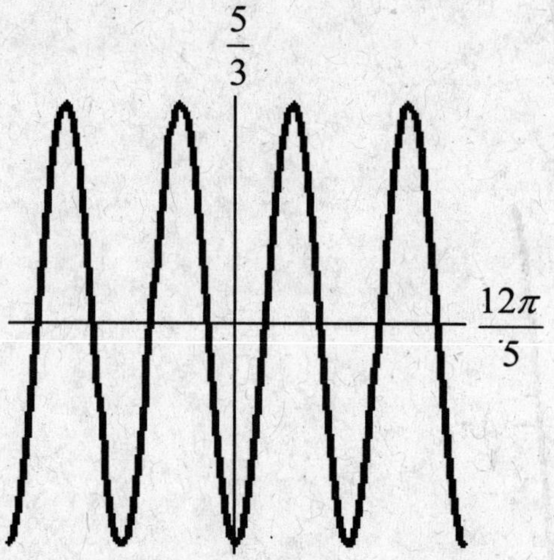

14. Use trigonometric identities to transform the left side of the equation into the right side. Assume all angles are positive acute angles, and show all of your work.

$$(\csc x - \cot x)(\csc x + \cot x) = 1$$

15. Convert the given angle measure from radians to degrees. Round to three decimal places.

$$\frac{3\pi}{7}$$

A) 0.023°
B) 77.143°
C) 154.286°
D) 38.571°
E) 420.000°

16. Find the reference angle θ' for the given angle θ.
   $\theta = -343°$
   A) 107°
   B) 73°
   C) 27°
   D) 17°
   E) 7°

17. The point $(-3,-4)$ is on the terminal side of an angle in standard position. Determine the exact value of $\sin\theta$.
   A) $\sin\theta = -\frac{3}{5}$
   B) $\sin\theta = -\frac{2}{3}$
   C) $\sin\theta = -\frac{4}{5}$
   D) $\sin\theta = -\frac{1}{4}$
   E) $\sin\theta = -\frac{5}{3}$

18. Find the exact value of $\csc\theta$, using the triangle shown in the figure below, if $a = 24$ and $b = 7$.

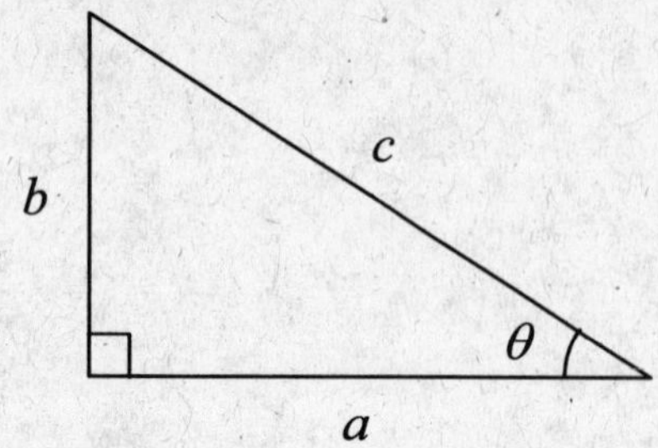

A) $\frac{25}{7}$
B) $\frac{25}{24}$
C) $\frac{24}{7}$
D) $\frac{7}{25}$
E) $\frac{24}{25}$

19. After leaving the runway, a plane's angle of ascent is $15°$ and its speed is 270 feet per second. How many minutes will it take for the airplane to climb to a height of 11,500 feet? Round answer to two decimal places.
A) 0.71 minutes
B) 2.74 minutes
C) 1.09 minutes
D) 1.51 minutes
E) 2.13 minutes

20. Sketch the graph of the function below, being sure to include at least two full periods.

$$y = \cos\left(x - \frac{\pi}{2}\right)$$

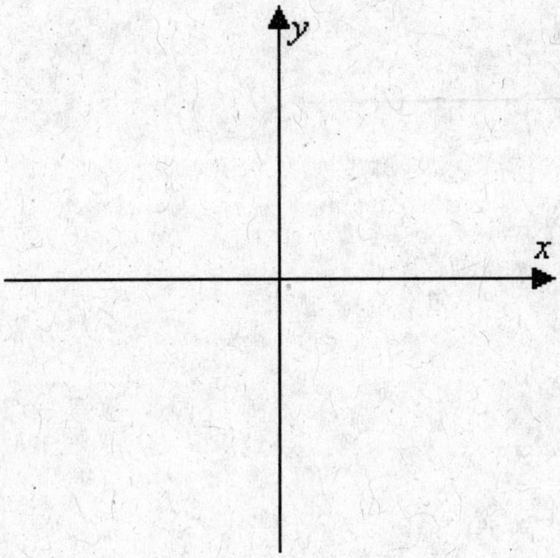

A)

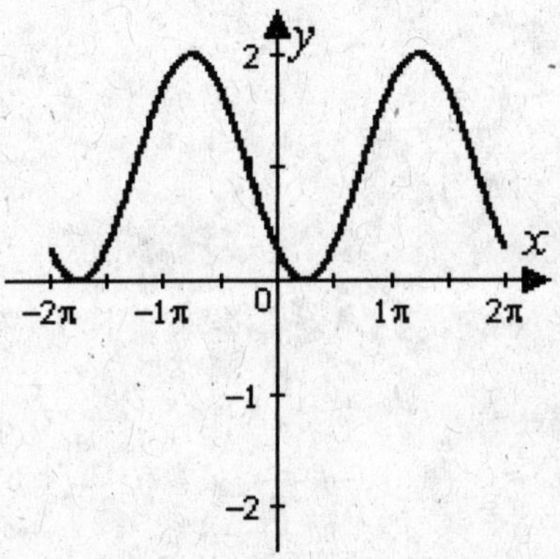

B)

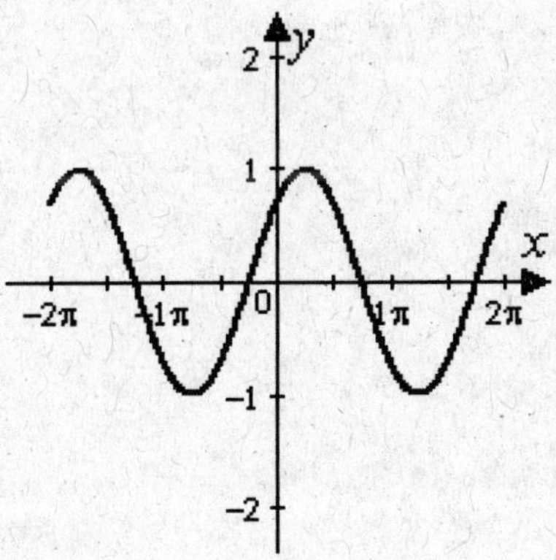

C)

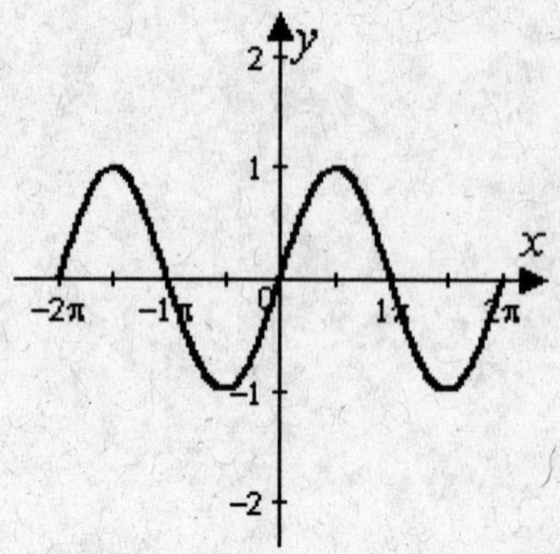

D)

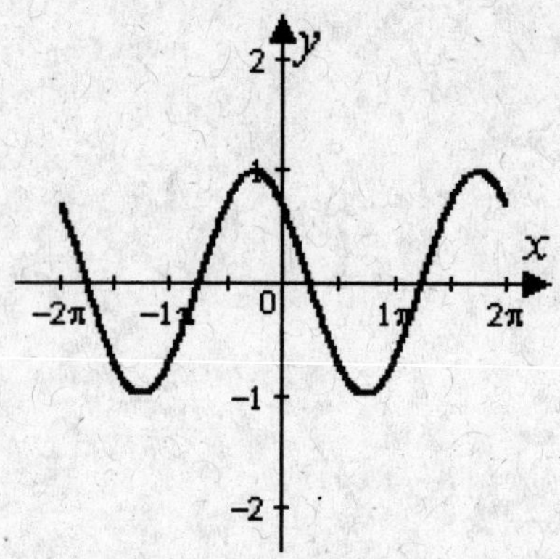

E)

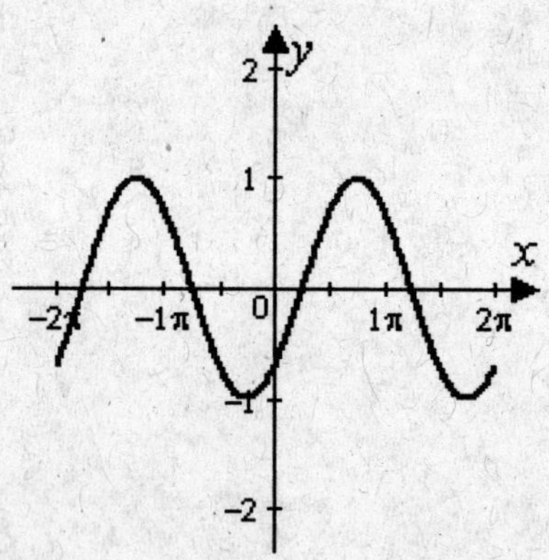

Answer Key

1. B
2. A
3. B
4. A
5. E
6. C
7. B
8. D
9. D
10. E
11. A
12. C
13. A
14.

$$\begin{aligned}(\csc x-\cot x)(\csc x+\cot x) &= \csc^2 x+\csc x\cot x-\csc x\cot x-\cot^2 x\\ &= \csc^2 x-\cot^2 x\\ &= \left(1+\cot^2 x\right)-\cot^2 x\\ &= 1\end{aligned}$$

15. B
16. D
17. C
18. A
19. B
20. C

Name: ______________________________ Date: _______________

1. Determine the quadrant in which the angle $-281.2°$ lies.
   A) Quadrant II
   B) Quadrant I
   C) Quadrant III
   D) Quadrant IV

2. Solve for y.

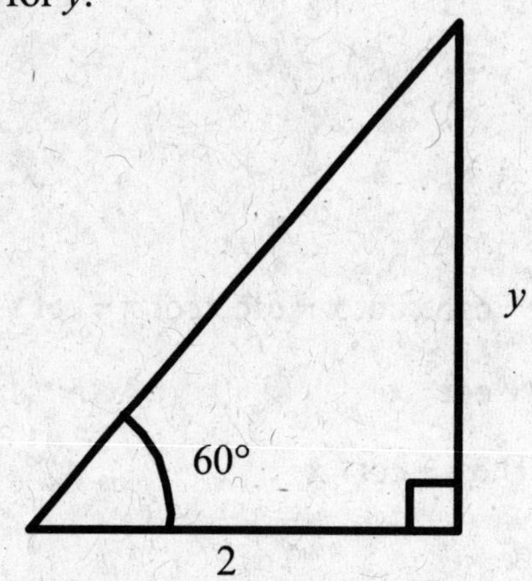

   A) $y = 2\sqrt{3}$
   B) $y = \dfrac{2\sqrt{2}}{3}$
   C) $y = \dfrac{\sqrt{3}}{2}$
   D) $y = \dfrac{2}{\sqrt{3}}$
   E) $y = 2\sqrt{2}$

3. Use a calculator to evaluate $\tan 4.8$. Round your answer to four decimal places.
   A) –3.5723
   B) –12.1349
   C) –11.3849
   D) 0.0840
   E) 0.5840

4. A certain satellite orbits 13,000 miles above Earth's surface (see figure). Find the angle of depression α from the satellite to the horizon. Assume the radius of the Earth is 4000 miles. Round your answer to the nearest hundredth of a degree.

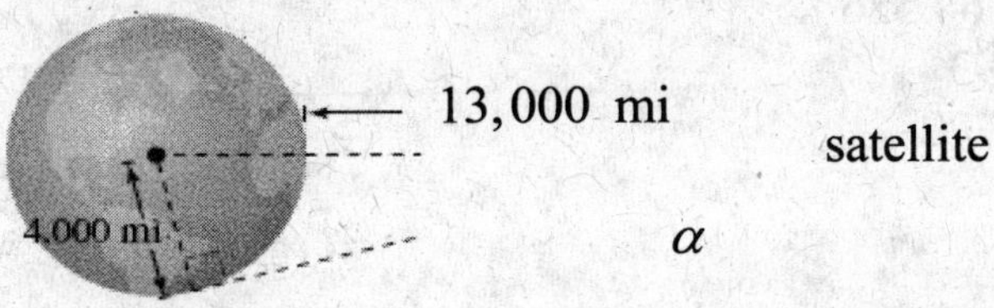

A) 13.61°
B) 17.92°
C) 72.08°
D) 76.39°
E) 13.24°

5. Find the indicated trigonometric value in the specified quadrant.

Function	*Quadrant*	*Trigonometric Value*
$\csc\theta = -\dfrac{11}{2}$	IV	$\sec\theta$

A) $\dfrac{2}{3\sqrt{13}}$
B) $\dfrac{3\sqrt{13}}{11}$
C) $\dfrac{11}{3\sqrt{13}}$
D) $\dfrac{2}{11}$
E) Undefined

6. Convert the given angle measure from degrees to radians. Round to three decimal places.

   $-365.9°$

   A) –6.386
   B) –5.863
   C) –20964.526
   D) –3.193
   E) –12.772

7. Determine the exact value of $\arctan(-1)$.

   A) $\frac{\pi}{4}$
   B) $-\frac{\pi}{2}$
   C) $-\frac{\pi}{4}$
   D) 0
   E) $\frac{\pi}{2}$

8. Approximate $\sin^{-1}(-0.51)$. Round your answer to four decimal places.

   A) –1.8685
   B) –2.0484
   C) –0.4882
   D) –0.5352
   E) –0.9249

9. The point $(-8,-15)$ is on the terminal side of an angle in standard position. Determine the exact value of $\sin\theta$.
   A) $\sin\theta=-\frac{8}{17}$
   B) $\sin\theta=-\frac{9}{8}$
   C) $\sin\theta=-\frac{15}{17}$
   D) $\sin\theta=-\frac{2}{15}$
   E) $\sin\theta=-\frac{17}{8}$

10. Determine a pair of coterminal angles (in radian measure) to the angle $\frac{\pi}{2}$.
   A) $\frac{3\pi}{2}, -\frac{\pi}{2}$
   B) $\frac{5\pi}{2}, \frac{3\pi}{2}$
   C) $\frac{7\pi}{2}, -\frac{\pi}{2}$
   D) $\frac{5\pi}{2}, -\frac{3\pi}{2}$
   E) $\frac{5\pi}{2}, -\frac{\pi}{2}$

11. Evaluate the tangent of the angle without using a calculator.
   $-300°$
   A) $\sqrt{3}$
   B) $-\frac{\sqrt{3}}{3}$
   C) $\frac{\sqrt{3}}{3}$
   D) $\frac{1}{2}$
   E) 0

12. Determine the exact value of the tangent of the quadrant angle $\frac{3\pi}{2}$.

A) undefined
B) 0
C) $\frac{\sqrt{3}}{2}$
D) $\frac{\sqrt{2}}{2}$
E) $\frac{1}{2}$

13. The point $(8,5)$ is on the terminal side of an angle in standard position. Determine the exact value of $\sin\theta$.

A) $\frac{5}{8}$
B) $\frac{5}{\sqrt{89}}$
C) $\frac{8}{5}$
D) $\frac{5}{13}$
E) $\frac{5}{\sqrt{13}}$

14. Determine the exact value, if it exists, of $\sec(-\pi)$.

A) $-\frac{\sqrt{3}}{2}$
B) -1
C) $\frac{\sqrt{3}}{2}$
D) 1
E) The value does not exist.

15. Downtown Sardis City is located due north of a straight segment of train track oriented in an east-west direction (see map below). A passenger on a train that is traveling from west to east notes that downtown Sardis City is visible at an angle $A = 36^\circ$ to the left of the tracks. After traveling a distance $d = 10$ kilometers, the passenger notes that the angle to Sardis City is $B = 41.5^\circ$. Estimate the distance from the track to downtown Sardis City. Round to the nearest kilometer.

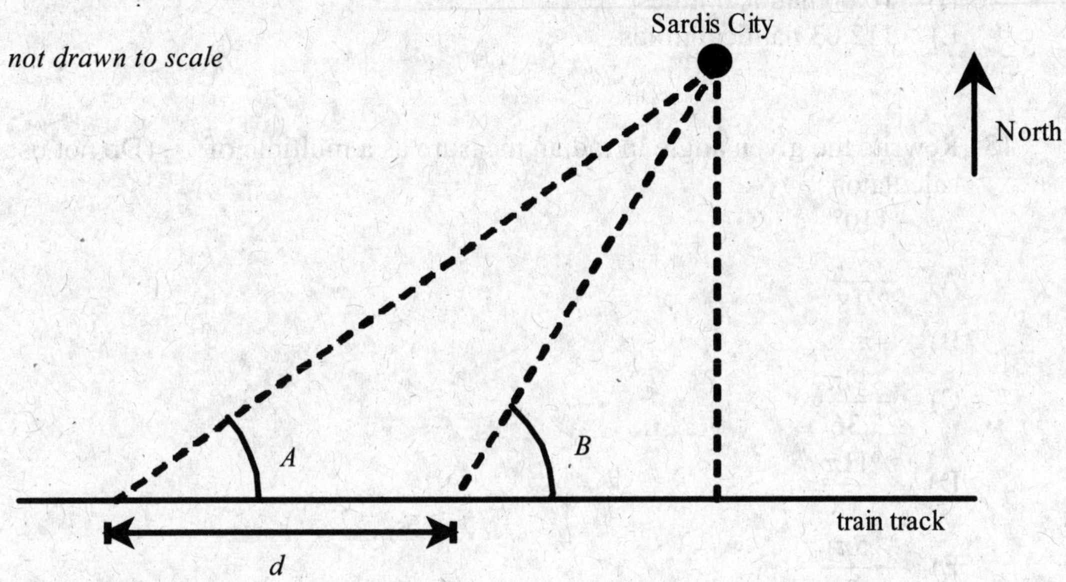

A) 43 km
B) 41 km
C) 44 km
D) 45 km
E) 46 km

16. Determine two coterminal angles (one positive and one negative) for the given angle. Give your answer in degrees.
$\theta = 290^\circ$

17. A ship leaves port at noon and has a bearing of S 35° W. The ship sails at 20 knots. How many nautical miles south will the ship have traveled by 8:00 P.M.? Round your answer to two decimals.
A) 91.77 nautical miles
B) 22.94 nautical miles
C) 131.06 nautical miles
D) 16.06 nautical miles
E) 112.03 nautical miles

18. Rewrite the given angle in radian measure as a multiple of π. (Do not use a calculator.)
-110°
A) $-\frac{7\pi}{18}$
B) $-\pi$
C) $-\frac{11\pi}{36}$
D) $-\frac{11\pi}{18}$
E) $-\frac{5\pi}{9}$

19. The circular blade of a saw has a diameter of 7 inches and rotates at 2500 revolutions per minute. Find the angular speed in radians per second.
A) $\frac{875\pi}{3}$ radians per second
B) $\frac{250\pi}{3}$ radians per second
C) $\frac{1750\pi}{3}$ radians per second
D) $\frac{125\pi}{3}$ radians per second
E) 14π radians per second

20. Find (if possible) the supplement of $\dfrac{9\pi}{11}$.

A) $\dfrac{2\pi}{11}$

B) $\dfrac{4\pi}{11}$

C) $\dfrac{10\pi}{11}$

D) $\dfrac{9\pi}{22}$

E) not possible

Answer Key

1. B
2. A
3. C
4. D
5. C
6. A
7. C
8. D
9. C
10. D
11. A
12. A
13. B
14. B
15. B
16. Answers may vary. One possible response is given below.
    $-70°, 650°$
17. C
18. D
19. B
20. A

Name: ______________________________ Date: ____________

1. Find the exact value of $\tan(u+v)$ given that $\sin u = -\frac{3}{5}$ and $\cos v = \frac{24}{25}$. (Both u and v are in Quadrant IV.)

   A) $\tan(u+v) = \frac{41}{75}$

   B) $\tan(u+v) = \frac{38}{75}$

   C) $\tan(u+v) = -\frac{4}{3}$

   D) $\tan(u+v) = \frac{89}{75}$

   E) $\tan(u+v) = -\frac{39}{25}$

2. Find all solutions of the following equation in the interval $[0, 2\pi)$.

   $\csc^2 x = \cot x + 1$

   A) $x = \frac{\pi}{4}, \frac{3\pi}{4}, \frac{4\pi}{3}$

   B) $x = \frac{\pi}{2}, \frac{3\pi}{2}$

   C) $x = 0, \frac{\pi}{2}, \pi, \frac{3\pi}{2}$

   D) $x = \frac{\pi}{4}, \frac{\pi}{2}, \frac{5\pi}{4}, \frac{3\pi}{2}$

   E) $x = \frac{\pi}{4}, \frac{4\pi}{3}, \frac{11\pi}{6}$

3. Use the figure below to find the exact value of the given trigonometric expression.

$\cot \frac{x}{2}$

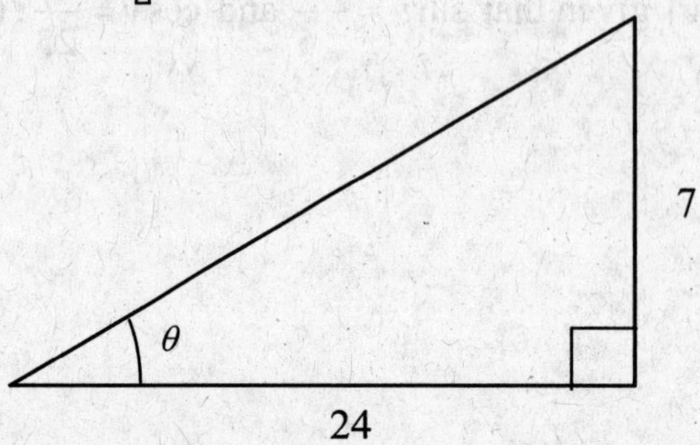

A) $\cot \frac{x}{2} = \frac{\sqrt{2}}{10}$

B) $\cot \frac{x}{2} = \frac{1}{7}$

C) $\cot \frac{x}{2} = 7$

D) $\cot \frac{x}{2} = \frac{7\sqrt{2}}{10}$

E) $\cot \frac{x}{2} = \frac{7}{12}$

4. Find the exact solutions of the given equation in the interval $[0, 2\pi)$.

$2\sin^2 x + \sin x = 1$

A) $x = 0, \frac{\pi}{3}, \frac{2\pi}{3}, \pi, \frac{4\pi}{3}$

B) $x = \frac{\pi}{6}, \frac{\pi}{2}, \frac{5\pi}{6}, \frac{3\pi}{2}$

C) $x = 0, \frac{\pi}{2}, \pi, \frac{3\pi}{2}$

D) $x = \frac{\pi}{6}, \frac{5\pi}{6}, \frac{3\pi}{2}$

E) $x = \frac{\pi}{4}, \frac{3\pi}{4}, \frac{5\pi}{4}, \frac{7\pi}{4}$

5. Find the exact value of $\cos(u-v)$ given that $\sin u = -\dfrac{9}{41}$ and $\cos v = \dfrac{15}{17}$. (Both u and v are in Quadrant IV.)

A) $\cos(u-v) = -\dfrac{194}{697}$

B) $\cos(u-v) = -\dfrac{39}{697}$

C) $\cos(u-v) = \dfrac{568}{697}$

D) $\cos(u-v) = -\dfrac{528}{697}$

E) $\cos(u-v) = \dfrac{672}{697}$

6. Use a double angle formula to rewrite the following expression.

$-16\sin(x)\cos(x)$

A) $\sin(-16x)$

B) $2\sin(-8x)$

C) $-8\cos(2x)$

D) $-8\sin(2x)$

E) $\cos(-16x)$

7. Find the exact value of $\cos(u+v)$ given that $\sin u = \dfrac{5}{13}$ and $\cos v = -\dfrac{4}{5}$. (Both u and v are in Quadrant II.)

A) $\cos(u+v) = \dfrac{28}{65}$

B) $\cos(u+v) = -\dfrac{16}{65}$

C) $\cos(u+v) = \dfrac{33}{65}$

D) $\cos(u+v) = -\dfrac{28}{65}$

E) $\cos(u+v) = \dfrac{56}{65}$

8. Use the figure below to determine the exact value of the given function.

$\sin 2\theta$

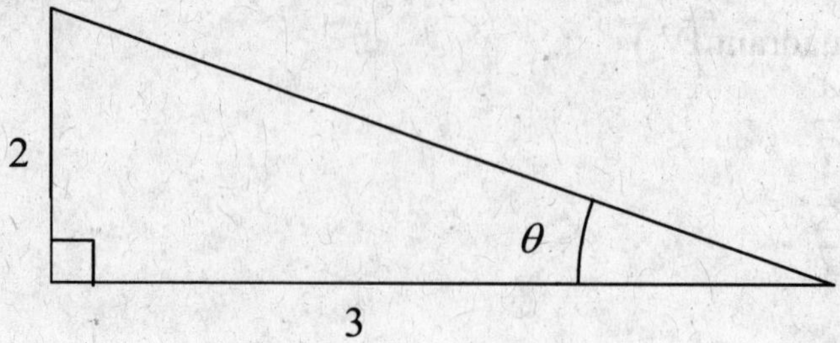

A) $\sin 2\theta = \frac{5}{13}$

B) $\sin 2\theta = \frac{7}{13}$

C) $\sin 2\theta = \frac{12}{5}$

D) $\sin 2\theta = \frac{9}{13}$

E) $\sin 2\theta = \frac{12}{13}$

9. Use the graph of the function $f(x) = -2\cos(x) + \sin(x)$ to approximate the maximum points of the graph in the interval $[0, 2\pi]$. Round your answer to one decimal.

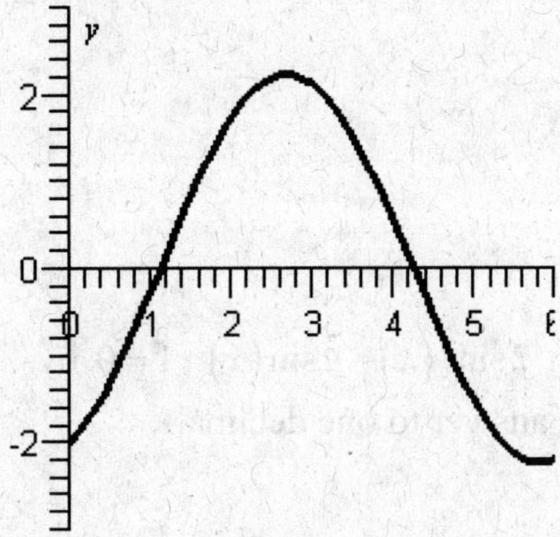

A) $(2.7, 2.2)$, $(6.3, -1.8)$
B) $(-1.8, 6.3)$, $(2.7, 2.2)$
C) $(2.7, -1.8)$, $(6.3, 2.2)$
D) $(-1.8, 6.3)$, $(2.2, 2.7)$
E) $(2.2, 2.7)$, $(6.3, -1.8)$

10. Expand the expression below and use fundamental trigonometric identities to simplify.

$$(\sin(\omega) + \cos(\omega))^2$$

A) $\sin^2(\omega) + \cos^2(\omega)$
B) $2\tan(\omega) + 1$
C) $2\sin(\omega)\cos(\omega) + 1$
D) 1
E) $2\cot(\omega) + 1$

11. Write the given expression as the sine of an angle.
$\sin 85° \cos 50° + \sin 50° \cos 85°$
A) $\sin(-100°)$
B) $\sin(135°)$
C) $\sin(35°)$
D) $\sin(85°)$
E) $\sin(50°)$

12. Approximate the solutions of the equation $2\sin^2(x) - 4\sin(x) + 1 = 0$ by considering its graph below. Round your answer to one decimal.

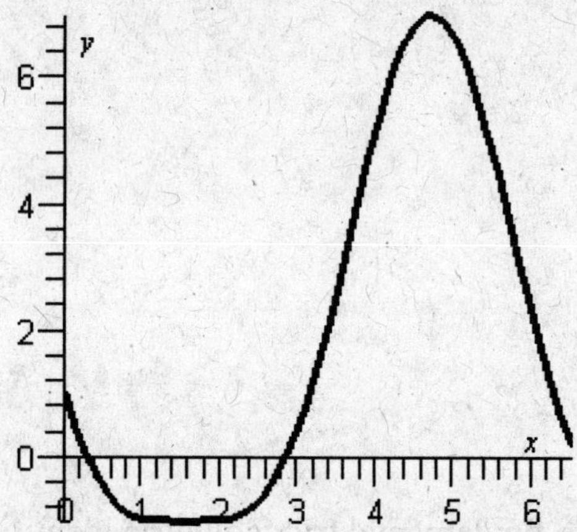

A) 0.3, 2.8
B) 0.3, 1.0
C) 1.0, 4.6
D) 0.3, 4.6
E) 2.8, 4.6

13. Which of the following is a solution to the given equation?

$2\cos x + \sqrt{3} = 0$

A) $x = \frac{2\pi}{3}$

B) $x = \frac{\pi}{4}$

C) $x = \frac{\pi}{6}$

D) $x = \frac{7\pi}{6}$

E) $x = \frac{7\pi}{4}$

14. Verify the identity shown below.

$$\frac{1-\sin\theta}{1+\sin\theta} = 2\sec^2\theta - 2\sec\theta\tan\theta - 1$$

15. Find the exact solutions of the given equation in the interval $[0, 2\pi)$.

$\sin 2x = \sin x$

A) $x = \frac{2\pi}{3}, \pi, \frac{4\pi}{3}$

B) $x = 0, \frac{\pi}{3}, \pi, \frac{5\pi}{3}$

C) $x = \frac{\pi}{2}, \frac{7\pi}{6}, \frac{11\pi}{6}$

D) $x = 0, \frac{\pi}{3}, \frac{2\pi}{3}, \pi, \frac{4\pi}{3}, \frac{5\pi}{3}$

E) $x = 0$

16. Solve the multi-angle equation below.

$$\sin(2x) = \frac{\sqrt{3}}{2}$$

A) $\frac{\pi}{8} + n\pi, \ \frac{\pi}{4} + n\pi$, where n is an integer

B) $\frac{\pi}{6} + n\pi, \ \frac{\pi}{4} + n\pi$, where n is an integer

C) $\frac{\pi}{6} + 2n\pi, \ \frac{\pi}{3} + 2n\pi$, where n is an integer

D) $\frac{\pi}{8} + 2n\pi, \ \frac{\pi}{4} + 2n\pi$, where n is an integer

E) $\frac{\pi}{6} + n\pi, \ \frac{\pi}{3} + n\pi$, where n is an integer

17. Which of the following is a solution to the given equation?

$$2\sin x - 1 = 0$$

A) $x = \frac{7\pi}{6}$

B) $x = \frac{2\pi}{3}$

C) $x = \frac{3\pi}{4}$

D) $x = \frac{5\pi}{6}$

E) $x = \frac{7\pi}{4}$

18. Simplify the given expression algebraically.

$$\cos(\pi + x)$$

A) $\sin x$

B) $-\sin x$

C) $\cos x$

D) $-\cos x$

E) 1

19. Determine which of the following are trigonometric identities.

I. $\dfrac{\cos(t)-\cos(s)}{\sin(t)+\sin(s)}+\dfrac{\sin(t)-\sin(s)}{\cos(t)+\cos(s)}=0$

II. $\dfrac{\cos(t)+\cos(s)}{\sin(t)+\sin(s)}+\dfrac{\sin(t)+\sin(s)}{\cos(t)+\cos(s)}=1$

III. $\dfrac{\cos(t)+\sin(s)}{\cos(t)\sin(s)}=\cos(s)+\sin(t)$

A) I is the only identity.
B) I, II, and III are identities.
C) II is the only identity.
D) II and II are the only identities.
E) I and II are the only identities.

20. Solve the following equation.

$3\csc^2(x)-4=0$

A) $\dfrac{\pi}{6}+\pi n,\ \dfrac{5\pi}{6}+\pi n,$ where n is an integer

B) $\dfrac{\pi}{6}+2\pi n,\ \dfrac{5\pi}{6}+2\pi n,$ where n is an integer

C) $\dfrac{\pi}{3}+\pi n,\ \dfrac{2\pi}{3}+\pi n,$ where n is an integer

D) $\dfrac{\pi}{3}+2\pi n,\ \dfrac{5\pi}{6}+2\pi n,$ where n is an integer

E) $\dfrac{\pi}{3}+2\pi n,\ \dfrac{2\pi}{3}+2\pi n,$ where n is an integer

Answer Key

1. C
2. D
3. C
4. D
5. E
6. D
7. C
8. E
9. A
10. C
11. B
12. A
13. D
14. $$\begin{aligned}\frac{1-\sin\theta}{1+\sin\theta} &= \frac{1-\sin\theta}{1+\sin\theta}\cdot\frac{1-\sin\theta}{1-\sin\theta}\\ &= \frac{(1-\sin\theta)^2}{1-\sin^2\theta}\\ &= \frac{1-2\sin\theta+\sin^2\theta}{\cos^2\theta}\\ &= \frac{1}{\cos^2\theta}-\frac{2\sin\theta}{\cos^2\theta}+\frac{\sin^2\theta}{\cos^2\theta}\\ &= \frac{1}{\cos^2\theta}-\frac{2\sin\theta}{\cos\theta}\cdot\frac{1}{\cos\theta}+\frac{\sin^2\theta}{\cos^2\theta}\\ &= \sec^2\theta-2\tan\theta\sec\theta+\tan^2\theta\\ &= \sec^2\theta-2\tan\theta\sec\theta+\sec^2\theta-1\\ &= 2\sec^2\theta-2\tan\theta\sec\theta-1\end{aligned}$$
15. B
16. E
17. D
18. D
19. A
20. C

Name: ____________________________ Date: _____________

1. Verify the identity shown below.

$$\frac{\tan\alpha+\cot\beta}{\tan\alpha\cot\beta}=\tan\beta+\cot\alpha$$

2. Which of the following is a solution to the given equation?

$\sec x-2=0$

A) $x=\frac{7\pi}{6}$

B) $x=\frac{\pi}{4}$

C) $x=\frac{5\pi}{6}$

D) $x=\frac{5\pi}{3}$

E) $x=\frac{7\pi}{4}$

3. Determine which of the following are trigonometric identities.

I. $\frac{\sin(y)-\sin(x)}{\cos(y)+\cos(x)}+\frac{\cos(y)-\cos(x)}{\sin(y)+\sin(x)}=0$

II. $\frac{\sin(y)+\sin(x)}{\cos(y)+\cos(x)}+\frac{\cos(y)+\cos(x)}{\sin(y)+\sin(x)}=1$

III. $\frac{\sin(y)+\cos(x)}{\sin(y)\cos(x)}=\sin(x)+\cos(y)$

A) I is the only identity.
B) II and II are the only identities.
C) I, II, and III are identities.
D) I and III are the only identities.
E) II is the only identity.

4. Verify the given identity.

$$\frac{\sin u + \sin v}{\cos u + \cos v} = \tan \frac{1}{2}(u + v)$$

5. Solve the following equation.

$$\csc^2(x) - 4 = 0$$

A) $\dfrac{\pi}{3} + \pi n,\ \dfrac{2\pi}{3} + \pi n,$ where n is an integer

B) $\dfrac{\pi}{3} + 2\pi n,\ \dfrac{2\pi}{3} + 2\pi n,$ where n is an integer

C) $\dfrac{\pi}{6} + \pi n,\ \dfrac{5\pi}{6} + \pi n,$ where n is an integer

D) $\dfrac{\pi}{6} + 2\pi n,\ \dfrac{2\pi}{3} + 2\pi n,$ where n is an integer

E) $\dfrac{\pi}{6} + 2\pi n,\ \dfrac{5\pi}{6} + 2\pi n,$ where n is an integer

6. Solve the multi-angle equation below.

$$\sin(2x) = -\frac{\sqrt{2}}{2}$$

A) $\dfrac{5\pi}{3} + n\pi,\ \dfrac{7\pi}{3} + n\pi,$ where n is an integer

B) $\dfrac{5\pi}{8} + n\pi,\ \dfrac{7\pi}{3} + n\pi,$ where n is an integer

C) $\dfrac{5\pi}{8} + 2n\pi,\ \dfrac{7\pi}{8} + 2n\pi,$ where n is an integer

D) $\dfrac{5\pi}{3} + 2n\pi,\ \dfrac{7\pi}{3} + 2n\pi,$ where n is an integer

E) $\dfrac{5\pi}{8} + n\pi,\ \dfrac{7\pi}{8} + n\pi,$ where n is an integer

7. Write the given expression as the cosine of an angle.
$\cos 30° \cos 50° + \sin 30° \sin 50°$
A) $\cos(50°)$
B) $\cos(-20°)$
C) $\cos(80°)$
D) $\cos(30°)$
E) $\cos(-100°)$

8. Write the given expression as the sine of an angle.
$\sin 75° \cos 35° - \sin 35° \cos 75°$
A) $\sin(-70°)$
B) $\sin(40°)$
C) $\sin(110°)$
D) $\sin(75°)$
E) $\sin(35°)$

9. Verify the identity shown below.
$$\sec^2\left(\frac{\pi}{2} - y\right) - 1 = \cot^2 y$$

10. Factor; then use fundamental identities to simplify the expression below and determine which of the following is *not* equivalent.
$\tan^3 x - \tan^2 x + \tan x - 1$
A) $\dfrac{\sin x - \cos x}{\cos^3 x}$
B) $\sec^2 x\left(\dfrac{\sin x - \cos x}{\cos x}\right)$
C) $\dfrac{\sin x}{\cos^3 x} - \sec^2 x$
D) $\tan x \sec^2 x - \sec^2 x$
E) $\sec^2 x - \tan^2 x$

11. Verify the identity shown below.

$$\left(1+\cot^2\theta\right)\tan^2\theta=\sec^2\theta$$

12. Find the exact value of $\cos(u+v)$ given that $\sin u=\frac{11}{61}$ and $\cos v=-\frac{40}{41}$. (Both u and v are in Quadrant II.)

A) $\cos(u+v)=\frac{160}{2501}$

B) $\cos(u+v)=-\frac{100}{2501}$

C) $\cos(u+v)=\frac{2301}{2501}$

D) $\cos(u+v)=-\frac{2290}{2501}$

E) $\cos(u+v)=\frac{980}{2501}$

13. Determine which of the following are trigonometric identities.

I. $\csc(\theta)\sec(\theta)=\tan(\theta)$

II. $\csc(\theta)\tan(\theta)=\sec(\theta)$

III. $\tan(\theta)\sec(\theta)=\csc(\theta)$

IV. $\csc(\theta)\sin(\theta)=1$

A) II is the only identity.
B) II and IV are the only identities.
C) I, III, and IV are the only identities.
D) I is the only identity.
E) II, III, and IV are the only identities.

14. Add or subtract as indicated; then use fundamental identities to simplify the expression below and determine which of the following is *not* equivalent.

$$\sin x + 1 + \frac{1}{\sin x - 1}$$

A) $\dfrac{\sin^2 x}{\sin x - 1}$

B) $\dfrac{1-\cos^2 x}{\sin x - 1}$

C) $\dfrac{1+\cos^2 x}{\sin x - 1}$

D) $\dfrac{\sec x - \cos x}{\tan x - \sec x}$

E) $\dfrac{\csc x - \cot x \cos x}{1 - \csc x}$

15. Use the product-to-sum formula to write the given product as a sum or difference.

$$10\sin\frac{\pi}{8}\cos\frac{\pi}{8}$$

A) $5\sin\dfrac{\pi}{8} + 5\cos\dfrac{\pi}{8}$

B) $5\sin\dfrac{\pi}{4}$

C) $5+5\cos\dfrac{\pi}{16}$

D) $5-5\cos\dfrac{\pi}{16}$

E) $-5\sin\dfrac{\pi}{16}$

16. Solve the following equation.

$\tan x - \sqrt{3} = 0$

A) $x = \frac{\pi}{6} + n\pi$ and $x = \frac{5\pi}{6} + n\pi$, where n is an integer

B) $x = \frac{\pi}{3} + n\pi$, where n is an integer

C) $x = \frac{\pi}{4} + n\pi$, where n is an integer

D) $x = \frac{\pi}{6} + n\pi$, where n is an integer

E) $x = \frac{2\pi}{3} + n\pi$ and $x = \frac{4\pi}{3} + n\pi$, where n is an integer

17. Use the half-angle formulas to determine the exact value of the following.

$\cos(22.5^\circ)$

A) $-\frac{\sqrt{2+\sqrt{3}}}{2}$

B) $\frac{\sqrt{2-\sqrt{2}}}{2}$

C) $-\frac{\sqrt{2-\sqrt{2}}}{2}$

D) $\frac{\sqrt{3-\sqrt{3}}}{2}$

E) $\frac{\sqrt{2+\sqrt{2}}}{2}$

18. Use the trigonometric substitution $x = 8\sec(\theta)$ to write the expression $\sqrt{x^2 - 64}$ as a trigonometric function of θ, where $0 < \theta < \frac{\pi}{2}$.

A) $8\tan(\theta)$

B) $64\tan(\theta)$

C) $64\sec(\theta)$

D) $8\sec(\theta)$

E) $8\sec(\theta) - 1$

19. If $x = 2\cot\theta$, use trigonometric substitution to write $\sqrt{4+x^2}$ as a trigonometric function of θ, where $0 < \theta < \pi$.
A) $2\cos\theta$
B) $2\csc\theta$
C) $2\cot\theta$
D) $2\sec\theta$
E) $2\sin\theta$

20. Use a double-angle formula to find the exact value of $\cos 2u$ when $\sin u = \frac{8}{17}$, where $\frac{\pi}{2} < u < \pi$.
A) $\cos 2u = -\frac{97}{289}$
B) $\cos 2u = \frac{120}{289}$
C) $\cos 2u = \frac{240}{289}$
D) $\cos 2u = \frac{161}{289}$
E) $\cos 2u = -\frac{450}{289}$

Answer Key

1. $$\frac{\tan\alpha+\cot\beta}{\tan\alpha\cot\beta} = \frac{\dfrac{\sin\alpha}{\cos\alpha}+\dfrac{\cos\beta}{\sin\beta}}{\dfrac{\sin\alpha}{\cos\alpha}\cdot\dfrac{\cos\beta}{\sin\beta}}$$
$$= \frac{\sin\alpha\sin\beta+\cos\alpha\cos\beta}{\cos\alpha\sin\beta}\cdot\frac{\cos\alpha\sin\beta}{\sin\alpha\cos\beta}$$
$$= \frac{\sin\alpha\sin\beta+\cos\alpha\cos\beta}{\sin\alpha\cos\beta}$$
$$= \frac{\sin\alpha\sin\beta}{\sin\alpha\cos\beta}+\frac{\cos\alpha\cos\beta}{\sin\alpha\cos\beta}$$
$$= \frac{\sin\beta}{\cos\beta}+\frac{\cos\alpha}{\sin\alpha}$$
$$= \tan\beta+\cot\alpha$$

2. D
3. A
4. $$\frac{\sin u+\sin v}{\cos u+\cos v} = \frac{2\sin\left(\dfrac{u+v}{2}\right)\cos\left(\dfrac{u-v}{2}\right)}{2\cos\left(\dfrac{u+v}{2}\right)\cos\left(\dfrac{u-v}{2}\right)}$$
$$= \frac{\sin\left(\dfrac{u+v}{2}\right)}{\cos\left(\dfrac{u+v}{2}\right)}$$
$$= \tan\left(\frac{u+v}{2}\right)$$
$$= \tan\frac{1}{2}(u+v)$$

5. C
6. E
7. B
8. B

9. $\sec^2\left(\dfrac{\pi}{2}-y\right)-1 = \csc^2 y - 1$

$= 1+\cot^2 y - 1$

$= \cot^2 y$

10. E

11. $\left(1+\cot^2\theta\right)\tan^2\theta = \left(1+\dfrac{\cos^2\theta}{\sin^2\theta}\right)\dfrac{\sin^2\theta}{\cos^2\theta}$

$= \dfrac{\sin^2\theta}{\cos^2\theta}+\dfrac{\cos^2\theta\sin^2\theta}{\sin^2\theta\cos^2\theta}$

$= \dfrac{\sin^2\theta}{\cos^2\theta}+\dfrac{\sin^2\theta\cos^2\theta}{\sin^2\theta\cos^2\theta}$

$= \tan^2\theta+1$

$= \sec^2\theta$

12. C
13. B
14. C
15. B
16. B
17. E
18. A
19. B
20. D

Name: ______________________________ Date: _____________

1. Verify the identity shown below.

$$\sqrt{\frac{1+\sin\theta}{1-\sin\theta}}=\frac{1+\sin\theta}{|\cos\theta|}$$

2. Verify the given identity.

$$\cos(x+y)\cos(x-y)=\cos^2 x-\sin^2 y$$

3. Solve the multi-angle equation below.

$$\sin(2x)=\frac{\sqrt{3}}{2}$$

A) $\frac{\pi}{8}+n\pi,\ \frac{\pi}{4}+n\pi,$ where n is an integer

B) $\frac{\pi}{6}+n\pi,\ \frac{\pi}{4}+n\pi,$ where n is an integer

C) $\frac{\pi}{6}+2n\pi,\ \frac{\pi}{3}+2n\pi,$ where n is an integer

D) $\frac{\pi}{8}+2n\pi,\ \frac{\pi}{4}+2n\pi,$ where n is an integer

E) $\frac{\pi}{6}+n\pi,\ \frac{\pi}{3}+n\pi,$ where n is an integer

4. Determine which of the following are trigonometric identities.

I. $\cot(\theta)\csc(\theta) = \sec(\theta)$

II. $\cot(\theta)\sec(\theta) = \csc(\theta)$

III. $\sec(\theta)\csc(\theta) = \cot(\theta)$

IV. $\cot(\theta)\sin(\theta) = 1$

A) II and IV are the only identities.
B) II is the only identity.
C) III is the only identity.
D) I, III, and IV are the only identities.
E) I is the only identity.

5. Simplify the given expression algebraically.

$$\cos\left(\frac{\pi}{2} + x\right)$$

A) $\sin x$

B) $-\sin x$

C) $\cos x$

D) $-\cos x$

E) 1

6. Use a graphing utility to approximate the solutions (to three decimal places) of the given equation in the interval $[0, 2\pi)$.

$$(\cos x)(15\cos x + 4) - 3 = 0$$

A) $x = 0.484, 1.799, 3.626, 4.490$
B) $x = 0.624, 1.932, 3.776, 4.435$
C) $x = 1.055, 2.078, 3.896, 4.721$
D) $x = 1.101, 2.118, 3.982, 5.104$
E) $x = 1.231, 2.214, 4.069, 5.052$

7. Factor; then use fundamental identities to simplify the expression below and determine which of the following is *not* equivalent.
$\csc^3 x - \csc^2 x - \csc x + 1$
A) $\csc x \cot^2 x - \cot^2 x$
B) $\cos^2 x\left(\dfrac{1-\sin x}{\sin^3 x}\right)$
C) $\dfrac{\cos^4 x}{\sin^3 x + \sin^4 x}$
D) $\dfrac{1-\sin x - \sin^2 x + \sin^3 x}{\sin^3 x}$
E) $\dfrac{\sec^2 x + \tan^2 x}{\csc x}$

8. Use the sum-to-product formulas to write the given expression as a product.
$\sin 9\theta - \sin 7\theta$
A) $2\sin 8\theta \cos\theta$
B) $2\cos 8\theta \cos\theta$
C) $-2\sin 8\theta \sin\theta$
D) $-2\cos 8\theta \cos\theta$
E) $2\cos 8\theta \sin\theta$

9. If $x = 8\cos\theta$, use trigonometric substitution to write $\sqrt{64 - x^2}$ as a trigonometric function of θ, where $0 < \theta < \pi$.
A) $8\sec\theta$
B) $8\sin\theta$
C) $8\cot\theta$
D) $8\cos\theta$
E) $8\csc\theta$

10. Verify the identity shown below.

$\tan^2\theta\sec^2\theta + \sec^2\theta = \sec^4\theta$

11. Determine which of the following are trigonometric identities.

I. $\dfrac{\sin(x)-\sin(y)}{\cos(x)+\cos(y)} + \dfrac{\cos(x)-\cos(y)}{\sin(x)+\sin(y)} = 0$

II. $\dfrac{\sin(x)+\sin(y)}{\cos(x)+\cos(y)} + \dfrac{\cos(x)+\cos(y)}{\sin(x)+\sin(y)} = 1$

III. $\dfrac{\sin(x)+\cos(y)}{\sin(x)\cos(y)} = \sin(y)+\cos(x)$

A) I is the only identity.
B) I and III are the only identities.
C) II and II are the only identities.
D) I and II are the only identities.
E) I, II, and III are identities.

12. If $\sin x = \dfrac{1}{2}$ and $\cos x = \dfrac{\sqrt{3}}{2}$, evaluate the following function.

$\csc x$

A) $\csc x = \dfrac{\sqrt{3}}{3}$

B) $\csc x = 2$

C) $\csc x = \sqrt{3}$

D) $\csc x = \dfrac{1}{3}$

E) $\csc x = \dfrac{2\sqrt{3}}{3}$

13. Find the exact value of $\tan(u+v)$ given that $\sin u = -\frac{7}{25}$ and $\cos v = \frac{12}{13}$. (Both u and v are in Quadrant IV.)

A) $\tan(u+v) = \frac{43}{253}$
B) $\tan(u+v) = \frac{316}{253}$
C) $\tan(u+v) = -\frac{204}{253}$
D) $\tan(u+v) = \frac{283}{253}$
E) $\tan(u+v) = -\frac{323}{253}$

14. Determine which of the following are trigonometric identities.

I. $\frac{\cos(4x)+\cos(2x)}{2\cot(3x)} = \cot(x)$

II. $\frac{\cos(4x)+\cos(x)}{\sin(3x)-\sin(x)} = \cot(2x)$

III. $\frac{\cos(6x)+\cos(2x)}{\sin(4x)+\sin(2x)} = \cot(3x)$

A) III is the only identity.
B) I, II, and III are identities.
C) I is the only identity.
D) None are identities.
E) II is the only identity.

15. Use the figure below to find the exact value of the given trigonometric expression.

$\cot\frac{x}{2}$

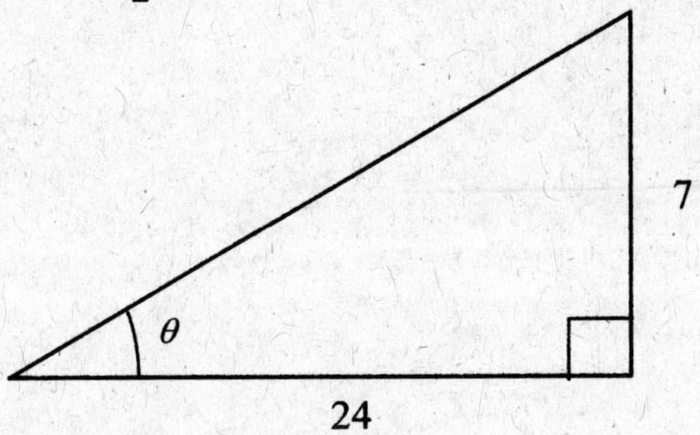

A) $\cot\frac{x}{2} = \frac{\sqrt{2}}{10}$

B) $\cot\frac{x}{2} = \frac{1}{7}$

C) $\cot\frac{x}{2} = 7$

D) $\cot\frac{x}{2} = \frac{7\sqrt{2}}{10}$

E) $\cot\frac{x}{2} = \frac{7}{12}$

16. Determine the exact value of the following expression.

$\cos(120^\circ - 0^\circ)$

A) $-\frac{\sqrt{3}}{2}$

B) $\frac{1}{2}$

C) $\frac{\sqrt{3}}{2}$

D) $-\frac{1}{2}$

E) $\frac{1}{2}-\frac{\sqrt{3}}{2}$

17. Use the half-angle formulas to determine the exact value of the following.

$\cos(-22.5^\circ)$

A) $-\frac{\sqrt{2+\sqrt{2}}}{2}$

B) $\frac{\sqrt{2-\sqrt{3}}}{2}$

C) $-\frac{\sqrt{2-\sqrt{3}}}{2}$

D) $\frac{\sqrt{3-\sqrt{2}}}{2}$

E) $\frac{\sqrt{2+\sqrt{3}}}{2}$

18. Use the graph below of the function to approximate the solutions to $4\cos(2x)-\cos(x)=0$ in the interval $[0,2\pi)$. Round your answers to one decimal.

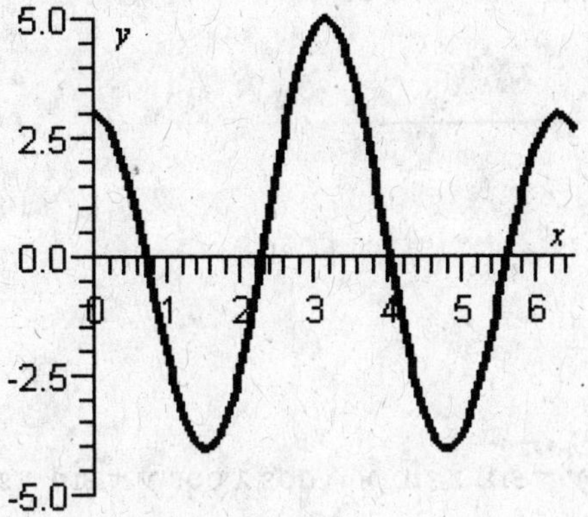

A) 0.7, 3.0, 1.5, 4.7
B) 3.0, 1.5, 4.7, 5.6
C) 3.0, 2.3, 4.0, 5.6
D) 0.7, 2.3, 4.0, 5.6
E) 0.7, 1.5, 4.0, 6.3

19. Use the cofunction identities to evaluate the expression below without the aid of a calculator.

$$\sin^2 62° + \sin^2 30° + \sin^2 28° + \sin^2 60°$$

A) 1
B) 2
C) –1
D) 0
E) $\frac{1}{2}$

20. Verify the given identity.

$$\frac{\sin 5x + \sin 3x}{\cos 5x + \cos 3x} = \tan 4x$$

Answer Key

1. $$\sqrt{\frac{1+\sin\theta}{1-\sin\theta}} = \sqrt{\frac{(1+\sin\theta)(1+\sin\theta)}{(1-\sin\theta)(1+\sin\theta)}} = \sqrt{\frac{(1+\sin\theta)^2}{1-\sin^2\theta}} = \frac{1+\sin\theta}{\sqrt{1-\sin^2\theta}} = \frac{1+\sin\theta}{|\cos\theta|}$$

2. $$\begin{aligned}\cos(x+y)\cos(x-y) &= (\cos x\cos y - \sin x\sin y)(\cos x\cos y + \sin x\sin y)\\ &= \cos^2 x\cos^2 y - \sin^2 x\sin^2 y\\ &= \cos^2 x(1-\sin^2 y) - (1-\cos^2 x)\sin^2 y\\ &= \cos^2 x - \cos^2 x\sin^2 y - \sin^2 y + \cos^2 x\sin^2 y\\ &= \cos^2 x - \sin^2 y\end{aligned}$$

3. E
4. B
5. B
6. E
7. E
8. E
9. B
10. $$\begin{aligned}\tan^2\theta\sec^2\theta + \sec^2\theta &= \sec^2\theta(\tan^2\theta+1)\\ &= \sec^2\theta\cdot\sec^2\theta\\ &= \sec^4\theta\end{aligned}$$

11. A
12. B
13. C
14. A
15. C
16. D
17. E
18. D
19. B

20. $$\frac{\sin 5x+\sin 3x}{\cos 5x+\cos 3x} = \frac{2\sin\left(\frac{5x+3x}{2}\right)\cos\left(\frac{5x-3x}{2}\right)}{2\cos\left(\frac{5x+3x}{2}\right)\cos\left(\frac{5x-3x}{2}\right)}$$
$$= \frac{2\sin 4x\cos x}{2\cos 4x\cos x}$$
$$= \tan 4x$$

Name: ______________________________ Date: ______________

1. Solve the multi-angle equation below.

$$\sin(2x) = -\frac{\sqrt{2}}{2}$$

A) $\frac{5\pi}{3}+n\pi,\ \frac{7\pi}{3}+n\pi,$ where n is an integer

B) $\frac{5\pi}{8}+n\pi,\ \frac{7\pi}{3}+n\pi,$ where n is an integer

C) $\frac{5\pi}{8}+2n\pi,\ \frac{7\pi}{8}+2n\pi,$ where n is an integer

D) $\frac{5\pi}{3}+2n\pi,\ \frac{7\pi}{3}+2n\pi,$ where n is an integer

E) $\frac{5\pi}{8}+n\pi,\ \frac{7\pi}{8}+n\pi,$ where n is an integer

2. Verify the given identity.

$$\frac{\cos u - \cos v}{\cos u + \cos v} = -\tan\frac{1}{2}(u+v)\tan\frac{1}{2}(u-v)$$

3. Determine which of the following are trigonometric identities.

I. $\dfrac{\cos(4x)-\cos(2x)}{2\tan(3x)}=-\tan(x)$

II. $\dfrac{\cos(3x)-\cos(x)}{\sin(3x)-\sin(x)}=-\tan(2x)$

III. $\dfrac{\cos(6x)-\cos(2x)}{\sin(4x)+\sin(2x)}=-\tan(3x$

A) II is the only identity.
B) I is the only identity.
C) I, II, and III are identities.
D) None are identities.
E) I and III are the only identities.

4. Approximate the solutions of the equation $2\sin^2(x)=2\cos(x)+1$ by considering its graph below. Round your answer to one decimal.

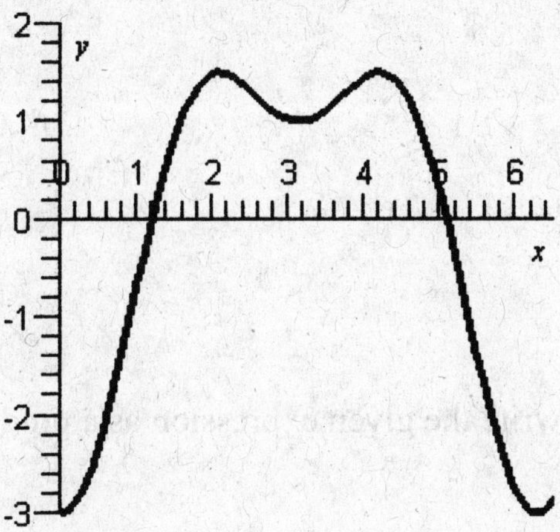

A) 2.0, 4.2
B) 2.0, 3.1
C) 3.1, 4.2
D) 1.2, 5.1
E) 1.2, 2.0

5. Which of the following is a solution to the given equation?

$\cot x + 1 = 0$

A) $x = \frac{2\pi}{3}$

B) $x = \frac{\pi}{4}$

C) $x = \frac{\pi}{6}$

D) $x = \frac{7\pi}{4}$

E) $x = \frac{7\pi}{6}$

6. Use the half-angle formula to simplify the given expression.

$\sqrt{\frac{1+\cos 16x}{2}}$

A) $\cos 32x$
B) $\cos 8x$
C) $\cos 16x$
D) $\cos 64x$
E) $\cos 4x$

7. Use the sum-to-product formulas to write the given expression as a product.

$\cos 8\theta - \cos 6\theta$

A) $-2\sin 7\theta \sin\theta$
B) $2\cos 7\theta \cos\theta$
C) $2\cos 7\theta \sin\theta$
D) $2\sin 7\theta \cos\theta$
E) $-2\cos 7\theta \cos\theta$

8. Solve the following trigonometric equation on the interval $[0, 2\pi)$.

$$\sin(x) + \cos(x) = 0$$

A) $\frac{5\pi}{4}, \frac{7\pi}{4}$
B) $\frac{3\pi}{4}, \frac{7\pi}{4}$
C) $\frac{\pi}{4}, \frac{7\pi}{4}$
D) $\frac{\pi}{4}, \frac{5\pi}{4}$
E) $\frac{3\pi}{4}, \frac{5\pi}{4}$

9. Use the graph below of the function to approximate the solutions to $3\cos(2x) - \cos(x) = 0$ in the interval $[0, 2\pi)$. Round your answers to one decimal.

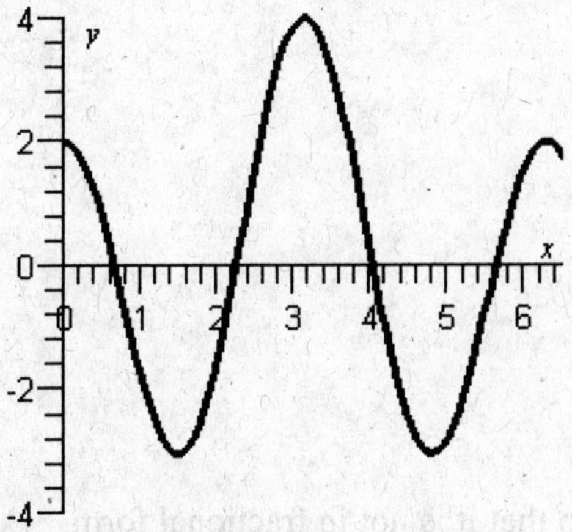

A) 0.7, 2.0, 1.5, 4.8
B) 2.0, 1.5, 4.8, 5.6
C) 2.0, 2.2, 4.0, 5.6
D) 0.7, 2.2, 4.0, 5.6
E) 0.7, 1.5, 4.0, 6.3

10. Multiply; then use fundamental identities to simplify the expression below and determine which of the following is *not* equivalent.

$$(\sin x + \cos x)(\sin x - \cos x)$$

A) $2\sin^2 x - \sec^2 x - \tan^2 x$

B) $\sin^2 x - \cos^2 x$

C) $1 - 2\cos^2 x$

D) $\csc^2 x - \cot^2 x - 2\cos^2 x$

E) $1 - 2\sin\left(\frac{\pi}{2} - x\right)\cos x$

11. Expand the expression below and use fundamental trigonometric identities to simplify.

$$(\sin(\omega) + \cos(\omega))^2$$

A) $\sin^2(\omega) + \cos^2(\omega)$

B) $2\tan(\omega) + 1$

C) $2\sin(\omega)\cos(\omega) + 1$

D) 1

E) $2\cot(\omega) + 1$

12. Rewrite the expression $\dfrac{\sin(y)}{1 - \cos(y)}$ so that it is not in fractional form.

A) $\sin^2 - \sin(y)\tan(y)$

B) $1 - \sin(y)\tan(y)$

C) $\csc(y) + \cot(y)$

D) $\sin^2 + \sin(y)\tan(y)$

E) $1 - \cos(y)$

13. Determine which of the following are trigonometric identities.

I. $\sin(\theta)+\cot(\theta)\cos(\theta)=\csc(\theta)$

II. $\cot(\theta)-\sin(\theta)\cos(\theta)=0$

III. $\sin(\theta)+\sin(\theta)\cos(\theta)=\csc(\theta)$

A) I is the only identity.
B) I and II are the only identities.
C) III is the only identity.
D) I, II, and III are identities.
E) II is the only identity.

14. Find all solutions of the following equation in the interval $[0,2\pi)$.

$2\cos^2 x = 2 + \sin x$

A) $x = 0, \frac{\pi}{4}, \frac{3\pi}{4}$

B) $x = 0, \frac{3\pi}{4}, \frac{4\pi}{3}, \frac{11\pi}{6}$

C) $x = \frac{\pi}{2}, \frac{2\pi}{3}, \frac{4\pi}{3}, \frac{3\pi}{2}$

D) $x = 0, \pi, \frac{7\pi}{6}, \frac{11\pi}{6}$

E) $x = \frac{\pi}{2}, \frac{3\pi}{4}, \frac{3\pi}{2}, \frac{7\pi}{6}$

15. If $x = 10\sin\theta$, use trigonometric substitution to write $\sqrt{100-x^2}$ as a trigonometric function of θ, where $-\frac{\pi}{2}<\theta<\frac{\pi}{2}$.

A) $10\sin\theta$
B) $10\cos\theta$
C) $10\tan\theta$
D) $10\csc\theta$
E) $10\sec\theta$

16. Use the figure below to determine the exact value of the given function.

$\sin 2\theta$

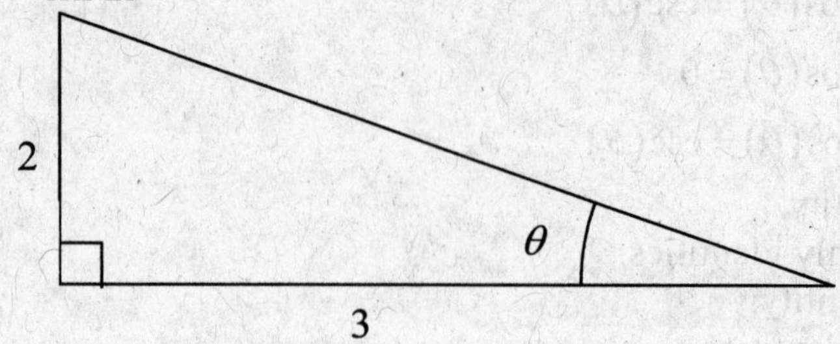

A) $\sin 2\theta = \dfrac{5}{13}$

B) $\sin 2\theta = \dfrac{7}{13}$

C) $\sin 2\theta = \dfrac{12}{5}$

D) $\sin 2\theta = \dfrac{9}{13}$

E) $\sin 2\theta = \dfrac{12}{13}$

17. Verify the identity shown below.

$$\frac{\sin^2\theta - 1}{\cos^2\theta - 1} = \csc^2\theta - 1$$

18. Solve the multiple-angle equation in the interval $[0, 2\pi)$.

$\sec 2x = 2$

A) $x = \frac{\pi}{8}, \frac{7\pi}{8}, \frac{9\pi}{8}, \frac{15\pi}{8}$

B) $x = \frac{\pi}{12}, \frac{5\pi}{12}, \frac{13\pi}{12}, \frac{17\pi}{12}$

C) $x = \frac{\pi}{6}, \frac{5\pi}{6}, \frac{7\pi}{6}, \frac{11\pi}{6}$

D) $x = \frac{\pi}{4}, \frac{3\pi}{4}, \frac{5\pi}{4}, \frac{7\pi}{4}$

E) $x = \frac{\pi}{3}, \frac{2\pi}{3}, \frac{4\pi}{3}, \frac{5\pi}{3}$

19. Verify the identity shown below.

$$\frac{\tan\theta + 1}{\sec\theta + \csc\theta} = \sin\theta$$

20. Solve the following equation.

$\tan^2 x + \tan x = 0$

A) $x = \pi + 2n\pi$ and $x = \frac{3\pi}{2} + 2n\pi$, where n is an :

B) $x = n\pi$ and $x = \frac{3\pi}{4} + n\pi$, where n is an integer

C) $x = \frac{2\pi}{3} + 2n\pi$ and $x = \frac{5\pi}{3} + 2n\pi$, where n is ar

D) $x = n\pi$ and $x = \frac{\pi}{2} + n\pi$, where n is an integer

E) $x = n\pi$ and $x = \frac{3\pi}{2} + 2n\pi$, where n is an integ

Answer Key

1. E

2. $$\frac{\cos u - \cos v}{\cos u + \cos v} = \frac{-2\sin\left(\frac{u+v}{2}\right)\sin\left(\frac{u-v}{2}\right)}{2\cos\left(\frac{u+v}{2}\right)\cos\left(\frac{u-v}{2}\right)}$$
$$= -\tan\left(\frac{u+v}{2}\right)\tan\left(\frac{u-v}{2}\right)$$
$$= -\tan\frac{1}{2}(u+v)\tan\frac{1}{2}(u-v)$$

3. A
4. D
5. D
6. B
7. A
8. B
9. D
10. A
11. C
12. C
13. A
14. D
15. B
16. E
17. $$\frac{\sin^2\theta - 1}{\cos^2\theta - 1} = \frac{\sin^2\theta - (\sin^2\theta + \cos^2\theta)}{\cos^2\theta - (\sin^2\theta + \cos^2\theta)}$$
$$= \frac{\sin^2\theta - \sin^2\theta - \cos^2\theta}{\cos^2\theta - \sin^2\theta - \cos^2\theta}$$
$$= \frac{-\cos^2\theta}{-\sin^2\theta}$$
$$= \cot^2\theta$$
$$= \csc^2\theta - 1$$

18. C

19. $$\frac{\tan\theta+1}{\sec\theta+\csc\theta} = \frac{\dfrac{\sin\theta}{\cos\theta}+1}{\dfrac{1}{\cos\theta}+\dfrac{1}{\sin\theta}}$$
$$= \frac{\dfrac{\sin\theta+\cos\theta}{\cos\theta}}{\dfrac{\sin\theta+\cos\theta}{\cos\theta\sin\theta}}$$
$$= \frac{\sin\theta+\cos\theta}{\cos\theta}\cdot\frac{\cos\theta\sin\theta}{\sin\theta+\cos\theta}$$
$$= \frac{\cos\theta\sin\theta}{\cos\theta}$$
$$= \sin\theta$$

20. B

Name: ______________________________ Date: _____________

1. Find the exact value of the given expression using a sum or difference formula.
   $\sin 165°$
   A) $\dfrac{-\sqrt{3}-1}{2\sqrt{2}}$
   B) $\dfrac{-\sqrt{3}+1}{2\sqrt{2}}$
   C) $\dfrac{\sqrt{3}-1}{2\sqrt{2}}$
   D) $\dfrac{\sqrt{3}+1}{2\sqrt{2}}$

2. Use the sum-to-product formulas to find the exact value of the given expression.
   $\cos 150° - \cos 30°$
   A) 1
   B) -1
   C) $-\sqrt{3}$
   D) $-\dfrac{\sqrt{3}}{2}$
   E) 0

3. Verify the identity shown below.
   $\dfrac{1+\tan\theta}{1+\cot\theta} = \tan\theta$

4. Find all solutions of the following equation on the interval $[0, 2\pi)$.

$$\cot(x) + \sqrt{3} = 0$$

A) $\frac{5\pi}{6}, \frac{\pi}{6}, \frac{11\pi}{6}, \frac{7\pi}{6}$
B) $\frac{\pi}{3}, \frac{4\pi}{3}$
C) $\frac{5\pi}{6}, \frac{11\pi}{6}$
D) $\frac{2\pi}{3}, \frac{5\pi}{3}$
E) $\frac{\pi}{6}, \frac{7\pi}{6}$

5. Approximate the solutions of the equation $2\sin^2(x) + 5\sin(x) - 2 = 0$ by considering its graph below. Round your answer to one decimal.

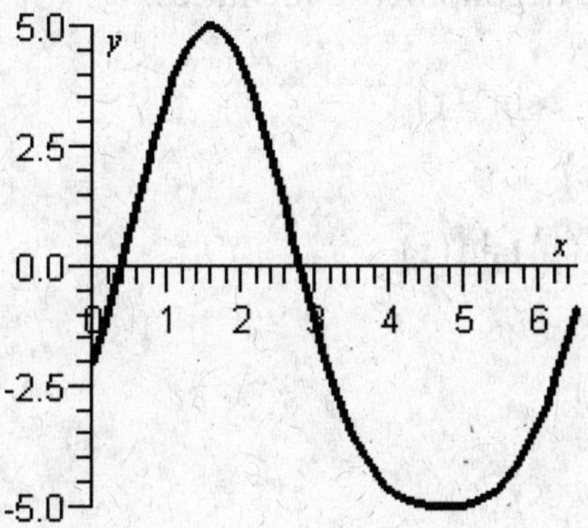

A) 0.4, 2.8
B) 0.4, 1.0
C) 1.0, 1.8
D) 0.4, 1.8
E) 1 8. 2 8.

6. Use fundamental identities to simplify the expression below and then determine which of the following is *not* equivalent.

   $\sin\alpha(\csc\alpha - \sin\alpha)$

   A) $1-\sin^2\alpha$

   B) $\dfrac{\csc^2\alpha - 1}{\csc^2\alpha}$

   C) $\dfrac{\csc^2\alpha - \sec^2\alpha + \tan^2\alpha}{\csc^2\alpha}$

   D) $1-\cot^2\alpha$

   E) $\cos^2\alpha$

7. Determine which of the following are trigonometric identities.

   I. $\cos^4(t)+\sin^4(t)=1-2\sin^2(t)+2\sin^4(t)$
   II. $\sin^5(t)=\sin^3(t)\cos^2(t)-\sin^3(t)$
   III. $\sin^3(t)\cos^2(t)=\left(\cos^2(t)-\cos^4(t)\right)\sin(t)$

   A) I is the only identity.
   B) I and III are the only identities.
   C) III is the only identity.
   D) I, II, and III are identities.
   E) I and II are the only identities.

8. Find the exact value of $\cos(u+v)$ given that $\sin u = \frac{11}{61}$ and $\cos v = -\frac{40}{41}$.

(Both u and v are in Quadrant II.)

A) $\cos(u+v) = \frac{160}{2501}$

B) $\cos(u+v) = -\frac{100}{2501}$

C) $\cos(u+v) = \frac{2301}{2501}$

D) $\cos(u+v) = -\frac{2290}{2501}$

E) $\cos(u+v) = \frac{980}{2501}$

9. Find all solutions of the following equation in the interval $[0, 2\pi)$.

$2\cos x - \sec x = 0$

A) $x = \frac{\pi}{4}, \frac{3\pi}{4}, \frac{5\pi}{4}, \frac{7\pi}{4}$

B) $x = \frac{\pi}{6}, \frac{5\pi}{6}, \frac{7\pi}{6}, \frac{11\pi}{6}$

C) $x = \frac{\pi}{3}, \frac{2\pi}{3}, \frac{4\pi}{3}, \frac{5\pi}{3}$

D) $x = 0, \frac{\pi}{2}, \pi, \frac{3\pi}{2}$

E) $x = \frac{\pi}{3}, \frac{3\pi}{4}, \frac{3\pi}{2}, \frac{7\pi}{6}$

10. Use the product-to-sum formula to write the given product as a sum or difference.

$$8\sin\frac{\pi}{8}\sin\frac{\pi}{8}$$

A) $4\sin\frac{\pi}{16}$

B) $4-4\cos\frac{\pi}{4}$

C) $4+4\cos\frac{\pi}{16}$

D) $-4\sin\frac{\pi}{16}$

E) $4\sin\frac{\pi}{8}+4\cos\frac{\pi}{8}$

11. Solve the multi-angle equation below.

$$\cos\left(\frac{x}{2}\right)=-\frac{\sqrt{3}}{2}$$

A) $\frac{10\pi}{3}+n\pi,\ \frac{14\pi}{3}+n\pi,$ where n is an integer

B) $\frac{10\pi}{3}+2n\pi,\ \frac{14\pi}{3}+2n\pi,$ where n is an integer

C) $\frac{5\pi}{3}+n\pi,\ \frac{14\pi}{3}+n\pi,$ where n is an integer

D) $\frac{5\pi}{3}+n\pi,\ \frac{7\pi}{3}+n\pi,$ where n is an integer

E) $\frac{5\pi}{3}+4n\pi,\ \frac{7\pi}{3}+4n\pi,$ where n is an integer

12. Use the product-to-sum formulas to write the expression below as a sum or difference.

$\sin(8\theta)\cos(6\theta)$

A) $\frac{1}{2}(\cos(2\theta)+\cos(14\theta))$

B) $\frac{1}{2}(\sin(2\theta)+\cos(14\theta))$

C) $\frac{1}{2}(\sin(14\theta)+\sin(2\theta))$

D) $\frac{1}{2}(\cos(2\theta)-\cos(14\theta))$

E) $\frac{1}{2}(\sin(14\theta)-\sin(2\theta))$

13. Verify the identity shown below.

$$\frac{\tan\theta+1}{\sec\theta+\csc\theta}=\sin\theta$$

14. Find all solutions of the following equation on the interval $[0,2\pi)$.

$\csc^2(x)-2=0$

A) $\frac{\pi}{4}, \frac{3\pi}{4}$

B) $\frac{\pi}{4}, \frac{3\pi}{4}, \frac{5\pi}{4}, \frac{7\pi}{4}$

C) $\frac{\pi}{4}, \frac{7\pi}{4}$

D) $\frac{5\pi}{4}, \frac{7\pi}{4}$

E) $\frac{3\pi}{4}, \frac{5\pi}{4}$

15. Verify the identity shown below.

$$\sec^2 \mu - \cot^2\left(\frac{\pi}{2} - \mu\right) = 1$$

16. Find the exact value of $\tan(u+v)$ given that $\sin u = -\frac{11}{61}$ and $\cos v = \frac{40}{41}$. (Both u and v are in Quadrant IV.)

A) $\tan(u+v) = \frac{37}{767}$

B) $\tan(u+v) = \frac{2488}{2301}$

C) $\tan(u+v) = -\frac{980}{2301}$

D) $\tan(u+v) = \frac{797}{767}$

E) $\tan(u+v) = -\frac{833}{767}$

17. Factor; then use fundamental identities to simplify the expression below and determine which of the following is *not* equivalent.

$$\cot^2 \alpha - \cot^2 \alpha \cos^2 \alpha$$

A) $\cos^2 \alpha$

B) $1 - \sin^2 \alpha$

C) $\tan^2 \alpha$

D) $\sin^2\left(\frac{\pi}{2} - \alpha\right)$

E) $\frac{1}{\sec^2 \alpha}$

18. If $x = 6\sin\theta$, use trigonometric substitution to write $\sqrt{36 - x^2}$ as a trigonometric function of θ, where $0 < \theta < \frac{\pi}{2}$.
    A) $6\csc\theta$
    B) $6\sec\theta$
    C) $6\sin\theta$
    D) $6\cos\theta$
    E) $6\tan\theta$

19. Which of the following is equivalent to the given expression?

    $\frac{\cot^2 x}{\csc x + 1}$

    A) $\csc x + 2\sec x$

    B) $\csc x - 1$

    C) $\cot x + 2$

    D) $\tan^2 x - \cot^2 x$

    E) $-2\sec x \csc^2 x$

20. Find the exact solutions of the given equation in the interval $[0, 2\pi)$.

$\sin x \cos 2x + \cos x \sin 2x = 0$

A) $x = \frac{\pi}{2}, \frac{7\pi}{6}, \frac{11\pi}{6}$

B) $x = 0, \frac{\pi}{3}, \frac{2\pi}{3}, \pi, \frac{4\pi}{3}, \frac{5\pi}{3}$

C) $x = 0, \frac{\pi}{3}, \frac{2\pi}{3}, \pi, \frac{4\pi}{3}$

D) $x = 0, \frac{\pi}{3}, \pi, \frac{5\pi}{3}$

E) $x = 0$

Answer Key

1. C
2. C
3. $$\frac{1+\tan\theta}{1+\cot\theta} = \frac{1+\dfrac{\sin\theta}{\cos\theta}}{1+\dfrac{\cos\theta}{\sin\theta}}$$
$$= \frac{\dfrac{\cos\theta+\sin\theta}{\cos\theta}}{\dfrac{\sin\theta+\cos\theta}{\sin\theta}}$$
$$= \frac{\cos\theta+\sin\theta}{\cos\theta}\cdot\frac{\sin\theta}{\sin\theta+\cos\theta}$$
$$= \frac{\sin\theta}{\cos\theta}$$
$$= \tan\theta$$
4. C
5. A
6. D
7. B
8. C
9. A
10. B
11. E
12. C
13. $$\frac{\tan\theta+1}{\sec\theta+\csc\theta} = \frac{\dfrac{\sin\theta}{\cos\theta}+1}{\dfrac{1}{\cos\theta}+\dfrac{1}{\sin\theta}}$$
$$= \frac{\dfrac{\sin\theta+\cos\theta}{\cos\theta}}{\dfrac{\sin\theta+\cos\theta}{\cos\theta\sin\theta}}$$
$$= \frac{\sin\theta+\cos\theta}{\cos\theta}\cdot\frac{\cos\theta\sin\theta}{\sin\theta+\cos\theta}$$
$$= \frac{\cos\theta\sin\theta}{\cos\theta}$$
$$= \sin\theta$$
14. B

15. $\sec^2 \mu - \cot^2\left(\frac{\pi}{2} - \mu\right) = \sec^2 \mu - \tan^2 \mu$

$$= \frac{1}{\cos^2 \mu} - \frac{\sin^2 \mu}{\cos^2 \mu}$$

$$= \frac{1 - \sin^2 \mu}{\cos^2 \mu}$$

$$= \frac{\cos^2 \mu}{\cos^2 \mu}$$

$$= 1$$

16. C
17. C
18. D
19. B
20. B

Name: ______________________________ Date: _______________

1. Factor; then use fundamental identities to simplify the expression below and determine which of the following is *not* equivalent.

$\sin^3 x - \sin^2 x - \sin x + 1$

A) $\cos^2 x - \cos^2 x \sin x$

B) $\dfrac{1 - \sin x}{\sec^2 x}$

C) $\dfrac{\cos^4 x}{1 + \sin x}$

D) $\cos^2 x(1 - \sin x)$

E) $\dfrac{1 + \cos^2 x}{\sin x}$

2. Solve the multi-angle equation below.

$\cos\left(\dfrac{x}{2}\right) = \dfrac{\sqrt{2}}{2}$

A) $\dfrac{\pi}{3} + n\pi,\ \dfrac{7\pi}{3} + n\pi,$ where n is an integer

B) $\dfrac{\pi}{3} + 2n\pi,\ \dfrac{7\pi}{3} + 2n\pi,$ where n is an integer

C) $\dfrac{\pi}{2} + n\pi,\ \dfrac{7\pi}{3} + n\pi,$ where n is an integer

D) $\dfrac{\pi}{2} + n\pi,\ \dfrac{7\pi}{2} + n\pi,$ where n is an integer

E) $\dfrac{\pi}{2} + 4n\pi,\ \dfrac{7\pi}{2} + 4n\pi,$ where n is an integer

3. Find the exact value of the given expression using a sum or difference formula.

$\cos\frac{7\pi}{12}$

A) $\frac{\sqrt{3}-1}{2\sqrt{2}}$

B) $\frac{-\sqrt{3}-1}{2\sqrt{2}}$

C) $\frac{\sqrt{3}+1}{2\sqrt{2}}$

D) $\frac{-\sqrt{3}+1}{2\sqrt{2}}$

4. Verify the identity shown below.

$\csc\theta - \cos\theta\cot\theta = \sin\theta$

5. Determine which of the following are trigonometric identities.

I. $\frac{\cos(4x)+\cos(2x)}{2\cot(3x)} = \cot(x)$

II. $\frac{\cos(4x)+\cos(x)}{\sin(3x)-\sin(x)} = \cot(2x)$

III. $\frac{\cos(6x)+\cos(2x)}{\sin(4x)+\sin(2x)} = \cot(3x)$

A) III is the only identity.
B) I and III are the only identities.
C) None are identities.
D) II is the only idcntity.
E) I and II are the only identities.

6. Solve the multiple-angle equation in the interval $[0, 2\pi)$.

$\tan 2x = -1$

A) $x = \frac{\pi}{4}, \frac{3\pi}{4}, \frac{5\pi}{4}, \frac{7\pi}{4}$

B) $x = \frac{\pi}{3}, \frac{2\pi}{3}, \frac{4\pi}{3}, \frac{5\pi}{3}$

C) $x = \frac{3\pi}{8}, \frac{7\pi}{8}, \frac{11\pi}{8}, \frac{15\pi}{8}$

D) $x = \frac{\pi}{6}, \frac{5\pi}{6}, \frac{7\pi}{6}, \frac{11\pi}{6}$

E) $x = \frac{\pi}{6}, \frac{5\pi}{6}, \frac{7\pi}{6}, \frac{11\pi}{6}$

7. Write the given expression as an algebraic expression.

$\cos(\arccos x - \arcsin x)$

A) 1

B) $\frac{x\sqrt{1-x^2} - x}{\sqrt{x^2+1}}$

C) 0

D) $2x\sqrt{1-x^2}$

E) $\frac{x\sqrt{1-x^2} + x}{\sqrt{x^2+1}}$

8. Find the exact value of $\cos(u+v)$ given that $\sin u = \frac{9}{41}$ and $\cos v = -\frac{15}{17}$. (Both u and v are in Quadrant II.)

A) $\cos(u+v) = \frac{225}{697}$

B) $\cos(u+v) = -\frac{185}{697}$

C) $\cos(u+v) = \frac{528}{697}$

D) $\cos(u+v) = -\frac{519}{697}$

E) $\cos(u+v) = \frac{455}{697}$

9. Use a double-angle formula to find the exact value of $\cos 2u$ when $\sin u = \frac{5}{13}$, where $\frac{\pi}{2} < u < \pi$.

A) $\cos 2u = -\frac{94}{169}$

B) $\cos 2u = \frac{60}{169}$

C) $\cos 2u = \frac{120}{169}$

D) $\cos 2u = \frac{119}{169}$

E) $\cos 2u = -\frac{288}{169}$

10. Use fundamental identities to simplify the expression below and then determine which of the following is *not* equivalent.

$\cot\beta\sec\beta$

A) $\dfrac{1}{\sin\beta}$

B) $\dfrac{\sec\beta}{\tan\beta}$

C) $\dfrac{1}{\cos\beta\tan\beta}$

D) $\sec\beta$

E) $\csc\beta$

11. Find the exact value of the given expression using a sum or difference formula.

$\sin 345°$

A) $\dfrac{\sqrt{3}+1}{2\sqrt{2}}$

B) $\dfrac{\sqrt{3}-1}{2\sqrt{2}}$

C) $\dfrac{-\sqrt{3}+1}{2\sqrt{2}}$

D) $\dfrac{-\sqrt{3}-1}{2\sqrt{2}}$

12. Verify the identity shown below.

$$\frac{1}{\sec\theta-\tan\theta}=\sec\theta+\tan\theta$$

13. Use the graph below to approximate the solutions of the equation $-2\cos(x) - \sin(x) = 0$ on the interval $[0, 2\pi)$. Round your answer to one decimal.

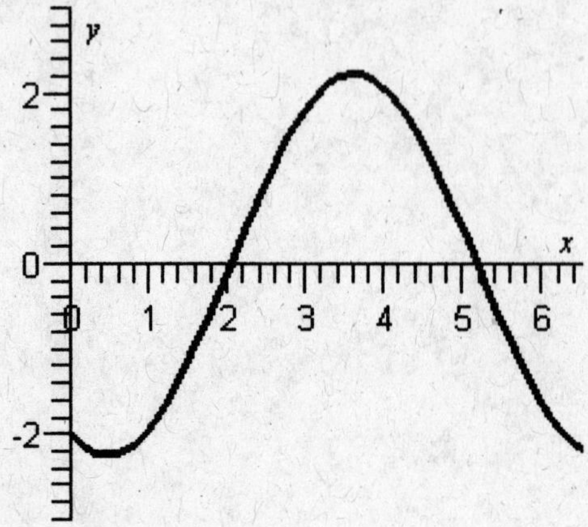

A) –2.0, 5.9
B) 2.0, 5.9
C) 5.2, 5.9
D) 2, 5.2
E) –2, 5.2

14. Determine which of the following are trigonometric identities.

I. $\dfrac{\sin(t)+\sin(s)}{\cos(t)-\cos(s)} + \dfrac{\cos(t)+\cos(s)}{\sin(t)-\sin(s)} = 0$

II. $\dfrac{\sin(t)+\sin(s)}{\cos(t)+\cos(s)} + \dfrac{\cos(t)+\cos(s)}{\sin(t)+\sin(s)} = 1$

III. $\dfrac{\sin(t)+\cos(s)}{\sin(t)\cos(s)} = \sin(s)+\cos(t)$

A) I is the only identity.
B) I and III are the only identities.
C) II and II are the only identities.
D) I and II are the only identities.
E) I, II, and III are identities.

15. If $\csc x = \dfrac{4\sqrt{3}}{3}$ and $\cos x < 0$, evaluate the function below.

$\sec x$

A) $\sec x = \dfrac{4}{\sqrt{3}}$

B) $\sec x = -\dfrac{\sqrt{3}}{4}$

C) $\sec x = -\dfrac{4\sqrt{13}}{13}$

D) $\sec x = -\dfrac{\sqrt{39}}{3}$

E) $\sec x = \dfrac{\sqrt{39}}{3}$

16. Use the graph below of the function to approximate the solutions to $4\cos(2x)-\cos(x)=0$ in the interval $[0,2\pi)$. Round your answers to one decimal.

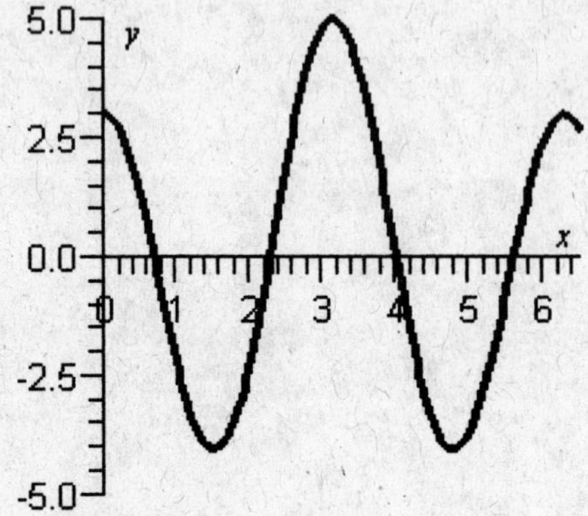

A) 0.7, 3.0, 1.5, 4.7
B) 3.0, 1.5, 4.7, 5.6
C) 3.0, 2.3, 4.0, 5.6
D) 0.7, 2.3, 4.0, 5.6
E) 0.7, 1.5, 4.0, 6.3

17. The rate of change of the function
$f(x) = -\csc x - \sin x$
is given by the expression
$\csc x \cot x - \cos x$.
Which of the following is its simplification?
A) $-\csc x \sec^2 x$

B) $\tan^2 x \cos x$

C) $\cot^2 x \cos x$

D) $-\cot x \cos^2 x$

E) $\sec^2 x$

18. Use the cofunction identities to evaluate the expression below without the aid of a calculator.

$$\cos^2 50° + \cos^2 58° + \cos^2 40° + \cos^2 32°$$

A) 1
B) 2
C) –1
D) 0
E) $\frac{1}{2}$

19. Use the product-to-sum formula to write the given product as a sum or difference.

$$6\sin\frac{\pi}{8}\cos\frac{\pi}{8}$$

A) $3\sin\frac{\pi}{8}+3\cos\frac{\pi}{8}$

B) $3\sin\frac{\pi}{4}$

C) $3+3\cos\frac{\pi}{16}$

D) $3-3\cos\frac{\pi}{16}$

E) $-3\sin\frac{\pi}{16}$

20. Multiply; then use fundamental identities to simplify the expression below and determine which of the following is *not* equivalent.

$$(\sin x + \cos x)(\sin x - \cos x)$$

A) $2\sin^2 x - \sec^2 x - \tan^2 x$

B) $\sin^2 x - \cos^2 x$

C) $1 - 2\cos^2 x$

D) $\csc^2 x - \cot^2 x - 2\cos^2 x$

E) $1 - 2\sin\left(\frac{\pi}{2} - x\right)\cos x$

Answer Key

1. E
2. E
3. D
4. $$\begin{aligned}\csc\theta - \cos\theta\cot\theta &= \frac{1}{\sin\theta} - \cos\theta\cdot\frac{\cos\theta}{\sin\theta}\\ &= \frac{1}{\sin\theta} - \frac{\cos^2\theta}{\sin\theta}\\ &= \frac{1-\cos^2\theta}{\sin\theta}\\ &= \frac{\sin^2\theta}{\sin\theta}\\ &= \sin\theta\end{aligned}$$
5. A
6. C
7. D
8. C
9. D
10. D
11. C

12.
$$\begin{aligned}
\frac{1}{\sec\theta-\tan\theta} &= \frac{1}{\dfrac{1}{\cos\theta}-\dfrac{\sin\theta}{\cos\theta}} \\
&= \frac{1}{\dfrac{1-\sin\theta}{\cos\theta}} \\
&= 1\cdot\frac{\cos\theta}{1-\sin\theta} \\
&= \frac{\cos\theta}{1-\sin\theta}\cdot\frac{1+\sin\theta}{1+\sin\theta} \\
&= \frac{\cos\theta(1+\sin\theta)}{1-\sin^2\theta} \\
&= \frac{\cos\theta(1+\sin\theta)}{\cos^2\theta} \\
&= \frac{1+\sin\theta}{\cos\theta} \\
&= \frac{1}{\cos\theta}+\frac{\sin\theta}{\cos\theta} \\
&= \sec\theta+\tan\theta
\end{aligned}$$

13. D
14. A
15. C
16. D
17. C
18. B
19. B
20. A

Name: ______________________________ Date: _______________

1. In the following triangle, $A = 58^\circ$, $b = 105$, and $a = 126$. Use the Law of Sines to find the measure of angle C, in degrees. Round your answer to two decimals.

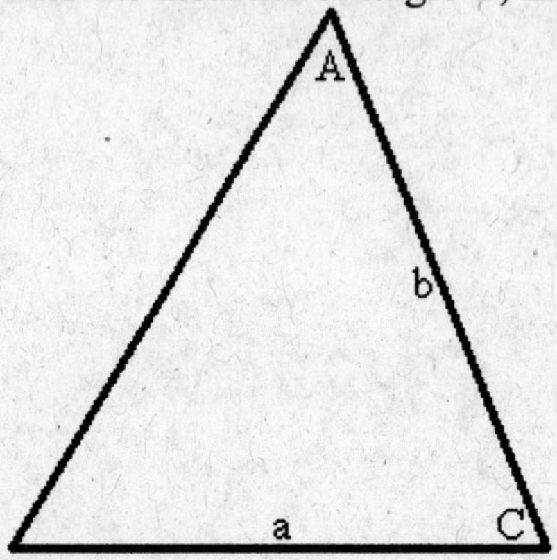

A) 121.03°
B) 135.03°
C) 45.03°
D) 77.03°
E) 44.97°

2. Given $A = 52^\circ$, $B = 65^\circ$, and $a = 7.1$, use the Law of Sines to solve the triangle for the value of b. Round answer to two decimal places.

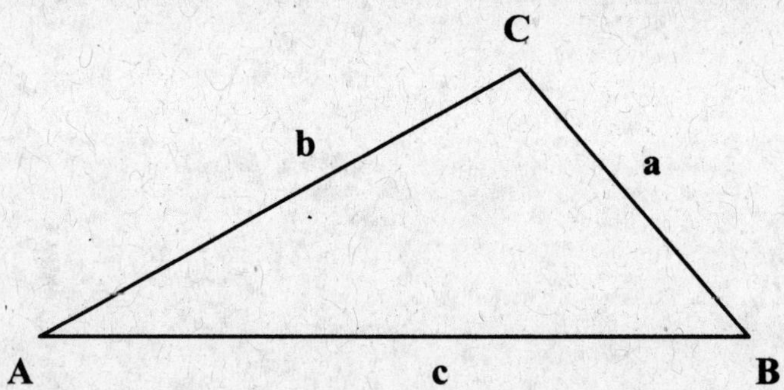

A) $b = 6.98$
B) $b = 8.03$
C) $b = 6.28$
D) $b = 8.17$
E) $b = 6.17$

3. A plane flies 710 miles from City A to City B with a bearing of 28° (clockwise from north). Then it flies 672 miles from City B to City C with a bearing of 45°. Find the straight-line distance from City C to City A. Round your answer to two decimals.
   A) 207.70 miles
   B) 1238.36 miles
   C) 1111.16 miles
   D) 955.99 miles
   E) 1366.83 miles

4. Find the angle between the vectors **u** and **v**.

$$\mathbf{u} = \cos\left(\frac{\pi}{3}\right)\mathbf{i} + \sin\left(\frac{\pi}{3}\right)\mathbf{j}, \quad \mathbf{v} = \cos\left(\frac{5\pi}{4}\right)\mathbf{i} + \sin\left(\frac{5\pi}{4}\right)\mathbf{j}$$

   A) 40°
   B) 75°
   C) 105°
   D) 165°
   E) 175°

5. Let $\mathbf{u} = \langle 6,5 \rangle$ and $\mathbf{v} = \langle 12,7 \rangle$. Find $\mathbf{u} - \mathbf{v}$.
   A) $\langle 13,17 \rangle$
   B) $\langle -6,-2 \rangle$
   C) $\langle 18,12 \rangle$
   D) $\langle -1,-7 \rangle$
   E) $\langle 17,13 \rangle$

6. Use the figure to determine the tension in the cable CA supporting the load. Round your answer to two decimals.

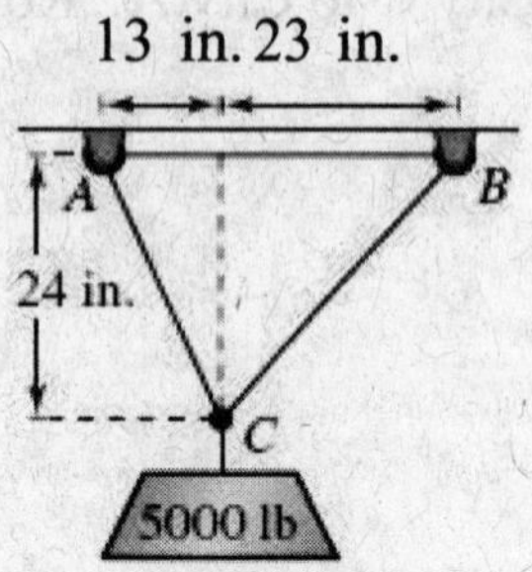

A) 3194.44 lbs.
B) 9996.76 lbs.
C) 6881.42 lbs.
D) 2500.81 lbs.
E) 3632.97 lbs.

7. Plot the following complex number.

   $3i$

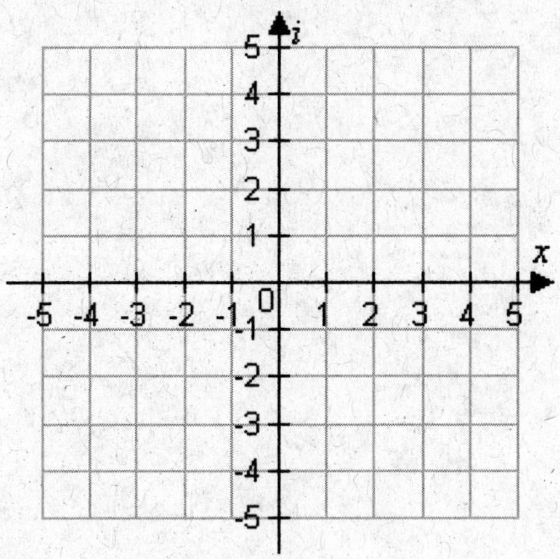

A)

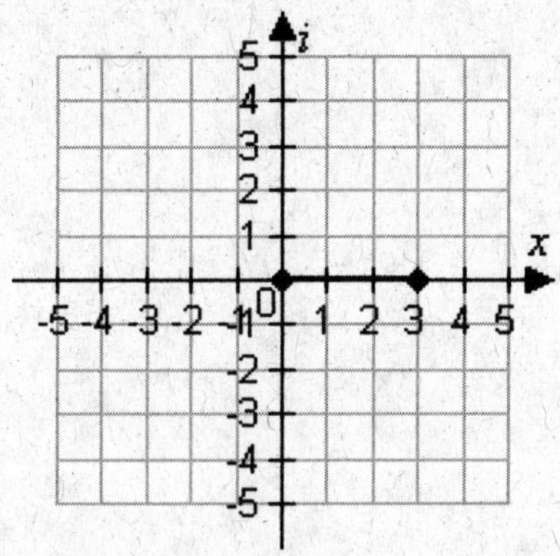

B)

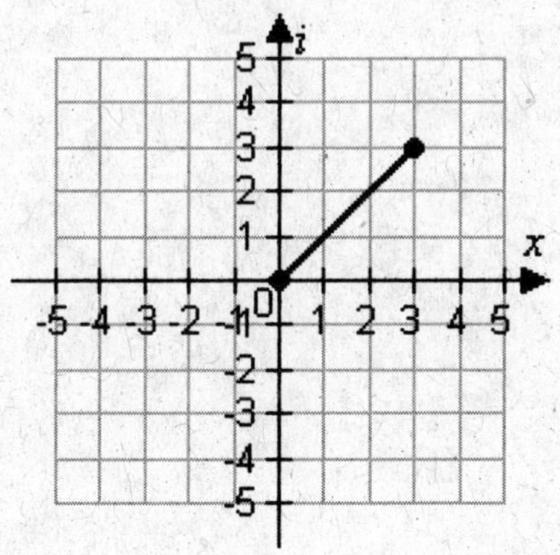

C)

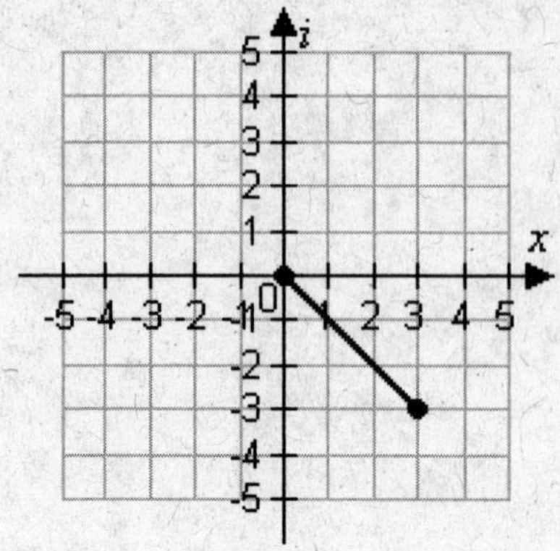

D)

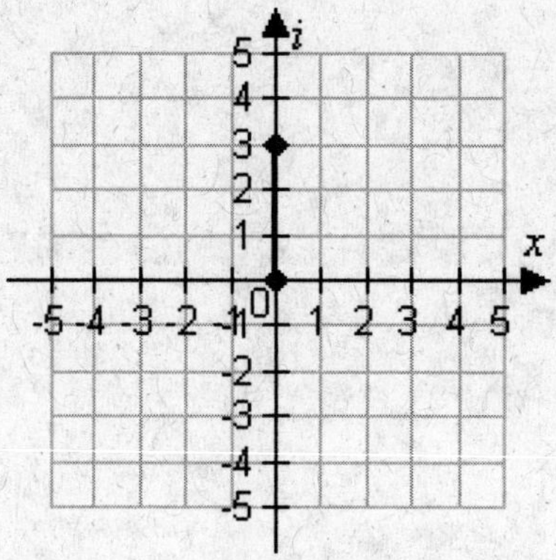

E)

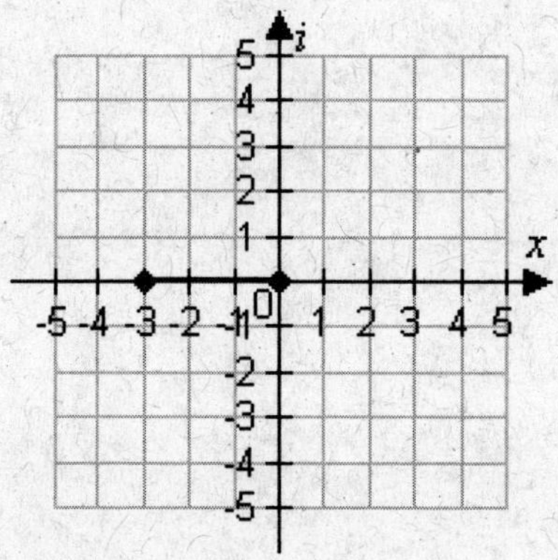

8. Using the figure below, sketch a graph of the given vector. [The graphs in the answer choices are drawn to the same scale as the graph below.]

2u + v

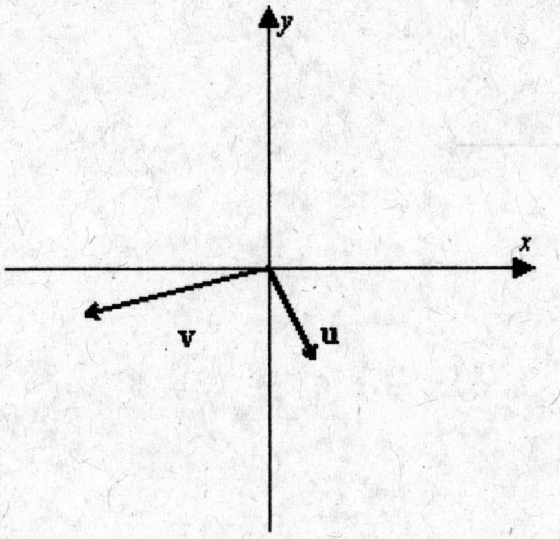

A)

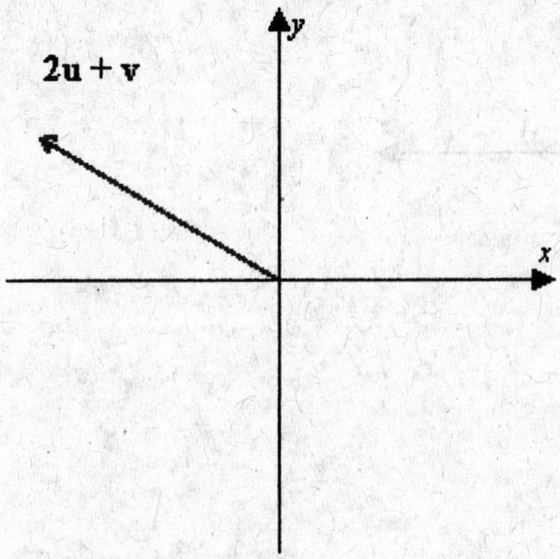

B)

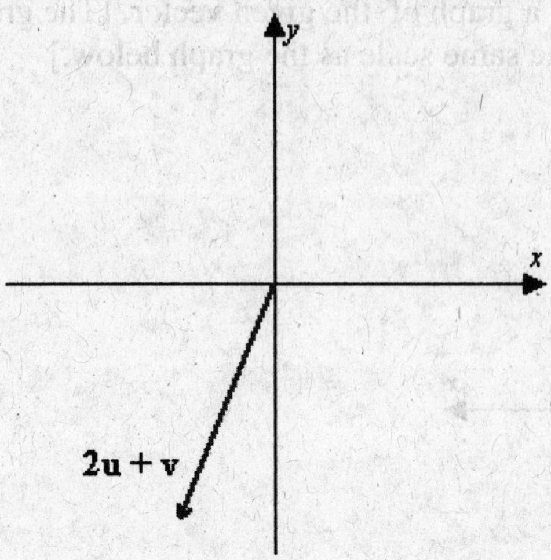

C)

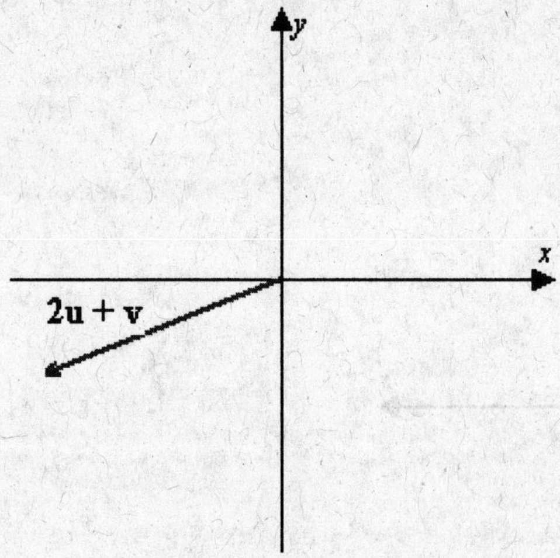

D)

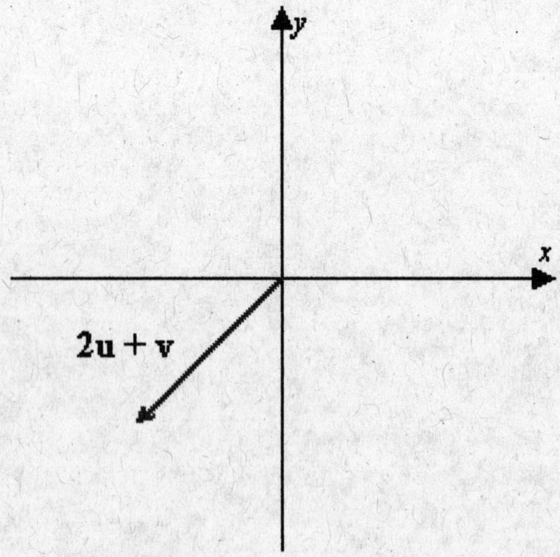

E) none of these

9. Find the vector **v** that has a magnitude of 6 and is in the same direction as **u**, where $\mathbf{u} = \langle -6, -5 \rangle$.

A) $\mathbf{v} = \left\langle \dfrac{-36}{\sqrt{61}}, \dfrac{-30}{\sqrt{61}} \right\rangle$

B) $\mathbf{v} = \left\langle \dfrac{-6}{\sqrt{61}}, \dfrac{-5}{\sqrt{61}} \right\rangle$

C) $\mathbf{v} = \left\langle \dfrac{-1}{\sqrt{61}}, \dfrac{-6}{\sqrt{61}} \right\rangle$

D) $\mathbf{v} = \left\langle \dfrac{-1}{\sqrt{61}}, \dfrac{1}{\sqrt{61}} \right\rangle$

E) $\mathbf{v} = \left\langle \dfrac{5}{\sqrt{61}}, \dfrac{6}{\sqrt{61}} \right\rangle$

10. Use DeMoivre's Theorem to find the indicated power of the following complex number.

$$\left(5 + 5\sqrt{3}\,i\right)^6$$

A) 1,953,125
B) 390,625
C) 78,125
D) 1,000,000
E) 3,125

11. Given vectors $\mathbf{u} = \langle -2, -3 \rangle$ and $\mathbf{v} = \langle 4, -4 \rangle$, determine the quantity indicated below.

$$-4\mathbf{u} \cdot 3\mathbf{v}$$

A) –216
B) –240
C) 68
D) –48
E) 4

12. Find all solutions to the following equation.

$$x^3 - 216 = 0$$

A) $x = 6,\ \ 3 + 3\sqrt{3}i,\ \ 3 - 3\sqrt{3}i$
B) $x = 6,\ \ -6$
C) $x = -6,\ \ 3 - 3\sqrt{3}i,\ \ 3 + 3\sqrt{3}i$
D) $x = 6$
E) $x = 6,\ \ -3 + 3\sqrt{3}i,\ \ -3 - 3\sqrt{3}i$

13. Find the standard form of the complex number shown below. Round answers to two decimal places.

$$6(\cos 190° + i \sin 190°)$$

A) $-5.91 - 1.04i$
B) $-6.14 - 1.43i$
C) $-6.66 - 2.14i$
D) $-7.15 - 2.39i$
E) $-7.33 - 2.57i$

14. Find the trigonometric form of the complex number shown below.

$$-7i$$

A) $7(\cos 0 + i \sin 0)$
B) $7(\cos \pi + i \sin \pi)$
C) $7\left(\cos \frac{3\pi}{2} + i \sin \frac{3\pi}{2}\right)$
D) $7\left(\cos \frac{\pi}{4} + i \sin \frac{\pi}{4}\right)$
E) $7\left(\cos \frac{\pi}{2} + i \sin \frac{\pi}{2}\right)$

15. Find the absolute value of the complex number $-4 + 6i$.

A) $\sqrt{10}$
B) $3\sqrt{10}$
C) 52
D) $5\sqrt{13}$
E) $2\sqrt{13}$

16. Given $a = 12$, $b = 8$, and $c = 11$, use the Law of Cosines to solve the triangle for the value of B. Round answer to two decimal places.
   A) 60.33°
   B) 76.53°
   C) 40.42°
   D) 80.44°
   E) 63.06°

17. A triangular parcel of land has sides of lengths 100, 580, and 600 feet. Approximate the area of the land. Round answer to nearest foot.
   A) 29,432 ft^2
   B) 28,800 ft^2
   C) 28,195 ft^2
   D) 28,182 ft^2
   E) 28,787 ft^2

18. Find the component form of vector **v** with initial point $(-5,1)$ and terminal point $(-3,-2)$.
   A) $\mathbf{v} = \langle 2,3 \rangle$
   B) $\mathbf{v} = \langle -3,4 \rangle$
   C) $\mathbf{v} = \langle 2,-3 \rangle$
   D) $\mathbf{v} = \langle 3,2 \rangle$
   E) $\mathbf{v} = \langle -6,-1 \rangle$

19. Find the fourth roots of $-\frac{1}{2}+\frac{\sqrt{3}}{2}i$. Write the roots in trigonometric form.

A) $w_1 = \cos(20°) + i\sin(20°)$
$w_2 = \cos(110°) + i\sin(110°)$
$w_3 = \cos(200°) + i\sin(200°)$
$w_4 = \cos(290°) + i\sin(290°)$

B) $w_1 = \cos(30°) + i\sin(30°)$
$w_2 = \cos(120°) + i\sin(120°)$
$w_3 = \cos(210°) + i\sin(210°)$
$w_4 = \cos(300°) + i\sin(300°)$

C) $w_1 = \cos(35°) + i\sin(35°)$
$w_2 = \cos(125°) + i\sin(125°)$
$w_3 = \cos(215°) + i\sin(215°)$
$w_4 = \cos(305°) + i\sin(305°)$

D) $w_1 = \cos(25°) + i\sin(25°)$
$w_2 = \cos(115°) + i\sin(115°)$
$w_3 = \cos(205°) + i\sin(205°)$
$w_4 = \cos(295°) + i\sin(295°)$

E) $w_1 = \cos(40°) + i\sin(40°)$
$w_2 = \cos(130°) + i\sin(130°)$
$w_3 = \cos(220°) + i\sin(220°)$
$w_4 = \cos(310°) + i\sin(310°)$

20. Given $a = 5$, $b = 9$, and $c = 12$, use Heron's Area Formula to find the area of triangle ABC. Round answer to two decimal places.

A) 20.40 sq. units
B) 16.97 sq. units
C) 19.60 sq. units
D) 12.65 sq. units
E) 17.40 sq. units

Answer Key

1. D
2. D
3. E
4. D
5. B
6. D
7. D
8. B
9. A
10. D
11. D
12. E
13. A
14. C
15. E
16. C
17. B
18. C
19. B
20. A

Name: ____________________________ Date: ______________

1. A boat race runs along a triangular course marked by buoys *A*, *B*, and *C*. The race begins with the boats headed west for 3900 meters. The other two sides of the course lie to the north of the first side, and their lengths are 1900 meters and 2500 meters, respectively. Find the bearing for the last leg of the race. Round your answer to two decimals.
   A) $-\text{inf}^\circ$
   B) 113.74°
   C) 180.01°
   D) 156.26°
   E) 90.01°

2. Find the fourth roots of $-\frac{1}{2}-\frac{\sqrt{3}}{2}i$. Write the roots in trigonometric form.

A) $w_1 = \cos(50°) + i\sin(50°)$
$w_2 = \cos(140°) + i\sin(140°)$
$w_3 = \cos(230°) + i\sin(230°)$
$w_4 = \cos(320°) + i\sin(320°)$

B) $w_1 = \cos(60°) + i\sin(60°)$
$w_2 = \cos(150°) + i\sin(150°)$
$w_3 = \cos(240°) + i\sin(240°)$
$w_4 = \cos(330°) + i\sin(330°)$

C) $w_1 = \cos(65°) + i\sin(65°)$
$w_2 = \cos(155°) + i\sin(155°)$
$w_3 = \cos(245°) + i\sin(245°)$
$w_4 = \cos(335°) + i\sin(335°)$

D) $w_1 = \cos(55°) + i\sin(55°)$
$w_2 = \cos(145°) + i\sin(145°)$
$w_3 = \cos(235°) + i\sin(235°)$
$w_4 = \cos(325°) + i\sin(325°)$

E) $w_1 = \cos(70°) + i\sin(70°)$
$w_2 = \cos(160°) + i\sin(160°)$
$w_3 = \cos(250°) + i\sin(250°)$
$w_4 = \cos(340°) + i\sin(340°)$

3. Perform the operation shown below and leave the result in trigonometric form.

$$\left[6\left(\cos 90° + i\sin 90°\right)\right]\left[2\left(\cos 280° + i\sin 280°\right)\right]$$

A) $8\left(\cos 370° + i\sin 370°\right)$
B) $12\left(\cos 252° + i\sin 252°\right)$
C) $12\left(\cos 190° + i\sin 190°\right)$
D) $8\left(\cos 190° + i\sin 190°\right)$
E) $12\left(\cos 370° + i\sin 370°\right)$

4. Find the absolute value of the complex number $-2-6i$.
   A) $2\sqrt{2}$
   B) $4\sqrt{2}$
   C) 40
   D) $5\sqrt{10}$
   E) $2\sqrt{10}$

5. Use DeMoivre's Theorem to find $(8+8i)^3$. Write the result in standard form.
   A) $1024-1024i$
   B) $-1024+1024i$
   C) $-1024-1024i$
   D) $512+512i$
   E) $1024+1024i$

6. Use Heron's area formula to find the area of the triangle pictured below, if $a = 8$ inches, b = 10 inches, and c = 4 inches.

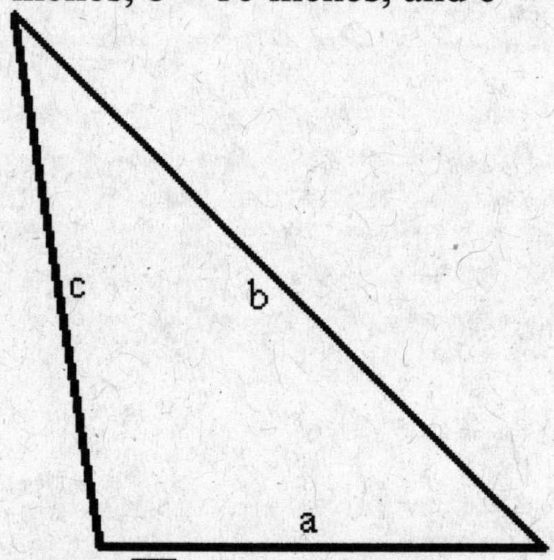

   A) $\sqrt{21}$
   B) 16
   C) 40
   D) $12\sqrt{21}$
   E) $\sqrt{231}$

7. A plane flies 740 miles from City A to City B with a bearing of 25° (clockwise from north). Then it flies 614 miles from City B to City C with a bearing of 49°. Find the straight-line distance from City C to City A. Round your answer to two decimals.
   A) 307.31 miles
   B) 1347.06 miles
   C) 1137.63 miles
   D) 911.87 miles
   E) 1324.67 miles

8. Determine the area of a triangle having the following measurements. Round your answer to two decimal places.
   $A = 128°,\ b = 8,\ \text{and}\ c = 15$
   A) 56.74 sq. units
   B) 47.28 sq. units
   C) 37.82 sq. units
   D) 42.55 sq. units
   E) 52.01 sq. units

9. Perform the indicated operation using trigonometric form. Leave answer in trigonometric form.
   $(8+8i)(2+2i)$
   A) $-16\left(\cos\frac{5\pi}{4}+i\sin\frac{5\pi}{4}\right)$
   B) $-32\left(\cos\frac{3\pi}{2}+i\sin\frac{3\pi}{2}\right)$
   C) $16\left(\cos\frac{\pi}{2}+i\sin\frac{\pi}{2}\right)$
   D) $32\left(\cos\frac{\pi}{2}+i\sin\frac{\pi}{2}\right)$
   E) $16\left(\cos\frac{3\pi}{4}+i\sin\frac{3\pi}{4}\right)$

10. Find the component form of **v** if $\|\mathbf{v}\| = 8$ and the angle it makes with the x-axis is $120°$.
   A) $\langle -4\sqrt{3}, 4 \rangle$
   B) $\langle -8, 8\sqrt{3} \rangle$
   C) $\langle -8\sqrt{3}, -8 \rangle$
   D) $\langle -4, 4\sqrt{3} \rangle$
   E) $\langle -4\sqrt{2}, 4\sqrt{2} \rangle$

11. Determine the area of a triangle having the following measurements. Round your answer to two decimal places.
   $C = 59°24'$, $a = 6$, and $b = 14.5$
   A) 37.44 sq. units
   B) 33.70 sq. units
   C) 29.95 sq. units
   D) 44.93 sq. units
   E) 41.19 sq. units

12. Determine whether **u** and **v** are orthogonal, parallel, or neither.
   $\mathbf{u} = \left\langle \frac{7}{3}, \frac{-1}{2} \right\rangle$, $\mathbf{v} = \langle 14, -3 \rangle$
   A) orthogonal
   B) parallel
   C) neither

13. The vector $\mathbf{u} = \langle 3700, 4600 \rangle$ gives the number of units of two models of laptops produced by a company. The vector $\mathbf{v} = \langle 1550, 1150 \rangle$ gives the prices (in dollars) of the two models of laptops, respectively. Identify the vector operation used to increase revenue by 3%.
   A) $\mathbf{u} \cdot (1.03)\|\mathbf{v}\|$
   B) $1.03(\mathbf{u} \cdot \mathbf{v})$
   C) $1.03\|\mathbf{u} \cdot \mathbf{v}\|$
   D) $1.03\|\mathbf{u}\| \cdot \mathbf{v}$
   E) $1.03(\mathbf{u} + \mathbf{v})$

14. After a severe storm, three sisters, April, May, and June, stood on their front porch and noticed that the tree in their front yard was leaning $5°$ from vertical toward the house. From the porch, which is 100 feet away from the base of the tree, they noticed that the angle of elevation to the top of the tree was $29°$. Approximate the height of the tree. Round answer to two decimal places.
    A) 57.19 feet
    B) 58.48 feet
    C) 68.48 feet
    D) 53.07 feet
    E) 60.01 feet

15. Write the complex number shown below in trigonometric form.

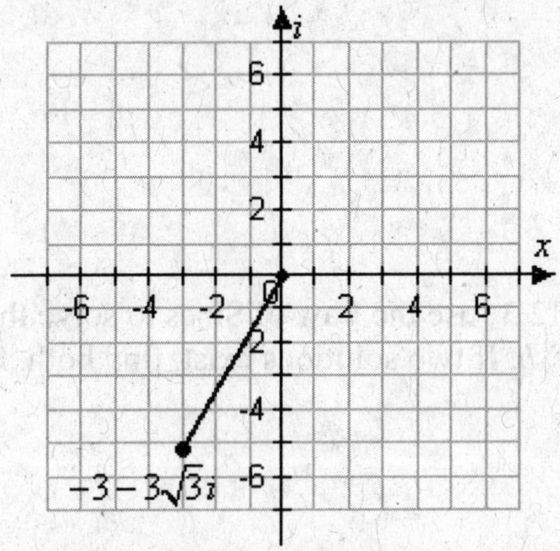

A) $6\left(\cos\frac{7\pi}{6}+i\sin\frac{7\pi}{6}\right)$

B) $6\left(\cos\frac{4\pi}{3}+i\sin\frac{4\pi}{3}\right)$

C) $6\left(\cos\frac{5\pi}{3}+i\sin\frac{5\pi}{3}\right)$

D) $6\left(\cos\frac{5\pi}{4}+i\sin\frac{5\pi}{4}\right)$

E) $6\left(\cos\frac{11\pi}{6}+i\sin\frac{11\pi}{6}\right)$

16. Let $\mathbf{u}=\langle 10,6\rangle$ and $\mathbf{v}=\langle 11,15\rangle$. Find $\mathbf{u}-\mathbf{v}$.
   A) $\langle 25,17\rangle$
   B) $\langle -1,-9\rangle$
   C) $\langle 21,21\rangle$
   D) $\langle -5,-5\rangle$
   E) $\langle 17,25\rangle$

17. Given vectors $\mathbf{u}=\langle -3,1\rangle$ and $\mathbf{v}=\langle 2,-2\rangle$, determine the quantity indicated below.
   $\mathbf{u}\cdot -4\mathbf{v}$
   A) 20
   B) −16
   C) 2
   D) 32
   E) 14

18. Given $C=130°$, $a=16.9$, and $c=12.3$, use the Law of Sines to solve the triangle (if possible) for the value of b. If two solutions exist, find both. Round answer to two decimal places.
   A) $b=3.73$
   B) $b=1.29$ and 5.61
   C) $b=6.98$
   D) $b=0.36$ and 5.74
   E) not possible

19. If $\|\mathbf{u}\|=5$ and $\|\mathbf{v}\|=3$, and the vectors make angles of $150°$ and $110°$ with the x-axis respectively, find the component form of the sum of **u** and **v**. Round answers to two decimal places.
   A) $\langle -5.36,5.32\rangle$
   B) $\langle -3.30,-0.32\rangle$
   C) $\langle 1.47,-1.51\rangle$
   D) $\langle 3.53,-7.15\rangle$
   E) $\langle -4.31,6.20\rangle$

20. Determine whether **u** and **v** are orthogonal, parallel, or neither.

    $\mathbf{u} = \langle 1, -5 \rangle$, $\mathbf{v} = \langle 20, 3 \rangle$

    A) orthogonal
    B) parallel
    C) neither

Answer Key

1. B
2. B
3. E
4. E
5. B
6. E
7. E
8. B
9. D
10. D
11. A
12. B
13. B
14. D
15. B
16. B
17. D
18. E
19. A
20. C

Name: ____________________________ Date: ____________

1. A 400 -pound trailer is sitting on an exit ramp inclined at $31°$ on Highway 35. How much force is required to keep the trailer from rolling back down the exit ramp? Round answer to two decimal places.
   A) 206.02 pounds
   B) 342.87 pounds
   C) 274.44 pounds
   D) 308.65 pounds
   E) 325.76 pounds

2. Find the component form of vector **v**.

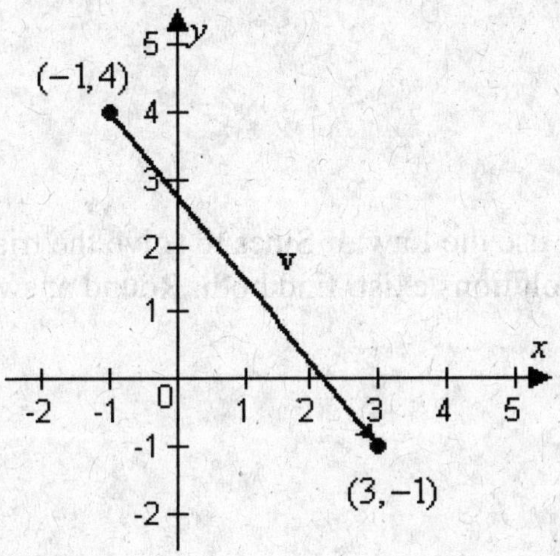

   A) $\mathbf{v} = \langle -5, 4 \rangle$

   B) $\mathbf{v} = \langle 4, -5 \rangle$

   C) $\mathbf{v} = \langle -4, -5 \rangle$

   D) $\mathbf{v} = \langle 4, -3 \rangle$

   E) $\mathbf{v} = \langle 5, 4 \rangle$

3. Given $a = 6$, $b = 9$, and $c = 7$, use the Law of Cosines to solve the triangle for the value of A. Round answer to two decimal places.

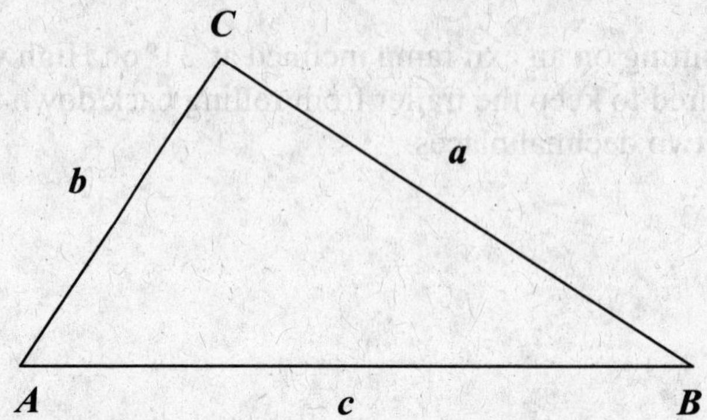

Figure not drawn to scale

A) 60.33°
B) 50.98°
C) 80.44°
D) 87.27°
E) 41.75°

4. Given $A = 16°$, $b = 12$, and $a = 10$, use the Law of Sines to solve the triangle (if possible) for the value of c. If two solutions exist, find both. Round answer to two decimal places.
A) $c = 18.62$
B) $c = 2.10$ and 20.97
C) $c = 19.75$
D) $c = 1.12$ and 17.85
E) not possible

5. Given $a = 7$, $b = 5$, and $c = 8$, use the Law of Cosines to solve the triangle for the value of B. Round answer to two decimal places.

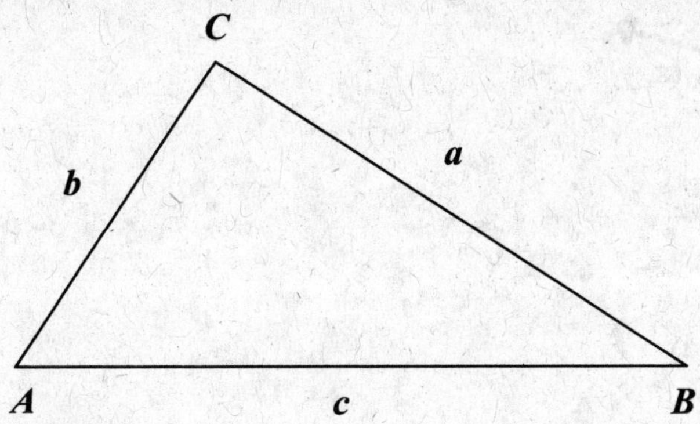

Figure not drawn to scale

A) 60.33°
B) 81.79°
C) 80.44°
D) 38.21°
E) 60.00°

6. Determine whether **u** and **v** are orthogonal, parallel, or neither.

$\mathbf{u} = \langle -7, 3 \rangle$, $\mathbf{v} = \langle 12, 35 \rangle$

A) orthogonal
B) parallel
C) neither

7. In the figure below, $(a,b)=(4,6)$ and $(c,d)=(6,5)$. Find (x,y) so that $\mathbf{u}=\mathbf{v}$.

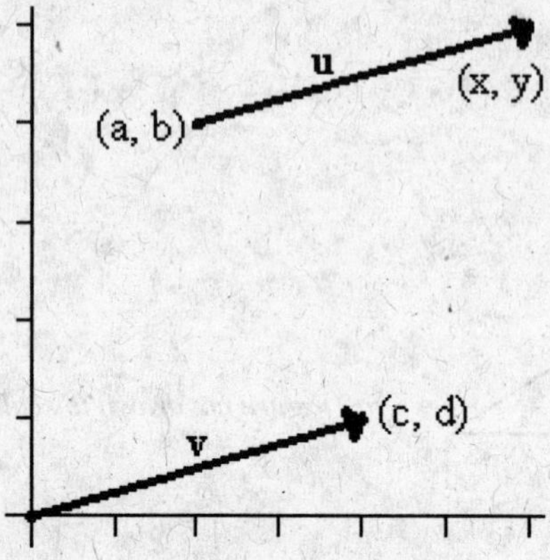

A) $(2,-1)$
B) $(12,9)$
C) $(0,1)$
D) $(9,12)$
E) $(10,11)$

8. A straight road makes an angle, *A*, of 20° with the horizontal. When the angle of elevation, *B*, of the sun is 60°, a vertical pole beside the road casts a shadow 8 feet long parallel to the road. Approximate the length of the pole. Round answer to two decimal places.

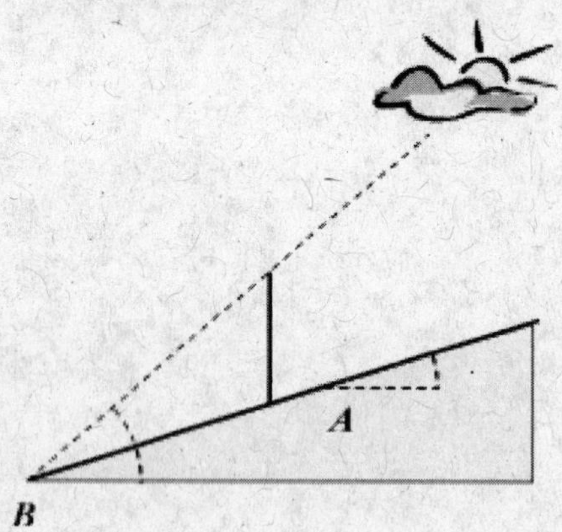

A) 5.47 feet
B) 6.22 feet
C) 10.28 feet
D) 13.86 feet
E) 5.94 feet

9. Given $A = 51°$, $b = 6$, and $c = 9$, use the Law of Cosines to solve the triangle for the value of a. Round answer to two decimal places.

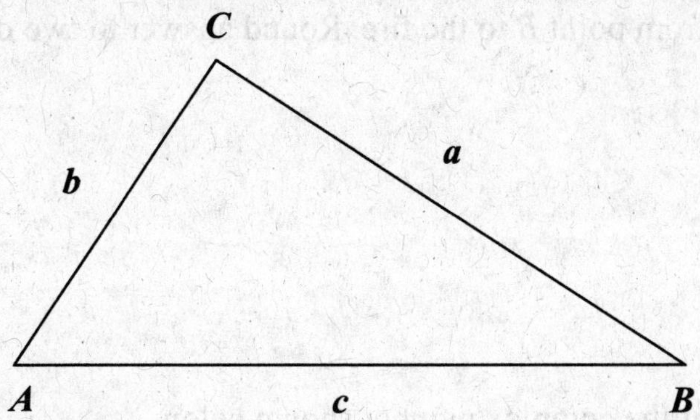

Figure not drawn to scale

A) 7.00
B) 10.30
C) 7.83
D) 13.60
E) 8.65

10. Use vectors to find the measure of the angle at vertex B of triangle ABC, when $A = (6,1), B = (-1,3)$, and $C = (-4,-5)$. Round answer to two decimal places.
A) 92.52
B) 94.61°
C) 95.95
D) 93.42
E) 97.13°

11. Determine the area of a triangle having the following measurements. Round your answer to two decimal places.
$A = 128°$, $b = 7$, and $c = 14$
A) 46.34 sq. units
B) 38.61 sq. units
C) 30.89 sq. units
D) 34.75 sq. units
E) 42.47 sq. units

12. A park ranger at point A observes a fire in the direction $N\,29°15'E$. Another ranger at point B, 9 miles due east of A, sites the same fire at $N\,50°39'W$. Determine the distance from point B to the fire. Round answer to two decimal places.
    A) 4.47 miles
    B) 4.72 miles
    C) 7.98 miles
    D) 7.07 miles
    E) 7.47 miles

13. Find the standard form of the complex number shown below.
    $$11\left(\cos\frac{5\pi}{6}+i\sin\frac{5\pi}{6}\right)$$
    A) $-\frac{11}{2}+\frac{11\sqrt{3}}{2}i$
    B) $\frac{11\sqrt{3}}{4}-\frac{11}{4}i$
    C) $-\frac{2}{11}+\frac{2\sqrt{3}}{11}i$
    D) $-\frac{11\sqrt{3}}{2}+\frac{11}{2}i$
    E) $\frac{4\sqrt{3}}{11}-\frac{4}{11}i$

14. Perform the operation shown below and leave the result in trigonometric form.
    $$\frac{8(\cos 2.3+i\sin 2.3)}{4(\cos 1.7+i\sin 1.7)}$$
    A) $2(\cos 0.6+i\sin 0.6)$
    B) $4(\cos 0.6+i\sin 0.6)$
    C) $4(\cos 1.4+i\sin 1.4)$
    D) $2(\cos 4+i\sin 4)$
    E) $2(\cos 3.91+i\sin 3.91)$

15. Use DeMoivre's Theorem to find the indicated power of the following complex number.

$$\left[2\left(\cos\frac{\pi}{3}+i\sin\frac{\pi}{3}\right)\right]^{12}$$

A) 2048
B) 4,096
C) $4,096+4096\sqrt{3}$
D) $2,048+2048\sqrt{3}$
E) –4,096

16. Determine the area of a triangle having the following measurements. Round your answer to two decimal places.

$B = 76°30'$, $a = 8$, and $c = 13$

A) 60.68 sq. units
B) 45.51 sq. units
C) 40.45 sq. units
D) 50.56 sq. units
E) 55.62 sq. units

17. Given $C = 126°$, $a = 10.10$, and $c = 18$, use the Law of Sines to solve the triangle (if possible) for the value of b. If two solutions exist, find both. Round answer to two decimal places.

A) $b = 3.39$
B) $b = 0.58$ and 2.73
C) $b = 10.10$
D) $b = 11.89$ and 4.03
E) not possible

18. Find the vector **v** that has a magnitude of 4 and is in the same direction as **u**, where $\mathbf{u}=\langle -5,4\rangle$.

A) $\mathbf{v}=\left\langle \dfrac{-20}{\sqrt{41}},\dfrac{16}{\sqrt{41}}\right\rangle$

B) $\mathbf{v}=\left\langle \dfrac{-5}{\sqrt{41}},\dfrac{4}{\sqrt{41}}\right\rangle$

C) $\mathbf{v}=\left\langle \dfrac{-4}{\sqrt{41}},\dfrac{1}{\sqrt{41}}\right\rangle$

D) $\mathbf{v}=\left\langle \dfrac{-1}{\sqrt{41}},\dfrac{1}{\sqrt{41}}\right\rangle$

E) $\mathbf{v}=\left\langle \dfrac{-4}{\sqrt{41}},\dfrac{5}{\sqrt{41}}\right\rangle$

19. If $\mathbf{u}=-\mathbf{i}-8\mathbf{j}$ and $\mathbf{v}=-4\mathbf{i}+\mathbf{j}$, find $\mathbf{w}=-3\mathbf{u}-4\mathbf{v}$.

A) $\mathbf{w}=15\mathbf{i}+28\mathbf{j}$
B) $\mathbf{w}=-\mathbf{i}+40\mathbf{j}$
C) $\mathbf{w}=16\mathbf{i}+20\mathbf{j}$
D) $\mathbf{w}=19\mathbf{i}+29\mathbf{j}$
E) $\mathbf{w}=19\mathbf{i}+20\mathbf{j}$

20. Given vectors $\mathbf{u}=\langle -2,5\rangle$ and $\mathbf{v}=\langle 2,-4\rangle$, determine the quantity indicated below.

$\mathbf{u}\cdot -\mathbf{v}$

A) 20
B) 2
C) 16
D) 24
E) 18

Answer Key

1. A
2. B
3. E
4. B
5. D
6. C
7. E
8. C
9. A
10. B
11. B
12. C
13. D
14. A
15. B
16. D
17. C
18. A
19. E
20. D

Name: ____________________________ Date: ______________

1. Given $C = 130°$, $a = 19.9$, and $c = 15.3$, use the Law of Sines to solve the triangle (if possible) for the value of b. If two solutions exist, find both. Round answer to two decimal places.
   A) $b = 4.20$
   B) $b = 1.76$ and 6.08
   C) $b = 7.45$
   D) $b = 0.83$ and 6.21
   E) not possible

2. Use DeMoivre's Theorem to find the indicated power of the following complex number.
   $\left(3+3\sqrt{3}\,i\right)^6$
   A) 19,683
   B) 6,561
   C) 2,187
   D) 46,656
   E) 243

3. Find the trigonometric form of the complex number shown below.
   $-3-3i$
   A) $3\left(\cos\frac{\pi}{4}+i\sin\frac{\pi}{4}\right)$
   B) $\sqrt{3}\left(\cos\frac{11\pi}{6}+i\sin\frac{11\pi}{6}\right)$
   C) $9\left(\cos\frac{7\pi}{6}+i\sin\frac{7\pi}{6}\right)$
   D) $3\sqrt{2}\left(\cos\frac{5\pi}{4}+i\sin\frac{5\pi}{4}\right)$
   E) $\sqrt{2}\left(\cos\frac{\pi}{4}+i\sin\frac{\pi}{4}\right)$

4. Find the standard form of the complex number shown below. Round answers to two decimal places.

   $2(\cos 340° + i \sin 340°)$

   A) $1.88 - 0.68i$
   B) $2.11 - 1.07i$
   C) $2.63 - 1.78i$
   D) $3.12 - 2.03i$
   E) $3.30 - 2.21i$

5. Given $a = 13$, $b = 11$, and $c = 6$, use Heron's Area Formula to find the area of triangle ABC. Round answer to two decimal places.

   A) 32.86 sq. units
   B) 28.14 sq. units
   C) 20.78 sq. units
   D) 30.59 sq. units
   E) 28.10 sq. units

6. Determine the area of a triangle having the following measurements. Round your answer to two decimal places.

   $B = 75°22'$, $a = 11$, and $c = 14$

   A) 89.40 sq. units
   B) 67.05 sq. units
   C) 59.60 sq. units
   D) 74.50 sq. units
   E) 81.95 sq. units

7. Determine whether **u** and **v** are orthogonal, parallel, or neither.

   $\mathbf{u} = \langle 1, 5 \rangle$, $\mathbf{v} = \langle -25, 2 \rangle$

   A) orthogonal
   B) parallel
   C) neither

8. Find the projection of **u** onto **v** if $\mathbf{u} = \langle 2, -1 \rangle,\ \mathbf{v} = \langle 5, 2 \rangle$.

A) $\left\langle \frac{40}{\sqrt{29}}, -\frac{8}{\sqrt{29}} \right\rangle$

B) $\left\langle \frac{16}{\sqrt{29}}, \frac{40}{\sqrt{29}} \right\rangle$

C) $\left\langle \frac{16}{\sqrt{29}}, -\frac{8}{\sqrt{29}} \right\rangle$

D) $\left\langle \frac{16}{\sqrt{29}}, \frac{16}{\sqrt{29}} \right\rangle$

E) $\left\langle \frac{40}{\sqrt{29}}, \frac{16}{\sqrt{29}} \right\rangle$

9. Perform the indicated operation using trigonometric form. Leave answer in trigonometric form.

$(-4-4i)(-3-3i)$

A) $-12\left(\cos\frac{5\pi}{4} + i\sin\frac{5\pi}{4}\right)$

B) $-24\left(\cos\frac{3\pi}{2} + i\sin\frac{3\pi}{2}\right)$

C) $12\left(\cos\frac{\pi}{2} + i\sin\frac{\pi}{2}\right)$

D) $24\left(\cos\frac{\pi}{2} + i\sin\frac{\pi}{2}\right)$

E) $12\left(\cos\frac{3\pi}{4} + i\sin\frac{3\pi}{4}\right)$

10. In the figure below, $a = 7$, $c = 20$, and $d = 14$. Use this information to solve the parallelogram for b. The diagonals of the parallelogram are represented by c and d. Round answer to two decimal places.

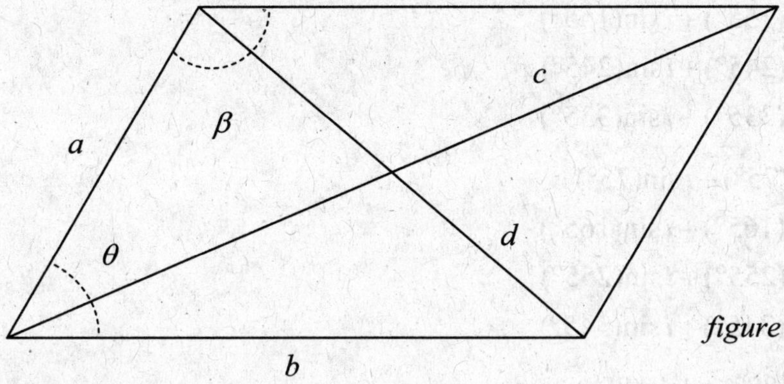

figure not drawn to scale

A) 7.14
B) 13.62
C) 15.78
D) 12.00
E) 11.46

11. Find the fourth roots of $\frac{1}{2}-\frac{\sqrt{3}}{2}i$. Write the roots in trigonometric form.

A) $w_1 = \cos(65°) + i\sin(65°)$
$w_2 = \cos(155°) + i\sin(155°)$
$w_3 = \cos(245°) + i\sin(245°)$
$w_4 = \cos(335°) + i\sin(335°)$

B) $w_1 = \cos(75°) + i\sin(75°)$
$w_2 = \cos(165°) + i\sin(165°)$
$w_3 = \cos(255°) + i\sin(255°)$
$w_4 = \cos(345°) + i\sin(345°)$

C) $w_1 = \cos(80°) + i\sin(80°)$
$w_2 = \cos(170°) + i\sin(170°)$
$w_3 = \cos(260°) + i\sin(260°)$
$w_4 = \cos(350°) + i\sin(350°)$

D) $w_1 = \cos(70°) + i\sin(70°)$
$w_2 = \cos(160°) + i\sin(160°)$
$w_3 = \cos(250°) + i\sin(250°)$
$w_4 = \cos(340°) + i\sin(340°)$

E) $w_1 = \cos(85°) + i\sin(85°)$
$w_2 = \cos(175°) + i\sin(175°)$
$w_3 = \cos(265°) + i\sin(265°)$
$w_4 = \cos(355°) + i\sin(355°)$

12. A plane flies 440 kilometers with a bearing of 320° from City A to City B (see figure below). The plane then flies 800 kilometers from City B to City C, which is directly west of City A. Find the bearing of the flight from City B to City C. Round your answer to two decimals.

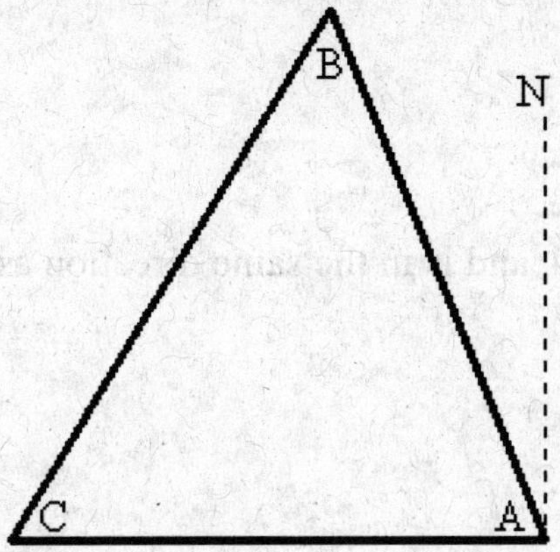

A) 245.08°
B) 335.08°
C) 24.92°
D) 20.70°
E) 249.30°

13. Given $\mathbf{u} = \langle -3, 7 \rangle$ and $\mathbf{v} = \langle 1, -5 \rangle$, find $\mathbf{u} \cdot \mathbf{v}$.

A) 32
B) –3
C) –38
D) 22
E) 8

14. Determine whether **u** are **v** and orthogonal, parallel, or neither.

$\mathbf{u} = \langle 6, 4 \rangle$, $\mathbf{v} = \langle -20, 30 \rangle$

A) orthogonal
B) parallel
C) neither

15. A triangular parcel of land has sides of lengths 900, 580, and 550 feet. Approximate the area of the land. Round answer to nearest foot.
   A) 154,345 ft^2
   B) 153,657 ft^2
   C) 152,972 ft^2
   D) 152,971 ft^2
   E) 153,656 ft^2

16. Find the vector **v** that has a magnitude of 4 and is in the same direction as **u,** where $\mathbf{u} = \langle -5, -3 \rangle$.
   A) $\mathbf{v} = \left\langle \dfrac{-20}{\sqrt{34}}, \dfrac{-12}{\sqrt{34}} \right\rangle$
   B) $\mathbf{v} = \left\langle \dfrac{-5}{\sqrt{34}}, \dfrac{-3}{\sqrt{34}} \right\rangle$
   C) $\mathbf{v} = \left\langle \dfrac{-4}{\sqrt{34}}, \dfrac{-4}{\sqrt{34}} \right\rangle$
   D) $\mathbf{v} = \left\langle \dfrac{-1}{\sqrt{34}}, \dfrac{1}{\sqrt{34}} \right\rangle$
   E) $\mathbf{v} = \left\langle \dfrac{3}{\sqrt{34}}, \dfrac{5}{\sqrt{34}} \right\rangle$

17. Use Heron's area formula to find the area of the triangle pictured below, if $a = 8$ inches, b = 14 inches, and c = 10 inches.

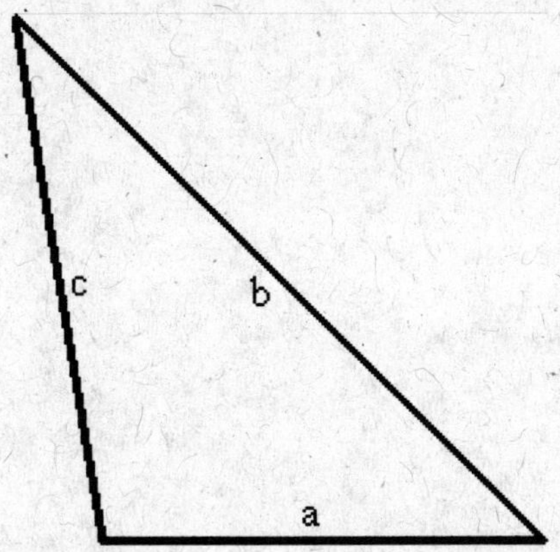

A) $4\sqrt{6}$
B) 40
C) 56
D) $12\sqrt{66}$
E) $16\sqrt{6}$

18. Find the magnitude of vector **v**.

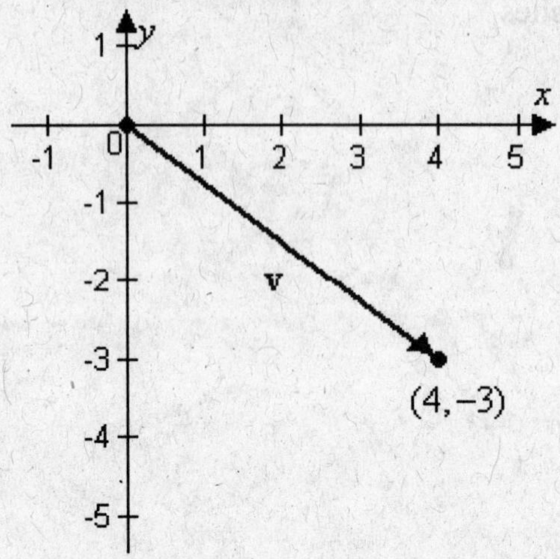

A) $\|\mathbf{v}\| = 4\sqrt{2}$

B) $\|\mathbf{v}\| = 2\sqrt{7}$

C) $\|\mathbf{v}\| = 6$

D) $\|\mathbf{v}\| = 6\sqrt{3}$

E) $\|\mathbf{v}\| = 5$

19. If $\|\mathbf{u}\| = 5$ and $\|\mathbf{v}\| = 4$, and the vectors make angles of 60° and 80° with the *x*-axis respectively, find the component form of the sum of **u** and **v**. Round answers to two decimal places.
A) $\langle 3.19, 8.27 \rangle$
B) $\langle 1.81, 0.39 \rangle$
C) $\langle 5.02, 6.44 \rangle$
D) $\langle 3.64, -1.44 \rangle$
E) $\langle 2.87, 8.39 \rangle$

20. Which of the following is a fifth root of $2(-1+i)$? 0 2

I. $\sqrt[10]{8}\left(\cos\left(\frac{3\pi}{20}\right)+i\sin\left(\frac{3\pi}{20}\right)\right)$

II. $\sqrt[10]{8}\left(\cos\left(\frac{19\pi}{20}\right)+i\sin\left(\frac{19\pi}{20}\right)\right)$

III. $\sqrt[5]{2}\left(\cos\left(\frac{3\pi}{20}\right)+i\sin\left(\frac{3\pi}{20}\right)\right)$

A) I is the only root.
B) None are roots.
C) III is the only root.
D) I and II are the only roots.
E) I and III are the only roots.

Answer Key

1. E
2. D
3. D
4. A
5. A
6. D
7. C
8. E
9. D
10. C
11. B
12. A
13. C
14. A
15. B
16. A
17. E
18. E
19. A
20. D

Name: ______________________________ Date: _______________

1. A straight road makes an angle, *A,* of $16°$ with the horizontal. When the angle of elevation, *B*, of the sun is $58°$, a vertical pole beside the road casts a shadow 6 feet long parallel to the road. Approximate the length of the pole. Round answer to two decimal places.

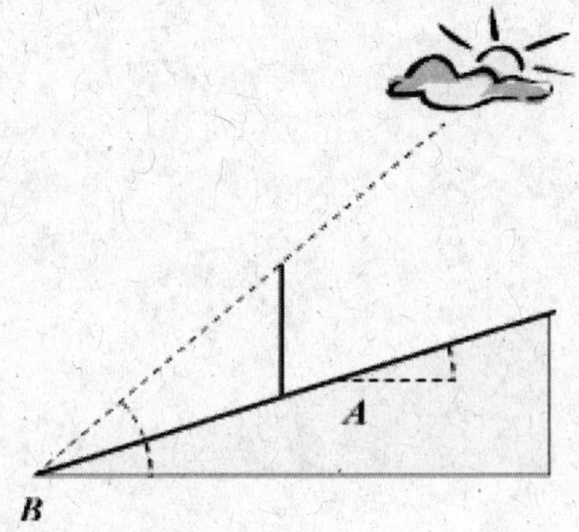

A) 3.12 feet
B) 4.75 feet
C) 7.58 feet
D) 9.60 feet
E) 4.73 feet

2. Given $A = 54°$, $B = 69°$, and $a = 7.1$, use the Law of Sines to solve the triangle for the value of b. Round answer to two decimal places.

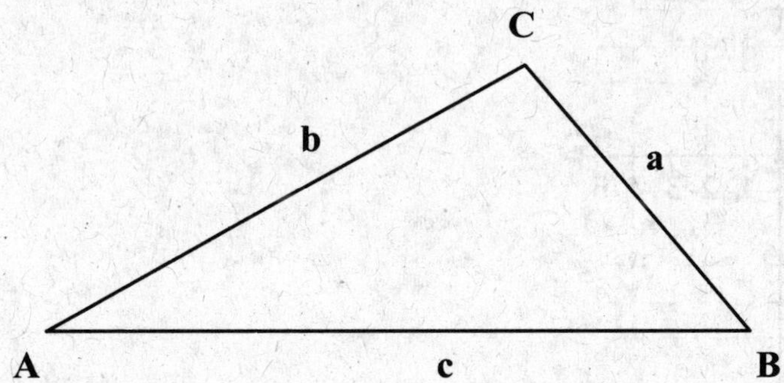

A) $b = 6.38$
B) $b = 7.36$
C) $b = 6.85$
D) $b = 8.19$
E) $b = 6.15$

3. Represent the complex number below graphically.

$\sqrt{3}-i$

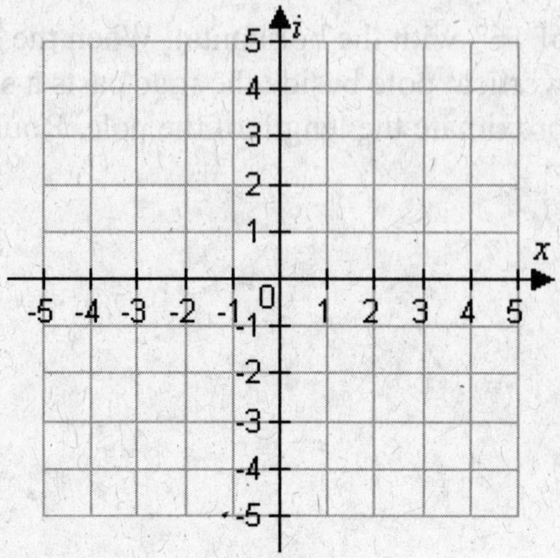

A)

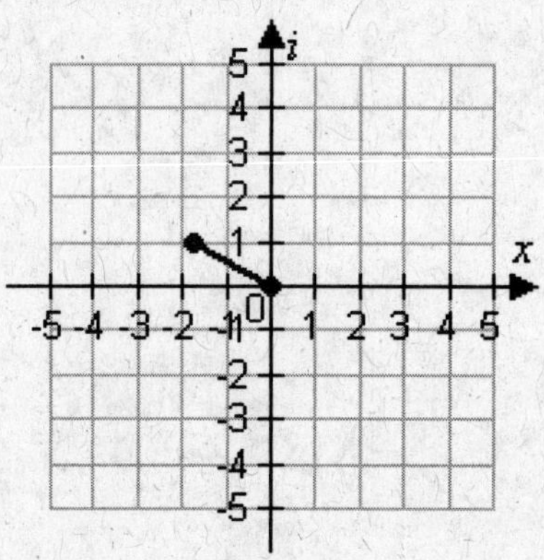

B)

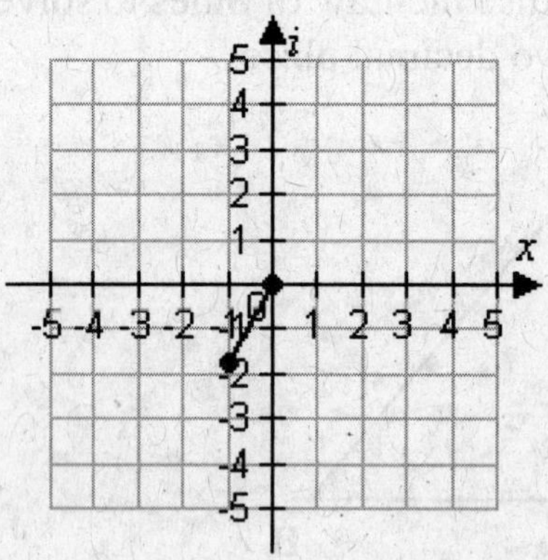

C)

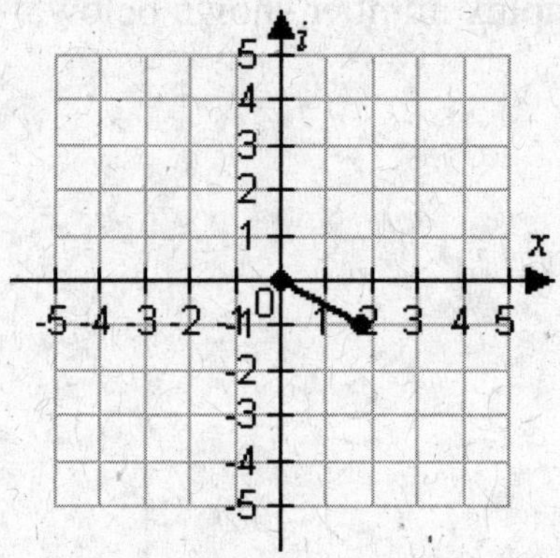

D)

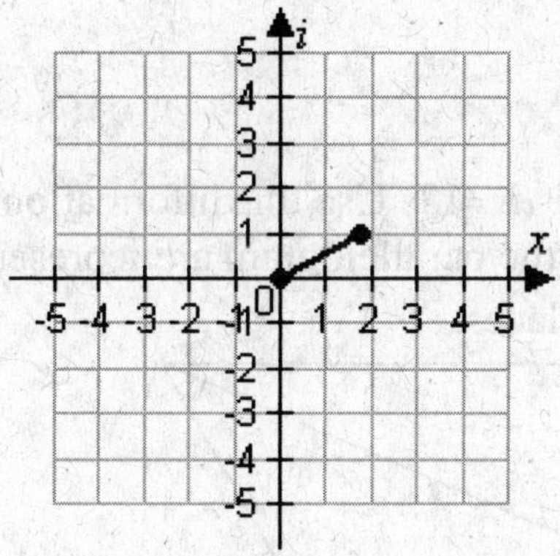

E)

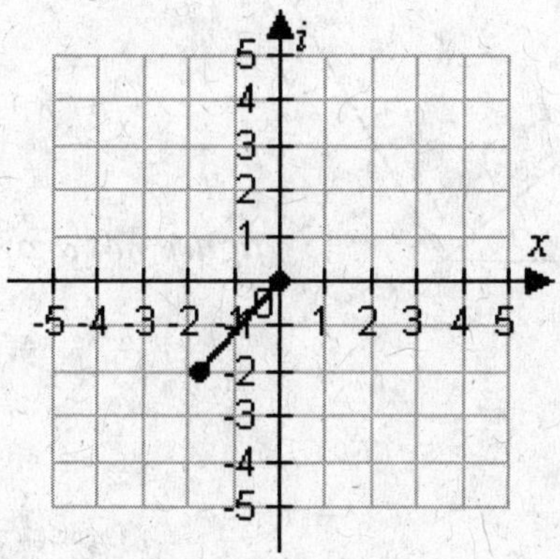

4. Find the trigonometric form of the complex number shown below.

   $-3i$

   A) $3(\cos 0 + i\sin 0)$
   B) $3(\cos \pi + i\sin \pi)$
   C) $3\left(\cos\frac{3\pi}{2} + i\sin\frac{3\pi}{2}\right)$
   D) $3\left(\cos\frac{\pi}{4} + i\sin\frac{\pi}{4}\right)$
   E) $3\left(\cos\frac{\pi}{2} + i\sin\frac{\pi}{2}\right)$

5. In the figure below, $a = 8$, $b = 12$, and $d = 13$. Use this information to solve the parallelogram for β. The diagonals of the parallelogram are represented by c and d. Round answer to two decimal places.

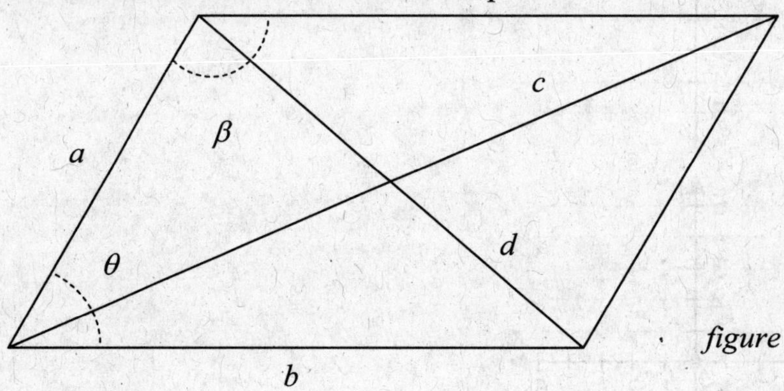

figure not drawn to scale

   A) 91.50°
   B) 78.28°
   C) 96.61°
   D) 101.72°
   E) 92.03°

6. Given vectors $\mathbf{u} = \langle u_1, u_2 \rangle$, $\mathbf{v} = \langle v_1, v_2 \rangle$, and $\mathbf{w} = \langle w_1, w_2 \rangle$, determine whether the result of the following expression is a vector or a scalar.

   $(\mathbf{w} \cdot \mathbf{w})\mathbf{w}$

   A) vector
   B) scalar

7. Find the trigonometric form of the complex number shown below.
   $8i$

   A) $8(\cos\pi + i\sin\pi)$
   B) $8\left(\cos\frac{\pi}{2} + i\sin\frac{\pi}{2}\right)$
   C) $8\left(\cos\frac{3\pi}{2} + i\sin\frac{3\pi}{2}\right)$
   D) $8\left(\cos\frac{\pi}{4} + i\sin\frac{\pi}{4}\right)$
   E) $8(\cos 0 + i\sin 0)$

8. In the figure below, $a = 8$, $b = 12$, and $c = 13$. Use this information to solve the parallelogram for θ. The diagonals of the parallelogram are represented by *c* and *d*. Round answer to two decimal places.

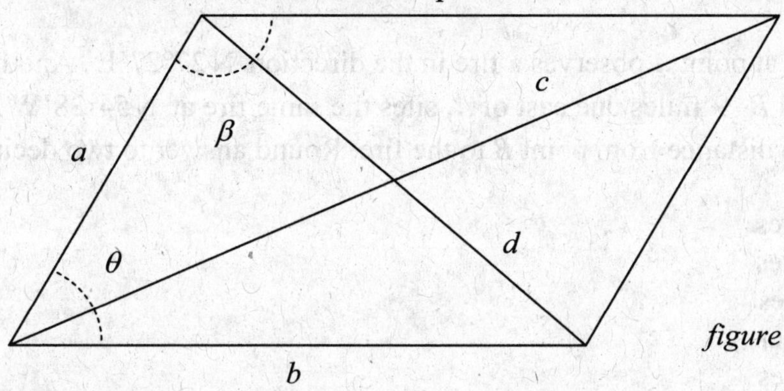

figure not drawn to scale

   A) 101.72°
   B) 95.86
   C) 90.00°
   D) 91.46
   E) 78.28°

9. A force of 50 pounds is exerted along a rope attached to a crate at an angle of 30° above the horizontal. The crate is moved 43 feet. How much work has been accomplished? Round answer to one decimal place.
   A) 2,150.0 foot-pounds
   B) 2,482.6 foot-pounds
   C) 2,006.0 foot-pounds
   D) 2,244.3 foot-pounds
   E) 1,862.0 foot-pounds

10. Perform the operation below and leave the result in trigonometric form.

$$\frac{17\left(\cos 59^\circ + i\sin 59^\circ\right)}{21\left(\cos 12^\circ + i\sin 12^\circ\right)}$$

A) $\frac{17}{21}\left(\cos 47^\circ + i\sin 47^\circ\right)$
B) $-4\left(\cos\frac{59^\circ}{12} + i\sin\frac{59^\circ}{12}\right)$
C) $38\left(\cos 71^\circ + i\sin 71^\circ\right)$
D) $357\left(\cos\frac{59^\circ}{12} + i\sin\frac{59^\circ}{12}\right)$
E) $\frac{17}{21}\left(\cos\frac{59^\circ}{12} + i\sin\frac{59^\circ}{12}\right)$

11. A park ranger at point *A* observes a fire in the direction $\text{N}\,27°27'\text{E}$. Another ranger at point *B*, 9 miles due east of *A*, sites the same fire at $\text{N}\,54°28'\text{W}$. Determine the distance from point *B* to the fire. Round answer to two decimal places.
A) 4.19 miles
B) 4.66 miles
C) 8.07 miles
D) 7.40 miles
E) 8.22 miles

12. Given $\mathbf{u} = -3\mathbf{i} - 2\mathbf{j}$ and $\mathbf{v} = -\mathbf{i} - 5\mathbf{j}$, determine $-9\mathbf{u} + 6\mathbf{v}$.
A) $-9\mathbf{u} + 6\mathbf{v} = 21\mathbf{i} - 12\mathbf{j}$
B) $-9\mathbf{u} + 6\mathbf{v} = 27\mathbf{i} - 21\mathbf{j}$
C) $-9\mathbf{u} + 6\mathbf{v} = -9\mathbf{i} - 3\mathbf{j}$
D) $-9\mathbf{u} + 6\mathbf{v} = 12\mathbf{i} - 3\mathbf{j}$
E) $-9\mathbf{u} + 6\mathbf{v} = -3\mathbf{i} + 3\mathbf{j}$

13. Find the component form of **v** if $\|\mathbf{v}\| = 4$ and the angle it makes with the *x*-axis is $120°$.

A) $\langle -2\sqrt{3}, 2 \rangle$
B) $\langle -4, 4\sqrt{3} \rangle$
C) $\langle -4\sqrt{3}, -4 \rangle$
D) $\langle -2, 2\sqrt{3} \rangle$
E) $\langle -2\sqrt{2}, 2\sqrt{2} \rangle$

14. Given vectors $\mathbf{u} = \langle u_1, u_2 \rangle$, $\mathbf{v} = \langle v_1, v_2 \rangle$, and $\mathbf{w} = \langle w_1, w_2 \rangle$, determine whether the result of the following expression is a vector or a scalar.

$\mathbf{u} \cdot \mathbf{w} + \mathbf{v} \cdot \mathbf{w}$

A) vector
B) scalar

15. Use Heron's area formula to find the area of the triangle pictured below, if $a = 10$ inches, b = 12 inches, and c = 4 inches.

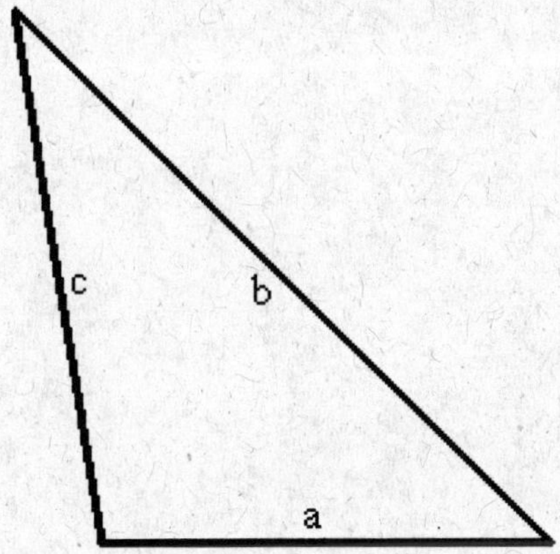

A) $3\sqrt{3}$
B) 20
C) 60
D) $8\sqrt{77}$
E) $3\sqrt{39}$

16. Given $a = 12$, $b = 5$, and $c = 15$, use the Law of Cosines to solve the triangle for the value of C. Round answer to two decimal places.

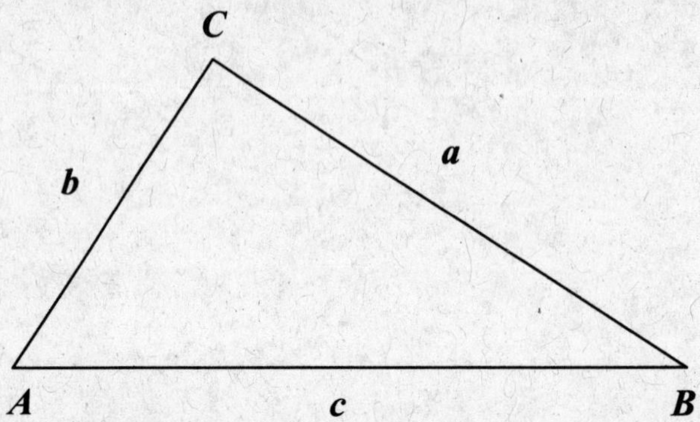

Figure not drawn to scale

A) 60.34°
B) 117.82°
C) 80.45°
D) 17.15°
E) 45.04°

17. Find the fourth roots of $-\frac{1}{2}+\frac{\sqrt{3}}{2}i$. Write the roots in trigonometric form.

A) $w_1 = \cos(20°) + i\sin(20°)$
$w_2 = \cos(110°) + i\sin(110°)$
$w_3 = \cos(200°) + i\sin(200°)$
$w_4 = \cos(290°) + i\sin(290°)$

B) $w_1 = \cos(30°) + i\sin(30°)$
$w_2 = \cos(120°) + i\sin(120°)$
$w_3 = \cos(210°) + i\sin(210°)$
$w_4 = \cos(300°) + i\sin(300°)$

C) $w_1 = \cos(35°) + i\sin(35°)$
$w_2 = \cos(125°) + i\sin(125°)$
$w_3 = \cos(215°) + i\sin(215°)$
$w_4 = \cos(305°) + i\sin(305°)$

D) $w_1 = \cos(25°) + i\sin(25°)$
$w_2 = \cos(115°) + i\sin(115°)$
$w_3 = \cos(205°) + i\sin(205°)$
$w_4 = \cos(295°) + i\sin(295°)$

E) $w_1 = \cos(40°) + i\sin(40°)$
$w_2 = \cos(130°) + i\sin(130°)$
$w_3 = \cos(220°) + i\sin(220°)$
$w_4 = \cos(310°) + i\sin(310°)$

18. Perform the indicated operation using trigonometric form. Leave answer in trigonometric form.

$$(3+3i)(5+5i)$$

A) $-15\left(\cos\frac{5\pi}{4}+i\sin\frac{5\pi}{4}\right)$

B) $-30\left(\cos\frac{3\pi}{2}+i\sin\frac{3\pi}{2}\right)$

C) $15\left(\cos\frac{\pi}{2}+i\sin\frac{\pi}{2}\right)$

D) $30\left(\cos\frac{\pi}{2}+i\sin\frac{\pi}{2}\right)$

E) $15\left(\cos\frac{3\pi}{4}+i\sin\frac{3\pi}{4}\right)$

19. Which of the following is a fourth root of 1?

I. $\cos\left(\frac{3\pi}{2}\right)+i\sin\left(\frac{3\pi}{2}\right)$

II. $\frac{1}{4}\left(\cos(0)+i\sin(0)\right)$

III. $\cos\left(\frac{\pi}{4}\right)+i\sin\left(\frac{\pi}{4}\right)$

A) None are roots.
B) II and III are the only roots.
C) I and II are the only roots.
D) I, II, and III are roots.
E) I is the only root.

20. After a severe storm, three sisters, April, May, and June, stood on their front porch and noticed that the tree in their front yard was leaning $6°$ from vertical toward the house. From the porch, which is 110 feet away from the base of the tree, they noticed that the angle of elevation to the top of the tree was $27°$. Approximate the height of the tree. Round answer to two decimal places.

A) 57.97 feet
B) 59.55 feet
C) 69.55 feet
D) 53.49 feet
E) 60.86 feet

Answer Key

1. C
2. D
3. C
4. C
5. D
6. A
7. B
8. A
9. E
10. A
11. C
12. A
13. D
14. B
15. E
16. B
17. B
18. D
19. E
20. D

Name: ____________________________ Date: _____________

1. Perform the operation shown below and leave the result in trigonometric form.
$$\frac{36(\cos 2.5 + i\sin 2.5)}{6(\cos 1.7 + i\sin 1.7)}$$
A) $6(\cos 0.8 + i\sin 0.8)$
B) $30(\cos 0.8 + i\sin 0.8)$
C) $30(\cos 1.5 + i\sin 1.5)$
D) $6(\cos 4.2 + i\sin 4.2)$
E) $6(\cos 4.25 + i\sin 4.25)$

2. A force of 45 pounds is exerted along a rope attached to a crate at an angle of 30° above the horizontal. The crate is moved 50 feet. How much work has been accomplished? Round answer to one decimal place.
A) 2,250.0 foot-pounds
B) 2,598.1 foot-pounds
C) 2,099.3 foot-pounds
D) 2,348.7 foot-pounds
E) 1,948.6 foot-pounds

3. Given vectors $\mathbf{u} = \langle u_1, u_2 \rangle$, $\mathbf{v} = \langle v_1, v_2 \rangle$, and $\mathbf{w} = \langle w_1, w_2 \rangle$, determine whether the result of the following expression is a vector or a scalar.
$\mathbf{u} \cdot 4\mathbf{v}$
A) vector
B) scalar

4. Use DeMoivre's Theorem to find the indicated power of the following complex number.
$$\left(3 + 3\sqrt{3}\,i\right)^6$$
A) 19,683
B) 6,561
C) 2,187
D) 46,656
E) 243

5. Given $a=3$, $b=8$, and $c=6$, use the Law of Cosines to solve the triangle for the value of C. Round answer to two decimal places.

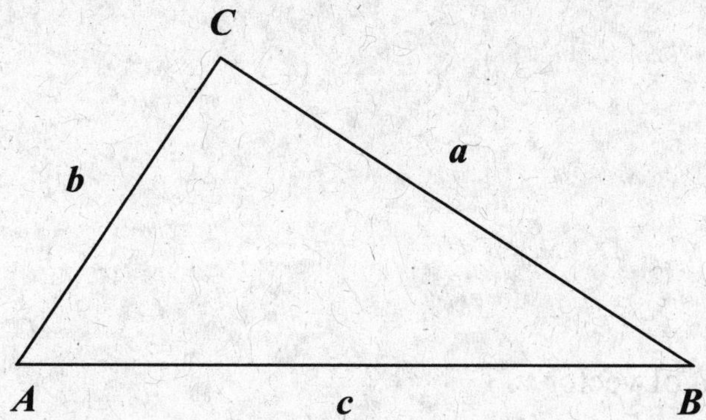

Figure not drawn to scale

A) 60.33°
B) 39.57°
C) 80.44°
D) 121.86°
E) 18.57°

6. Find the trigonometric form of the complex number shown below.
$-6i$

A) $6(\cos 0 + i\sin 0)$

B) $6(\cos \pi + i\sin \pi)$

C) $6\left(\cos\frac{3\pi}{2} + i\sin\frac{3\pi}{2}\right)$

D) $6\left(\cos\frac{\pi}{4} + i\sin\frac{\pi}{4}\right)$

E) $6\left(\cos\frac{\pi}{2} + i\sin\frac{\pi}{2}\right)$

7. Let **w** be a vector with initial point $(2,1)$ and terminal point $(9,-3)$. Write **w** as a linear combination of the standard unit vectors **i** and **j**.
A) $\mathbf{w} = \mathbf{i} + 12\mathbf{j}$
B) $\mathbf{w} = -5\mathbf{i} + 8\mathbf{j}$
C) $\mathbf{w} = 7\mathbf{i} - 4\mathbf{j}$
D) $\mathbf{w} = 11\mathbf{i} + 4\mathbf{j}$
E) $\mathbf{w} = 3\mathbf{i} - 12\mathbf{j}$

8. Let $\mathbf{u} = \langle 6,8 \rangle$ and $\mathbf{v} = \langle 14,10 \rangle$. Find $\mathbf{u} - \mathbf{v}$.
   A) $\langle 16,22 \rangle$
   B) $\langle -8,-2 \rangle$
   C) $\langle 20,18 \rangle$
   D) $\langle -4,-6 \rangle$
   E) $\langle 22,16 \rangle$

9. Find the component form of vector **v**.

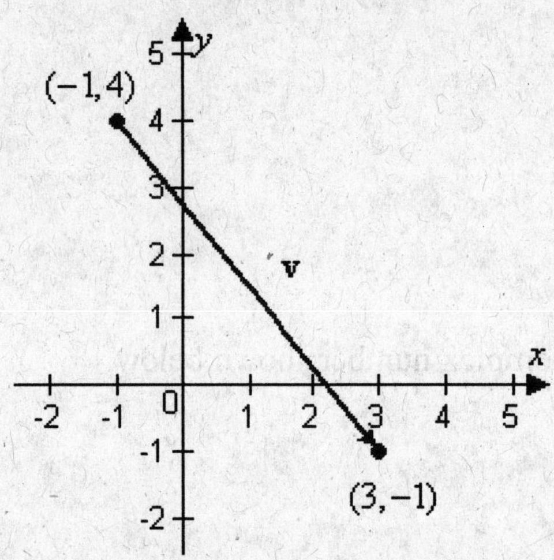

   A) $\mathbf{v} = \langle -5,4 \rangle$
   B) $\mathbf{v} = \langle 4,-5 \rangle$
   C) $\mathbf{v} = \langle -4,-5 \rangle$
   D) $\mathbf{v} = \langle 4,-3 \rangle$
   E) $\mathbf{v} = \langle 5,4 \rangle$

10. An airplane is flying at a bearing of 143° with an airspeed of 335 kilometers per hour. Because of the wind, its ground speed and direction are, respectively, 325 kilometers per hour and 138°. Find the direction and speed of the wind. Round your answer to two decimals.
    A) 31.36°, 30.47 kilometers per hour
    B) 121.36°, 25.71 kilometers per hour
    C) –37.01°, 30.47 kilometers per hour
    D) 31.36°, 25.71 kilometers per hour
    E) 121.36°, 25.69 kilometers per hour

11. Plot the following complex number.

$3i$

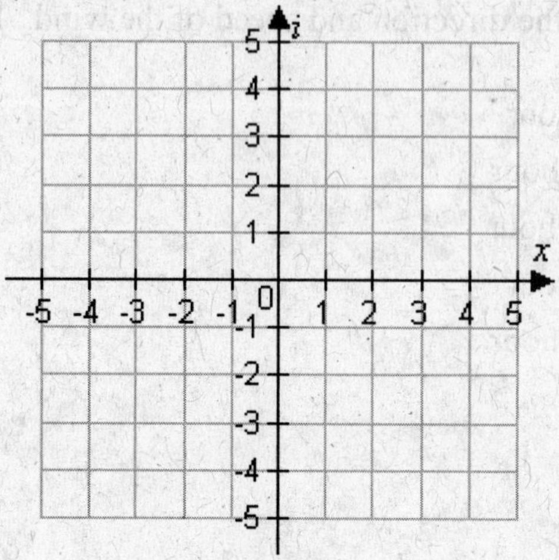

A)

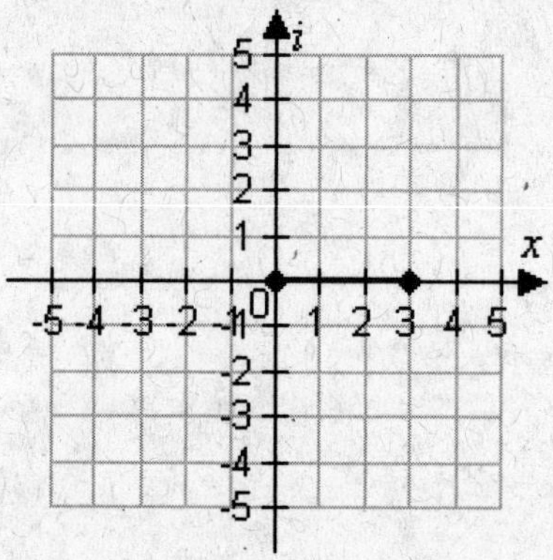

B)

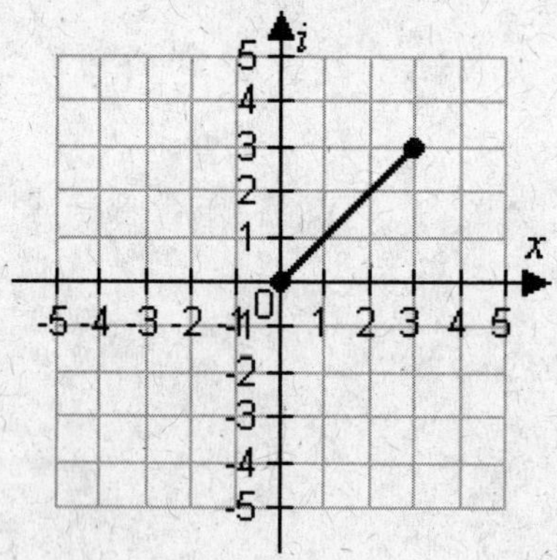

C)
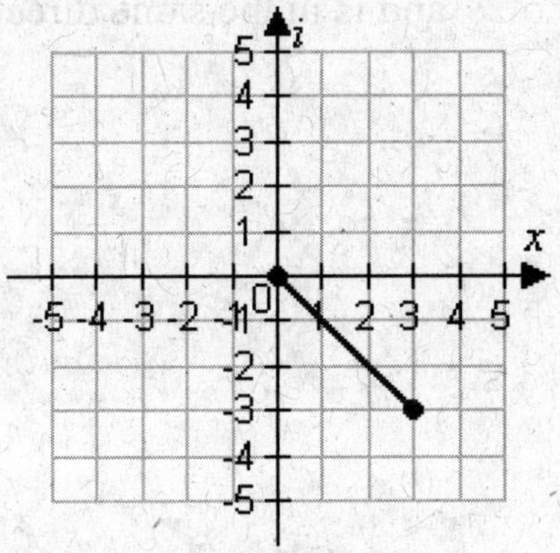

D)
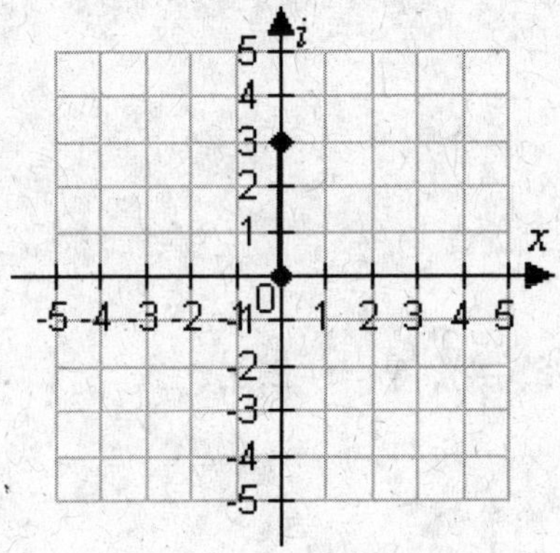

E)
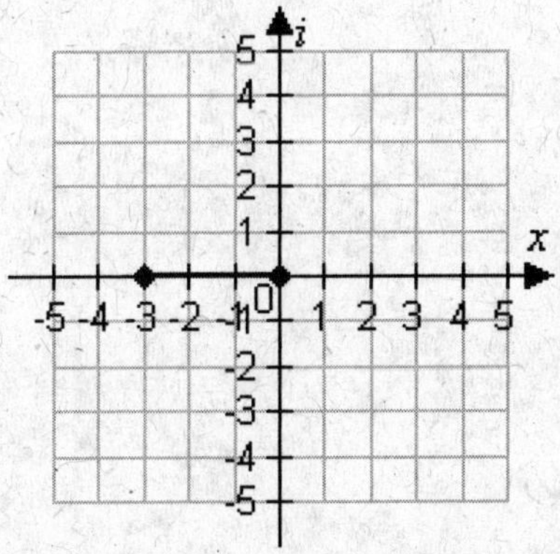

12. Use DeMoivre's Theorem to find the indicated power of the folllowing complex number.

$$(-6+6i)^4$$

A) –5,184
B) 46,656
C) 5,184
D) 186,624
E) 7,776

13. Find the vector **v** that has a magnitude of 8 and is in the same direction as **u**, where $\mathbf{u} = \langle -6, -4 \rangle$.

A) $\mathbf{v} = \left\langle \frac{-24}{\sqrt{13}}, \frac{-16}{\sqrt{13}} \right\rangle$

B) $\mathbf{v} = \left\langle \frac{-3}{\sqrt{13}}, \frac{-2}{\sqrt{13}} \right\rangle$

C) $\mathbf{v} = \left\langle \frac{-8}{\sqrt{13}}, \frac{-4}{\sqrt{13}} \right\rangle$

D) $\mathbf{v} = \left\langle \frac{-2}{\sqrt{13}}, \frac{2}{\sqrt{13}} \right\rangle$

E) $\mathbf{v} = \left\langle \frac{4}{\sqrt{13}}, \frac{6}{\sqrt{13}} \right\rangle$

14. Find the projection of **u** onto **v** if $\mathbf{u} = \langle 5, -4 \rangle$, $\mathbf{v} = \langle -2, 3 \rangle$.

A) $\left\langle \frac{44}{\sqrt{13}}, \frac{88}{\sqrt{13}} \right\rangle$

B) $\left\langle -\frac{110}{\sqrt{13}}, \frac{44}{\sqrt{13}} \right\rangle$

C) $\left\langle -\frac{110}{\sqrt{13}}, \frac{88}{\sqrt{13}} \right\rangle$

D) $\left\langle -\frac{110}{\sqrt{13}}, -\frac{66}{\sqrt{13}} \right\rangle$

E) $\left\langle \frac{44}{\sqrt{13}}, -\frac{66}{\sqrt{13}} \right\rangle$

15. Given $A = 52°$, $B = 79°$, and $a = 5.1$, use the Law of Sines to solve the triangle for the value of b. Round answer to two decimal places.

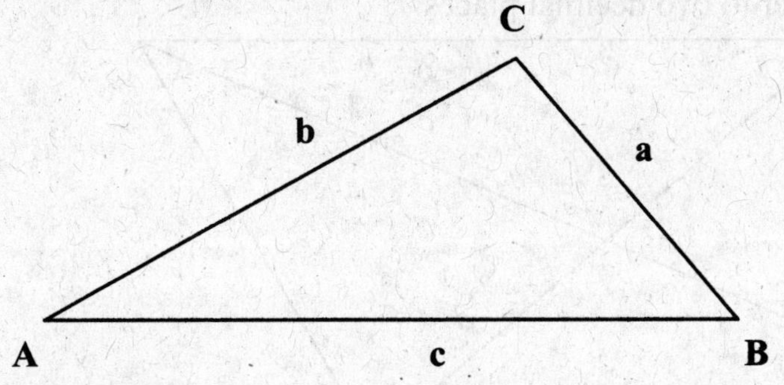

A) $b = 3.92$
B) $b = 4.88$
C) $b = 5.33$
D) $b = 6.35$
E) $b = 4.09$

16. In the figure below, $(a,b) = (2,7)$ and $(c,d) = (7,5)$. Find (x,y) so that $\mathbf{u} = \mathbf{v}$.

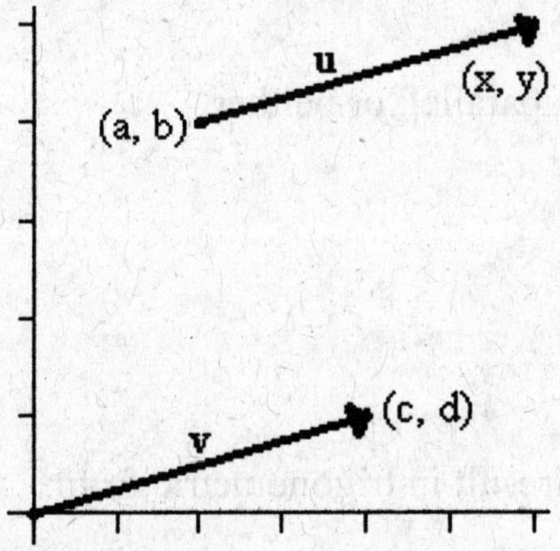

A) $(5,-2)$
B) $(14,7)$
C) $(0,3)$
D) $(7,14)$
E) $(9,12)$

17. In the figure below, $a = 9$, $b = 11$, and $c = 12$. Use this information to solve the parallelogram for θ. The diagonals of the parallelogram are represented by c and d. Round answer to two decimal places.

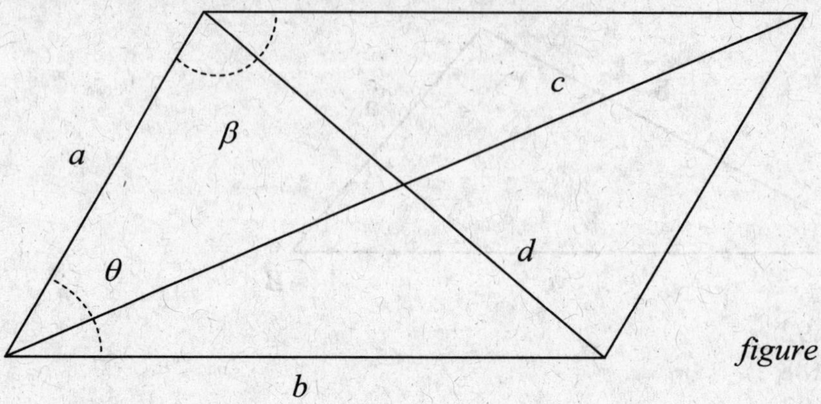

figure not drawn to scale

A) 107.03°
B) 98.52
C) 90.00°
D) 92.13
E) 72.97°

18. Determine whether **u** and **v** are orthogonal, parallel, or neither.
$\mathbf{u} = \langle 1,5 \rangle$, $\mathbf{v} = \langle -30,-5 \rangle$

A) orthogonal
B) parallel
C) neither

19. Perform the operation below and leave the result in trigonometric form.
$$\frac{4(\cos 58^\circ + i\sin 58^\circ)}{12(\cos 13^\circ + i\sin 13^\circ)}$$

A) $\frac{1}{3}(\cos 45^\circ + i\sin 45^\circ)$
B) $-8\left(\cos\frac{58^\circ}{13} + i\sin\frac{58^\circ}{13}\right)$
C) $16(\cos 71^\circ + i\sin 71^\circ)$
D) $48\left(\cos\frac{58^\circ}{13} + i\sin\frac{58^\circ}{13}\right)$
E) $\frac{1}{3}\left(\cos\frac{58^\circ}{13} + i\sin\frac{58^\circ}{13}\right)$

20. A plane flies 450 kilometers with a bearing of 319° from City A to City B (see figure below). The plane then flies 730 kilometers from City B to City C, which is directly west of City A. Find the bearing of the flight from City B to City C. Round your answer to two decimals.

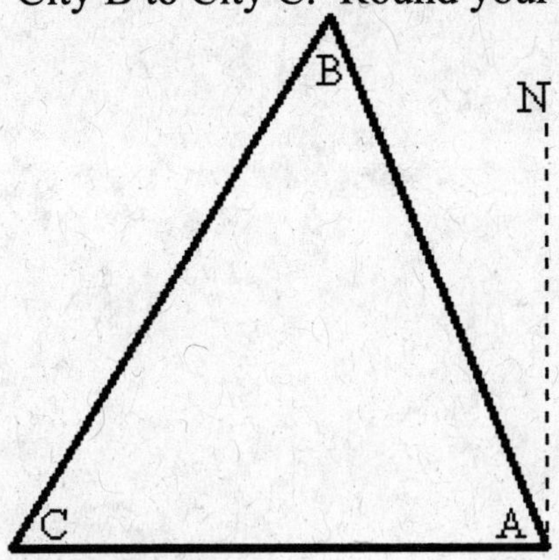

A) 242.27°
B) 332.27°
C) 27.73°
D) 23.85°
E) 246.15°

Answer Key

1. A
2. E
3. B
4. D
5. B
6. C
7. C
8. B
9. B
10. A
11. D
12. A
13. A
14. E
15. D
16. E
17. A
18. C
19. A
20. A

Name: ______________________ Date: ____________

1. If possible, find $3A + 4B$.

$$A = \begin{bmatrix} -4 & 3 & 9 \\ 3 & -2 & 5 \end{bmatrix}, B = \begin{bmatrix} -8 & 9 & 5 \\ 0 & 2 & 1 \end{bmatrix}$$

A) $\begin{bmatrix} 20 & -27 & 7 \\ 9 & -14 & 11 \end{bmatrix}$

B) $\begin{bmatrix} -12 & 12 & 14 \\ 3 & 0 & 6 \end{bmatrix}$

C) $\begin{bmatrix} 1 & 0 & 0 \\ 0 & 1 & 0 \end{bmatrix}$

D) $\begin{bmatrix} -44 & 45 & 47 \\ 9 & 2 & 19 \end{bmatrix}$

E) not possible

2. Use a system of equations to find the specified equation that passes through the points. Solve the system using matrices.
   Parabola: $y = ax^2 + bx + c$

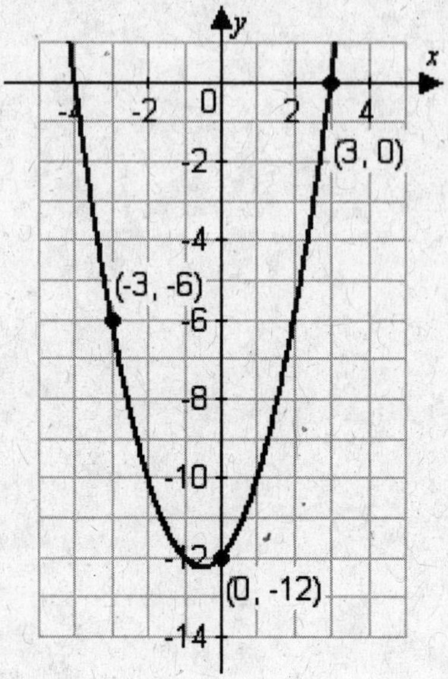

A) $y = -x^2 - 12$
B) $y = -x^2 - x - 12$
C) $y = x^2 + x - 12$
D) $y = x^2 + 9x - 36$
E) $y = x^2 - 12$

3. Use a determinant to determine whether the points below are collinear.
   $(3,-1),\ (0,-3),\ (27,15)$
   A) the points are collinear
   B) the points are not collinear

4. Solve the system by the method of elimination.

$$\begin{cases} 4x+5y=-71 \\ x-y=-2 \end{cases}$$

A) $(-6,-4)$
B) $(-9,-7)$
C) $\left(-\frac{61}{3},\frac{52}{3}\right)$
D) $\left(8,-\frac{111}{4}\right)$
E) inconsistent

5. Solve the system of linear equations.

$$\begin{cases} x+y+z+w=3 \\ -x-y+z+w=-5 \\ x \qquad +3z+w=-6 \\ -4x-4y-2z-w=-13 \end{cases}$$

A) $(-1,5,1,-2)$
B) $(-1,-2,5,1)$
C) $(1,5,-2,-1)$
D) $(-1,5,-2,1)$
E) inconsistent

6. Write the system of linear equations represented by the augmented matrix. (Use variables x, y, z, and w.)

$$\left[\begin{array}{cccc|c} -4 & 0 & 0 & -9 & -9 \\ 7 & -8 & 0 & 0 & 1 \\ 0 & -4 & -7 & -9 & -8 \\ 0 & 0 & -4 & 1 & -8 \end{array}\right]$$

A) $$\begin{cases} -4x - 9y = -9 \\ 7x - 8y = 1 \\ -4x - 7y - 9z = -8 \\ -4x + y = -8 \end{cases}$$

B) $$\begin{cases} -4x - 9z = -9 \\ 7x - 8y = 1 \\ -4y - -7z - 9w = -8 \\ -4z + w = -8 \end{cases}$$

C) $$\begin{cases} -4x - 9z = -9 \\ 7x - 8z = 1 \\ -4y - 7z - 9w = -8 \\ -4z + w = -8 \end{cases}$$

D) $$\begin{cases} -4x - 9w = -9 \\ 7x - 8y = 1 \\ -4y - 7z - 9w = -8 \\ -4y + z = -8 \end{cases}$$

E) $$\begin{cases} -4x - 9w = -9 \\ 7x - 8y = 1 \\ -4y - 7z - 9w = -8 \\ -4z + w = -8 \end{cases}$$

7. Solve for X in the equation given.

$$9A+6B=3X, A=\begin{bmatrix}6 & -5 & 3\\ 8 & -3 & -7\end{bmatrix} \text{ and } B=\begin{bmatrix}3 & 5 & -1\\ 1 & -2 & 5\end{bmatrix}$$

A) $\begin{bmatrix}72 & -15 & 21\\ 78 & -39 & -33\end{bmatrix}$

B) $\begin{bmatrix}9 & 0 & 2\\ 9 & -5 & -2\end{bmatrix}$

C) $\begin{bmatrix}24 & -5 & 7\\ 26 & -13 & -11\end{bmatrix}$

D) $\begin{bmatrix}12 & -25 & 11\\ 22 & -5 & -31\end{bmatrix}$

E) not possible

8. Write the partial fraction decomposition of the rational expression.

$$\frac{x^2}{x^4-21x^2-100}$$

A) $\frac{1}{29}\left(\frac{4}{x^2+4}+\frac{5}{2(x-5)}-\frac{5}{2(x+5)}\right)$

B) $\frac{1}{x^2}-\frac{1}{2}-\frac{x^2}{20}$

C) $\frac{1}{29}\left(\frac{4}{x^2+4}+\frac{5(x+4)}{2(x-5)}+\frac{5}{2(x-5)}\right)$

D) $\frac{1}{29}\left(\frac{4}{x^2+4}-\frac{5}{2(x-5)}+\frac{5}{2(x+5)}\right)$

E) $\frac{1}{29}\left(\frac{1}{x^2+4}+\frac{1}{2(x-5)}-\frac{1}{2(x+5)}\right)$

9. Write the system of linear equations represented by the augmented matrix. Then use back-substitution to solve. (Use variables x, y, and z.)

$$\left[\begin{array}{ccc|c} 1 & 4 & -3 & -23 \\ 0 & 1 & 4 & 19 \\ 0 & 0 & 1 & 5 \end{array}\right]$$

A) $x = -44, y = 3, z = -3$
B) $x = -1, y = -4, z = 5$
C) $x = -4, y = 1, z = -5$
D) $x = -4, y = -1, z = 5$
E) $x = 1, y = 5, z = 4$

10. Solve the system by the method of substitution.

$$\begin{cases} y = -3x \\ y = x^3 + 3x^2 + 3x \end{cases}$$

A) (0, 6)
B) (0, 0)
C) (6, 3), (–6, –3)
D) (6, 3)
E) no real solution

11. Solve for x given the following equation involving a determinant.

$$\begin{vmatrix} x+2 & 5 \\ -1 & x+8 \end{vmatrix} = 0$$

A) $x = 7, 3$
B) $x = -8, -2$
C) $x = 8, 2$
D) $x = -7, -3$
E) $x = 6, -5$

12. Solve the system of linear equations

$$\begin{cases} 4x_1 - 8x_2 - 4x_3 - 8x_4 = 0 \\ 12x_1 - 20x_2 - 8x_3 - 12x_4 = -9 \\ 8x_1 - 20x_2 - 8x_3 - 20x_4 = 6 \\ -4x_1 + 16x_2 + 16x_3 + 44x_4 = 0 \end{cases}$$

using the inverse matrix $\frac{1}{4}\begin{bmatrix} -24 & 7 & 1 & -2 \\ -10 & 3 & 0 & -1 \\ -29 & 7 & 3 & -2 \\ 12 & -3 & -1 & 1 \end{bmatrix}$.

A) $\begin{bmatrix} x_1 \\ x_2 \\ x_3 \\ x_4 \end{bmatrix} = \begin{bmatrix} -\frac{57}{4} \\ -\frac{27}{4} \\ -\frac{45}{4} \\ \frac{21}{4} \end{bmatrix}$

B) $\begin{bmatrix} x_1 \\ x_2 \\ x_3 \\ x_4 \end{bmatrix} = \begin{bmatrix} -\frac{21}{4} \\ -\frac{15}{4} \\ \frac{27}{4} \\ -\frac{27}{4} \end{bmatrix}$

C) $\begin{bmatrix} x_1 \\ x_2 \\ x_3 \\ x_4 \end{bmatrix} = \begin{bmatrix} -\frac{27}{4} \\ -\frac{57}{4} \\ 0 \\ \frac{21}{4} \end{bmatrix}$

D) $\begin{bmatrix} x_1 \\ x_2 \\ x_3 \\ x_4 \end{bmatrix} = \begin{bmatrix} -\frac{9}{4} \\ \frac{3}{4} \\ 0 \\ \frac{3}{2} \end{bmatrix}$

E) $\begin{bmatrix} x_1 \\ x_2 \\ x_3 \\ x_4 \end{bmatrix} = \begin{bmatrix} 0 \\ -\frac{3}{4} \\ 0 \\ -\frac{3}{2} \end{bmatrix}$

13. Find the equilibrium point of the demand and supply equations. (The equilibrium point is the price p and number of units x that satisfy both the demand and supply equations.)

Demand	Supply
$p = 130 - 0.09x$	$p = 0.4x - 115$

A) $(500, 85)$
B) $(200, 112)$
C) $(835, 219)$
D) $(300, 103)$
E) inconsistent

14. Match the system of linear equations with its graph.

$$\begin{cases} 3x + 6y - 12 = 0 \\ 15x - 5y - 10 = 0 \end{cases}$$

A)

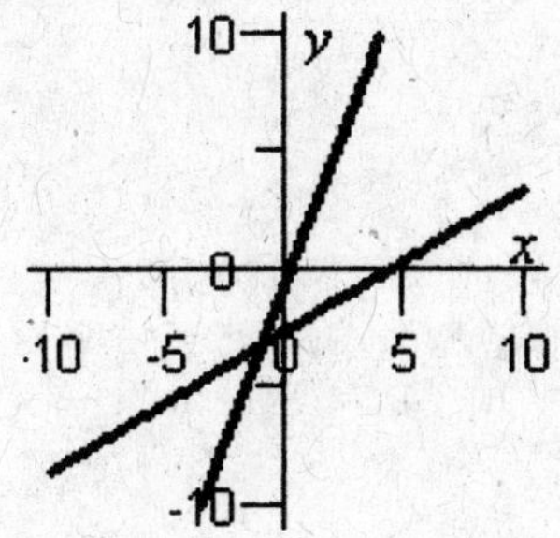

B)

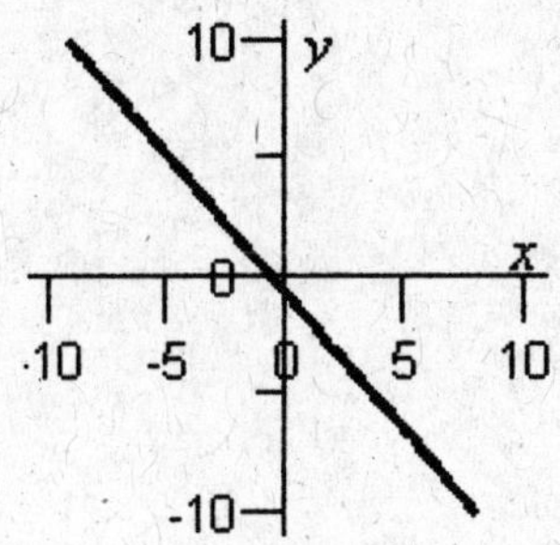

C)

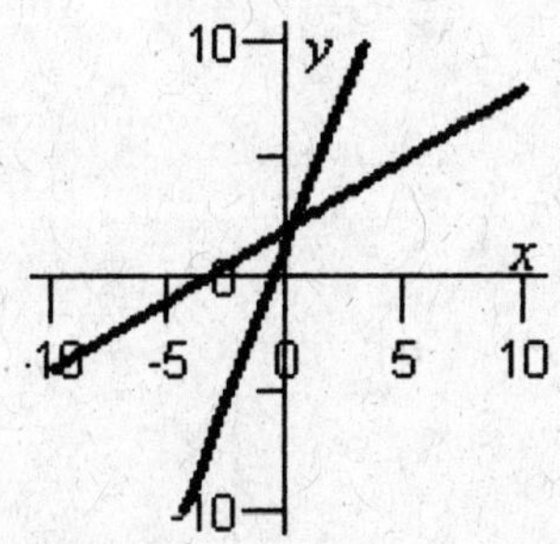

D)

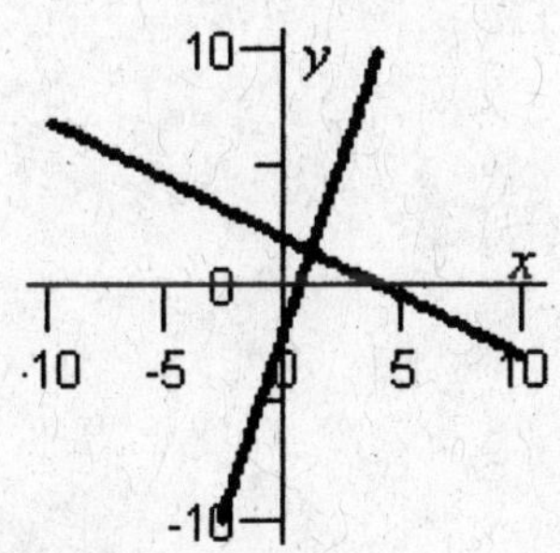

E)

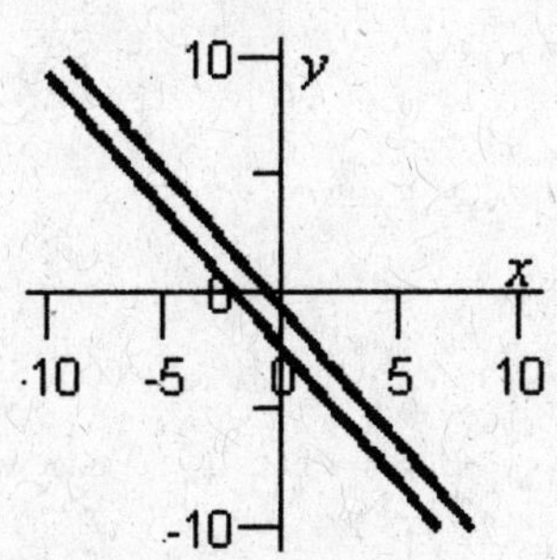

15. Write the system of linear equations as a matrix equation $AX = B$, and use Gauss-Jordan elimination on the augmented matrix $[A \vdots B]$ to solve for the matrix X.

$$\begin{cases} x_1 - 3x_2 - 9x_3 = 43 \\ x_1 + 4x_2 + 7x_3 = -51 \\ 7x_1 - 7x_2 + 3x_3 = -57 \end{cases}$$

A) $X = \begin{bmatrix} -16 \\ 10 \\ -40 \end{bmatrix}$

B) $X = \begin{bmatrix} 10 \\ 8 \\ -2 \end{bmatrix}$

C) $X = \begin{bmatrix} -8 \\ -2 \\ -5 \end{bmatrix}$

D) $X = \begin{bmatrix} -10 \\ 16 \\ 40 \end{bmatrix}$

E) $X = \begin{bmatrix} 8 \\ 5 \\ -2 \end{bmatrix}$

16. Solve using any method.

$$\begin{cases} -7x - 9y = -6 \\ y = x - 3 \end{cases}$$

A) $(2,-1)$
B) $\left(1,-\dfrac{3}{7}\right)$
C) $\left(-\dfrac{21}{16},-\dfrac{27}{16}\right)$
D) $\left(\dfrac{33}{16},-\dfrac{15}{16}\right)$
E) inconsistent

17. Evaluate the expression.

$$\begin{bmatrix} 5 & -1 \\ 4 & -4 \end{bmatrix}\begin{bmatrix} 0 & 3 \\ -3 & 0 \end{bmatrix}\begin{bmatrix} -5 & 0 \\ 1 & 4 \end{bmatrix}$$

A) $\begin{bmatrix} 0 & 60 \\ -48 & 48 \end{bmatrix}$
B) $\begin{bmatrix} 0 & -48 \\ 60 & 48 \end{bmatrix}$
C) $\begin{bmatrix} 0 & 0 \\ -12 & 0 \end{bmatrix}$
D) $\begin{bmatrix} 15 & -12 \\ 60 & 12 \end{bmatrix}$
E) not possible

18. Determine which ordered pair is a solution of the system.

$$\begin{cases} x - 2y^2 = -7 \\ -5x + 6y = 7 \end{cases}$$

A) (1, 2)
B) (1, –2)
C) (–1, 2)
D) (–1, –2)
E) (0, –2)

19. Solve system of equations by the method of substitution.

$$\begin{cases} 15x^3 - 25y = 0 \\ 15x - y = 0 \end{cases}$$

A) $(0,0)$, $(-5,-75)$, $(5,75)$
B) $(0,0)$, $(-5,75)$, $(5,75)$
C) $(0,0)$, $(5,-75)$, $(5,75)$
D) $(0,0)$, $(-5,76)$, $(5,76)$
E) $(0,0)$, $(-5,-74)$, $(5,74)$

20. Given:

$$A = \begin{bmatrix} 8 & -7 & 0 \\ -5 & -2 & 3 \end{bmatrix},\ B = \begin{bmatrix} 3 & 0 & 7 \\ 4 & 9 & -2 \end{bmatrix},\ c = 8 \text{ and } d = -3,$$

determine $cA + dB$.

A) $\begin{bmatrix} 55 & -7 & 7 \\ -52 & 7 & 1 \end{bmatrix}$

B) $\begin{bmatrix} 73 & -56 & 21 \\ -28 & 11 & 18 \end{bmatrix}$

C) $\begin{bmatrix} 55 & -56 & -21 \\ -52 & -43 & 30 \end{bmatrix}$

D) $\begin{bmatrix} 55 & -56 & -21 \\ -1 & 7 & 1 \end{bmatrix}$

E) not possible

Answer Key

1. D
2. C
3. A
4. B
5. D
6. E
7. C
8. A
9. D
10. B
11. D
12. A
13. A
14. D
15. C
16. D
17. A
18. A
19. A
20. C

Name: ______________________________ Date: ______________

1. Find the inverse of the matrix $\begin{bmatrix} 4 & -5 \\ 1 & 5 \end{bmatrix}$ (if it exists).

A) $-\dfrac{1}{25}\begin{bmatrix} 5 & 5 \\ -1 & 4 \end{bmatrix}$

B) $\dfrac{1}{25}\begin{bmatrix} 5 & 5 \\ -1 & 4 \end{bmatrix}$

C) $\dfrac{1}{25}\begin{bmatrix} -5 & -5 \\ 1 & -4 \end{bmatrix}$

D) $\dfrac{1}{25}\begin{bmatrix} 4 & -5 \\ 1 & 5 \end{bmatrix}$

E) does not exist

2. Solve system of equations by the method of substitution.

$$\begin{cases} 4x^3 - 25y = 0 \\ 4x - y = 0 \end{cases}$$

A) $(0,0)$, $(-5,-20)$, $(5,20)$
B) $(0,0)$, $(-5,20)$, $(5,20)$
C) $(0,0)$, $(5,-20)$, $(5,20)$
D) $(0,0)$, $(-5,21)$, $(5,21)$
E) $(0,0)$, $(-5,-19)$, $(5,19)$

3. Solve for X in the equation given.

$$3X = 2A - B, A = \begin{bmatrix} -6 & 2 \\ 8 & -5 \end{bmatrix} \text{ and } B = \begin{bmatrix} -3 & 10 \\ 7 & -1 \end{bmatrix}$$

A) $\begin{bmatrix} -3 & -2 \\ 3 & -3 \end{bmatrix}$

B) $\begin{bmatrix} -9 & -6 \\ 9 & -9 \end{bmatrix}$

C) $\begin{bmatrix} -\frac{1}{3} & -\frac{1}{2} \\ \frac{1}{3} & -\frac{1}{3} \end{bmatrix}$

D) $\begin{bmatrix} -3 & 2 \\ -3 & -3 \end{bmatrix}$

E) $\begin{bmatrix} 1 & 0 \\ 0 & 1 \end{bmatrix}$

4. Determine the order of the matrix.

$$\begin{bmatrix} -4 & -8 & 2 \\ 0 & -7 & -4 \end{bmatrix}$$

A) 2×3
B) 3×3
C) 3×1
D) 3×2
E) 2×2

5. Solve the system of linear equations

$$\begin{cases} 6x_1 - 12x_2 - 6x_3 - 12x_4 = 0 \\ 18x_1 - 30x_2 - 12x_3 - 18x_4 = -3 \\ 12x_1 - 30x_2 - 12x_3 - 30x_4 = 2 \\ -6x_1 + 24x_2 + 24x_3 + 66x_4 = 0 \end{cases}$$

using the inverse matrix $\frac{1}{6}\begin{bmatrix} -24 & 7 & 1 & -2 \\ -10 & 3 & 0 & -1 \\ -29 & 7 & 3 & -2 \\ 12 & -3 & -1 & 1 \end{bmatrix}$.

A) $\begin{bmatrix} x_1 \\ x_2 \\ x_3 \\ x_4 \end{bmatrix} = \begin{bmatrix} -\frac{19}{6} \\ -\frac{3}{2} \\ -\frac{5}{2} \\ \frac{7}{6} \end{bmatrix}$

B) $\begin{bmatrix} x_1 \\ x_2 \\ x_3 \\ x_4 \end{bmatrix} = \begin{bmatrix} -\frac{7}{6} \\ -\frac{5}{6} \\ \frac{3}{2} \\ -\frac{3}{2} \end{bmatrix}$

C) $\begin{bmatrix} x_1 \\ x_2 \\ x_3 \\ x_4 \end{bmatrix} = \begin{bmatrix} -\frac{3}{2} \\ -\frac{19}{6} \\ 0 \\ \frac{7}{6} \end{bmatrix}$

D) $\begin{bmatrix} x_1 \\ x_2 \\ x_3 \\ x_4 \end{bmatrix} = \begin{bmatrix} -\frac{1}{2} \\ \frac{1}{6} \\ 0 \\ \frac{1}{3} \end{bmatrix}$

E) $\begin{bmatrix} x_1 \\ x_2 \\ x_3 \\ x_4 \end{bmatrix} = \begin{bmatrix} 0 \\ -\frac{1}{6} \\ 0 \\ -\frac{1}{3} \end{bmatrix}$

6. Use a system of equations to find the cubic function $f(x) = ax^3 + bx^2 + cx + d$ that satisfies given equations. Solve the system using matrices.

$$f(2) = -26$$
$$f(-2) = 6$$
$$f(1) = 0$$
$$f(-1) = -2$$

A) $a = 2,\ b = -2,\ c = 1,\ d = -1$
B) $a = -26,\ b = 6,\ c = 0,\ d = -2$
C) $a = -3,\ b = -3,\ c = 4,\ d = 2$
D) $a = -26,\ b = 30,\ c = -12,\ d = 16$
E) $a = 50{,}598,\ b = -54{,}650\ c = -730,\ d = 518$

7. Determine whether the two systems of linear equations yield the same solutions. If so, find the solutions using matrices.

$$\begin{cases} x + y - 9z = -67 \\ \quad\; y + 5z = 31 \\ \qquad\quad z = 7 \end{cases}$$

$$\begin{cases} x - 9y + 6z = 76 \\ \qquad y - 9z = -67 \\ \qquad\qquad z = 7 \end{cases}$$

A) $x = 0, y = -4, z = 7$
B) $x = 0, y = 4, z = 7$
C) $x = 4, y = 7, z = 0$
D) $x = -4, y = 7, z = 0$
E) The systems yield different solutions.

8. Perform the indicated row operations on the matrix. Show the final result.

$$\begin{bmatrix} 1 & -2 & -3 \\ 2 & -3 & -1 \\ 1 & -8 & -33 \end{bmatrix}$$

Add –2 times R_1 to R_2.
Add –1 times R_1 to R_3.

A) $\begin{bmatrix} 1 & -9 & -36 \\ 2 & -3 & -1 \\ 1 & -8 & -33 \end{bmatrix}$

B) $\begin{bmatrix} 1 & -2 & -3 \\ 0 & 1 & -4 \\ 0 & -10 & -36 \end{bmatrix}$

C) $\begin{bmatrix} 1 & -2 & -3 \\ 0 & 1 & 5 \\ 0 & -6 & -30 \end{bmatrix}$

D) $\begin{bmatrix} 1 & -2 & -3 \\ -3 & 4 & -1 \\ 0 & 6 & 30 \end{bmatrix}$

E) $\begin{bmatrix} 1 & -2 & -3 \\ 0 & -1 & 2 \\ 0 & -4 & -29 \end{bmatrix}$

9. Find the minor M_{13} and its cofactor C_{13} of the matrix $\begin{bmatrix} 9 & -6 & 24 \\ -9 & 6 & -18 \\ 3 & -9 & 18 \end{bmatrix}$.

A) $M_{13} = \begin{vmatrix} 9 & 24 \\ 3 & 18 \end{vmatrix} = 90$
$C_{13} = -90$

B) $M_{13} = \begin{vmatrix} -6 & 24 \\ 6 & -18 \end{vmatrix} = -36$
$C_{13} = -36$

C) $M_{13} = \begin{vmatrix} -9 & 6 \\ 3 & -9 \end{vmatrix} = 63$
$C_{13} = -63$

D) $M_{13} = \begin{vmatrix} -6 & 24 \\ 6 & -18 \end{vmatrix} = -36$
$C_{13} = 36$

E) $M_{13} = \begin{vmatrix} -9 & 6 \\ 3 & -9 \end{vmatrix} = 63$
$C_{13} = 63$

10. Given:

$$A=\begin{bmatrix}-7 & -4 & 2\\ 4 & -5 & -8\end{bmatrix},\ B=\begin{bmatrix}-4 & 1 & -7\\ 8 & -8 & 3\end{bmatrix} \text{ and } c=-6,$$

determine cAB^2.

A) $\begin{bmatrix}108 & 1998\\ 4116 & 816\end{bmatrix}$

B) $\begin{bmatrix}-4704 & -96 & -1176\\ -6144 & -9600 & -3456\end{bmatrix}$

C) $\begin{bmatrix}672 & 24 & -588\\ -1536 & 1920 & 432\end{bmatrix}$

D) $\begin{bmatrix}-864 & -1494 & 1842\\ 2304 & 1944 & 1446\\ 2880 & 3060 & -156\end{bmatrix}$

E) not possible

11. Solve for X in the equation given.

$$16A+12B=-4X, A=\begin{bmatrix}5 & 6 & 3\\ -2 & 5 & -4\end{bmatrix} \text{ and } B=\begin{bmatrix}-5 & 8 & -1\\ -3 & -9 & -5\end{bmatrix}$$

A) $\begin{bmatrix}20 & 192 & 36\\ -68 & -28 & -124\end{bmatrix}$

B) $\begin{bmatrix}0 & 14 & 2\\ -5 & -4 & -9\end{bmatrix}$

C) $\begin{bmatrix}-5 & -48 & -9\\ 17 & 7 & 31\end{bmatrix}$

D) $\begin{bmatrix}-35 & 0 & -15\\ -1 & -47 & 1\end{bmatrix}$

E) not possible

12. Evaluate the expression.

$$\begin{bmatrix} -5 \\ -1 \\ 5 \end{bmatrix}\left(\begin{bmatrix} 9 & -7 \end{bmatrix}+\begin{bmatrix} -7 & -8 \end{bmatrix}\right)$$

A) $\begin{bmatrix} -10 \\ -2 \\ 10 \end{bmatrix}$

B) $\begin{bmatrix} -10 & 75 \\ -2 & 15 \\ 10 & -75 \end{bmatrix}$

C) $\begin{bmatrix} -10 & -2 & 10 \\ 75 & 15 & -75 \end{bmatrix}$

D) $\begin{bmatrix} -45 & 35 & 35 & 40 \\ -9 & 7 & 7 & 8 \\ 45 & -35 & -35 & -40 \end{bmatrix}$

E) not possible

13. Solve for x given the following equation involving a determinant.

$$\begin{vmatrix} x+1 & 7 \\ -1 & x+9 \end{vmatrix} = 0$$

A) $x = 8, 2$
B) $x = -9, -1$
C) $x = 9, 1$
D) $x = -8, -2$
E) $x = 8, -7$

14. Find the equation of the circle

$$x^2 + y^2 + Dx + Ey + F = 0$$

that passes through the points $(1,9),(-4,4),(6,4)$.

A) $x^2 + y^2 - 2x - 8y - 8 = 0$

B) $x^2 + y^2 - x - 4y - 8 = 0$

C) $x^2 + y^2 - 2x - 8y + 42 = 0$

D) $x^2 + y^2 - 2x - 8y - 25 = 0$

E) $x^2 + y^2 - x - 4y - 25 = 0$

15. Evaluate the expression.

$$\begin{bmatrix} 4 & -3 \\ -9 & 1 \end{bmatrix} + \begin{bmatrix} 2 & 8 \\ -2 & -4 \end{bmatrix} + \begin{bmatrix} -2 & -1 \\ -8 & 4 \end{bmatrix}$$

A) $\begin{bmatrix} 8 & 6 \\ -3 & -7 \end{bmatrix}$

B) $\begin{bmatrix} 4 & 4 \\ -19 & 1 \end{bmatrix}$

C) $\begin{bmatrix} 6 & 5 \\ -11 & -3 \end{bmatrix}$

D) $\begin{bmatrix} 1 & 0 \\ 0 & 1 \end{bmatrix}$

E) not possible

16. Given:

$$A=\begin{bmatrix}-6 & -10 & 7\\ -2 & 4 & 3\end{bmatrix},\ B=\begin{bmatrix}-1 & 7 & -3\\ 1 & 2 & -2\end{bmatrix},\ c=-4 \text{ and } d=2,$$

determine $cA+dB$.

A) $\begin{bmatrix}22 & -3 & 4\\ 10 & 6 & 1\end{bmatrix}$

B) $\begin{bmatrix}26 & 26 & -22\\ 6 & -20 & -8\end{bmatrix}$

C) $\begin{bmatrix}22 & 54 & -34\\ 10 & -12 & -16\end{bmatrix}$

D) $\begin{bmatrix}22 & 54 & -34\\ -1 & 6 & 1\end{bmatrix}$

E) not possible

17. Solve the system by the method of elimination. Round numbers to three decimal places.

$$\begin{cases} 5.2x+8.1y=6.9\\ -2.6x+1.1y=-0.7\end{cases}$$

A) $(2.020, 1.873)$
B) $(0.495, 0.534)$
C) $(-0.864, 1.407)$
D) $(0.192, 0.728)$
E) no solution

18. Determine whether the system of linear equations is consistent or inconsistent.

$$\begin{cases} -9x+6y=-1\\ -81x+54y=-10\end{cases}$$

A) inconsistent
B) consistent

19. Find $|AB|$, if

$$A=\begin{bmatrix}5 & -2 & -2\\ -3 & 4 & -1\\ -5 & 6 & 2\end{bmatrix},\ B=\begin{bmatrix}1 & -1 & -3\\ -5 & -3 & 4\\ -5 & -1 & -2\end{bmatrix}$$

A) 3080
B) 616
C) 150
D) 280
E) 1300

20. Evaluate the expression.

$$\frac{2}{9}\begin{bmatrix}5 & -2 & 1\end{bmatrix}+\begin{bmatrix}4 & 8 & -1\end{bmatrix}$$

A) $\begin{bmatrix}9 & 6 & 0\end{bmatrix}$
B) $\begin{bmatrix}-\frac{26}{9} & -\frac{76}{9} & \frac{11}{9}\end{bmatrix}$
C) $\begin{bmatrix}\frac{14}{9} & \frac{4}{9} & \frac{1}{9}\end{bmatrix}$
D) $\begin{bmatrix}\frac{46}{9} & \frac{68}{9} & -\frac{7}{9}\end{bmatrix}$
E) not possible

Answer Key

1. B
2. A
3. A
4. A
5. A
6. C
7. E
8. C
9. E
10. E
11. C
12. B
13. D
14. A
15. B
16. C
17. B
18. A
19. A
20. D

Name: ____________________ Date: __________

1. Find the inverse of the matrix $\begin{bmatrix} 6 & 12 \\ -18 & -30 \end{bmatrix}$.

A) $\frac{1}{7}\begin{bmatrix} 6 & 12 \\ -18 & -30 \end{bmatrix}$
B) $\frac{1}{6}\begin{bmatrix} -5 & -2 \\ 3 & 1 \end{bmatrix}$
C) $\begin{bmatrix} -6 & -12 \\ 30 & 18 \end{bmatrix}$
D) $\begin{bmatrix} 6 & -18 \\ 24 & -42 \end{bmatrix}$
E) $\frac{1}{13}\begin{bmatrix} -6 & -12 \\ -18 & 30 \end{bmatrix}$

2. An augmented matrix that represents a system of linear equations (in variables *x*, *y*, and *z*) has been reduced using Gauss-Jordan elimination. Write the solution represented by the augmented matrix.

$$\left[\begin{array}{ccc|c} 1 & 0 & 0 & 9 \\ 0 & 1 & 0 & 8 \\ 0 & 0 & 1 & -7 \end{array}\right]$$

A) $x = 9, y = 8, z = -7$
B) $x = -9, y = -8, z = 7$
C) $x = 0, y = 0, z = 0$
D) $x = 9x, y = -7y, z = 8z$
E) $x = 9, y = 0, z = 0$

3. Determine whether the two systems of linear equations yield the same solutions. If so, find the solutions using matrices.

$$\begin{cases} x + 9y + 9z = -22 \\ y - 6z = -23 \\ z = 3 \end{cases}$$

$$\begin{cases} x + 8y - 5z = -63 \\ y + 4z = 7 \\ z = 3 \end{cases}$$

A) $x = -4, y = -5, z = 3$
B) $x = 4, y = 5, z = 3$
C) $x = 5, y = 3, z = -4$
D) $x = -5, y = 3, z = -4$
E) The systems yield different solutions.

4. Solve the system of equations algebraically.

$$\begin{cases} xy + 25 = 0 \\ 49x - 100y - 700 = 0 \end{cases}$$

A) $\left(50, -\frac{1}{2}\right)$
B) $\left(-50, \frac{1}{2}\right), \left(50, -\frac{1}{2}\right)$
C) $\left(\frac{50}{7}, -\frac{7}{2}\right)$
D) $\left(-\frac{50}{7}, \frac{7}{2}\right), \left(\frac{50}{7}, -\frac{7}{2}\right)$
E) no solution

5. Find the determinant of $\begin{bmatrix} 0 & -6 & 0 \\ -9 & 6 & -18 \\ 3 & 0 & 18 \end{bmatrix}$ by the method of expansion by cofactors.

A) –648
B) 1296
C) –1296
D) 162
E) –594

6. Find the determinant of the matrix $\begin{bmatrix} -\frac{1}{6} & -\frac{1}{2} \\ -1 & -\frac{4}{5} \end{bmatrix}$.

A) $-\frac{7}{30}$
B) $\frac{19}{30}$
C) $\frac{1}{24}$
D) $-\frac{11}{30}$
E) $\frac{7}{24}$

7. Solve using any method.

$$\begin{cases} 7x + 6y = 15 \\ y = x - 7 \end{cases}$$

A) $(7,0)$
B) $\left(-6, \frac{51}{7}\right)$
C) $\left(-\frac{27}{13}, -\frac{64}{13}\right)$
D) $\left(\frac{57}{13}, -\frac{34}{13}\right)$
E) inconsistent

8. Solve the system of linear equations.

$$\begin{cases} 2x - 2y + 3z = -21 \\ 2x \qquad + z = -11 \\ -2x + 4y - 4z = 28 \end{cases}$$

A) $(3,-3,4)$
B) $(-4,2,-3)$
C) $(2,-3,-4)$
D) $(-3,4,3)$
E) $(-4,2,6)$

9. Solve the system of equations graphically or algebraically.

$$\begin{cases} y = 13x + 24 \\ y = \sqrt{3x + 576} \end{cases}$$

A) $\left(\frac{627}{169}, \frac{627}{13}\right)$
B) $\left(\frac{3}{13}, 27\right)$
C) $(0,24)$
D) $\left(\frac{3}{169}, \frac{315}{13}\right)$
E) $(1,37)$

10. If possible, find $4A + 2B$.

$$A = \begin{bmatrix} 0 & -4 & 4 \\ 1 & 0 & -3 \end{bmatrix}, B = \begin{bmatrix} -3 & 2 & 5 \\ -5 & 8 & -2 \end{bmatrix}$$

A) $\begin{bmatrix} 6 & -20 & 6 \\ 14 & -16 & -8 \end{bmatrix}$

B) $\begin{bmatrix} -3 & -2 & 9 \\ -4 & 8 & -5 \end{bmatrix}$

C) $\begin{bmatrix} 1 & 0 & 0 \\ 0 & 1 & 0 \end{bmatrix}$

D) $\begin{bmatrix} -6 & -12 & 26 \\ -6 & 16 & -16 \end{bmatrix}$

E) not possible

11. Solve the system by the method of elimination. Round numbers to three decimal places.

$$\begin{cases} 5.7x + 7.1y = 3.2 \\ -0.8x - 4.8y = -1.1 \end{cases}$$

A) $(2.872, 5.844)$
B) $(0.348, 0.171)$
C) $(0.229, 0.267)$
D) $(0.175, 0.310)$
E) no solution

12. Find x and y.

$$\begin{bmatrix} 1 & x \\ y & 3 \end{bmatrix} = \begin{bmatrix} 1 & -6 \\ -4 & 3 \end{bmatrix}$$

A) $x = 6, y = 4$
B) $x = 1, y = 3$
C) $x = -4, y = -6$
D) $x = -6, y = -4$
E) $x = -6, y = -6$

13. Solve the system of linear equations.

$$\begin{cases} x+y+z+w=11 \\ -2x-y-3z+4w=-5 \\ -2x \qquad -5z-w=-23 \\ 3x+5y+z-2w=16 \end{cases}$$

A) $(5,1,3,2)$
B) $(5,2,1,3)$
C) $(3,1,2,5)$
D) $(5,1,2,3)$
E) inconsistent

14. Fill in the blank using elementary row operations to form a row-equivalent matrix.

$$\begin{bmatrix} 2 & -1 & -9 \\ 8 & -6 & 4 \end{bmatrix}$$

$$\begin{bmatrix} 2 & -1 & -9 \\ 0 & \square & 40 \end{bmatrix}$$

A) $\begin{bmatrix} 2 & -1 & -9 \\ 0 & -2 & 40 \end{bmatrix}$
B) $\begin{bmatrix} 2 & -1 & -9 \\ 0 & 0 & 40 \end{bmatrix}$
C) $\begin{bmatrix} 2 & -1 & -9 \\ 0 & -10 & 40 \end{bmatrix}$
D) $\begin{bmatrix} 2 & -1 & -9 \\ 0 & -1 & 40 \end{bmatrix}$
E) $\begin{bmatrix} 2 & -1 & -9 \\ 0 & 2 & 40 \end{bmatrix}$

15. Solve system by the method of substitution and graph your solution.

$$\begin{cases} y = x^3 + 3x^2 + 2x \\ y = -x \end{cases}$$

A)

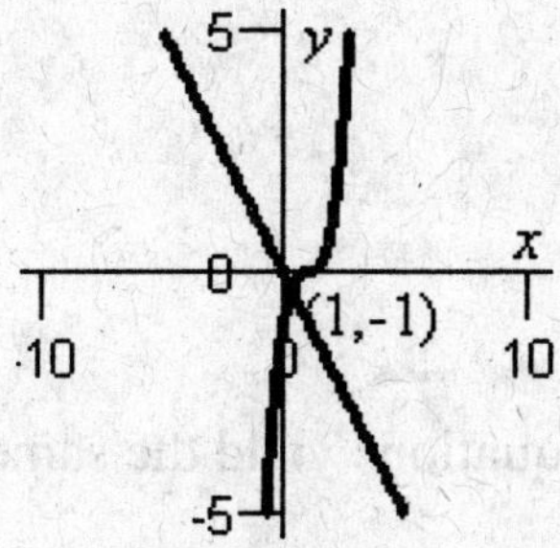

B)

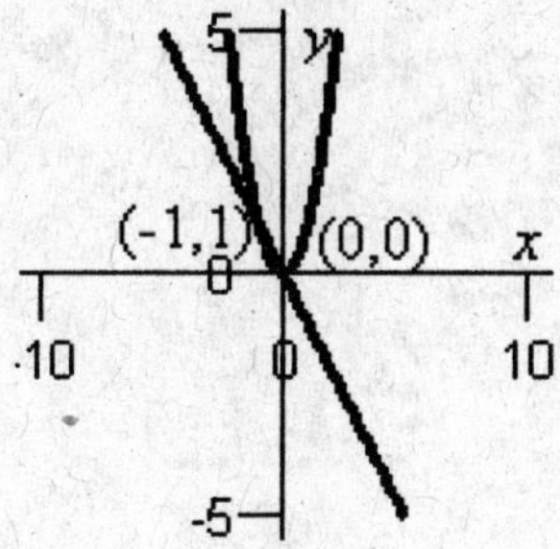

C)

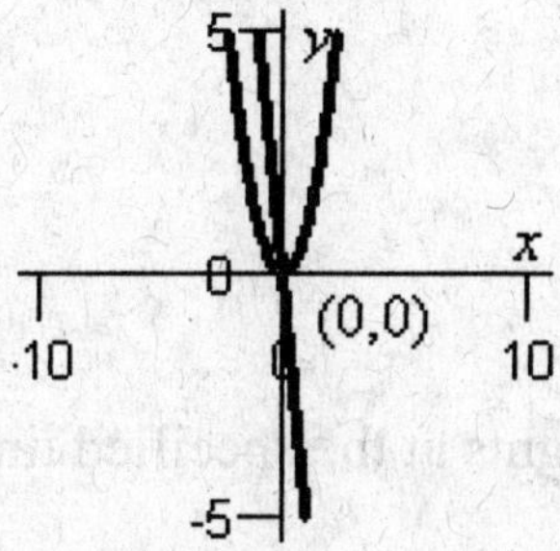

D)

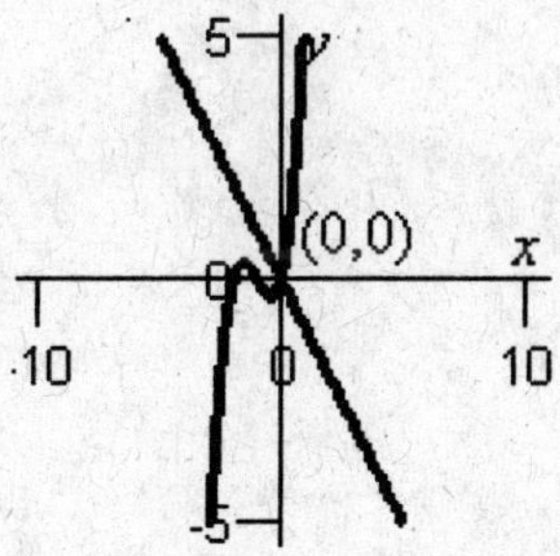

E) 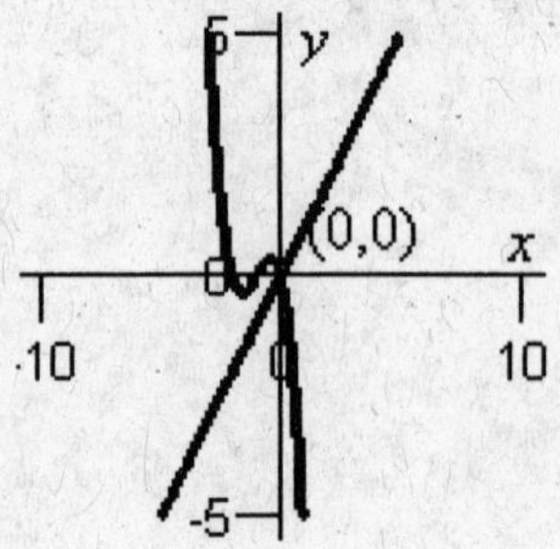

16. Determine whether the two systems of linear equations yield the same solutions. If so, find the solutions using matrices.

$$\begin{cases} x + 9y - 2z = 41 \\ y - 7z = 46 \\ z = -6 \end{cases}$$

$$\begin{cases} x + 4y + 2z = -3 \\ y + 5z = -26 \\ z = -6 \end{cases}$$

A) $x = -7, y = 4, z = -6$
B) $x = 7, y = -4, z = -6$
C) $x = -4, y = -6, z = -7$
D) $x = 4, y = -6, z = -7$
E) The systems yield different solutions.

17. An object moving vertically is at the given heights at the specified times. Find the position equation $s = \frac{1}{2}at^2 + v_0t + s_0$ for the object.
At $t = 1$ second, $s = 234$ feet
At $t = 2$ seconds, $s = 202$ feet
At $t = 3$ seconds, $s = 138$ feet
A) $s = -32t^2 + 16t + 234$
B) $s = -16t^2 + 16t - 234$
C) $s = -16t^2 + 16t + 234$
D) $s = -8t^2 + t - 234$
E) $s = -16t^2 - 16t + 234$

18. Write the partial fraction decomposition of the rational expression.

$$\frac{1}{100x^2-81}$$

A) $\frac{1}{18}\left(\frac{x-10}{10x-9}-\frac{x-10}{10x+9}\right)$

B) $\frac{1}{100x^2}-\frac{1}{81}$

C) $\frac{1}{18}\left(\frac{1}{10x-9}-\frac{1}{10x+9}\right)$

D) $\frac{1}{9}\left(\frac{1}{10x+9}-\frac{1}{10x-9}\right)$

E) $\frac{1}{9}\left(\frac{1}{10x-9}-\frac{1}{10x+9}\right)$

19. Solve the system of linear equations.

$$\begin{cases} x+y+z=-6 \\ x-6y-7z=76 \\ 8y-6z=-18 \end{cases}$$

A) $(4,-7,-3)$

B) $(-7,4,-3)$

C) $(5,-6,-5)$

D) $(-6,5,-5)$

E) $(8,-8,-6)$

20. Write the partial fraction decomposition of the improper rational expression.

$$\frac{x^3-4x^2+3x+2}{x^2+x-2}$$

A) $x-5+\frac{1}{3}\left(\frac{2}{x-1}+\frac{28}{x+2}\right)$

B) $x-5+\frac{1}{3}\left(\frac{18}{x-1}+\frac{28}{x+2}\right)$

C) $x-5+\frac{1}{3}\left(\frac{18}{x+1}+\frac{28}{x-2}\right)$

D) $x-5+\frac{1}{3}\left(\frac{2}{x-1}-\frac{28}{x+2}\right)$

E) $\frac{1}{3}\left(\frac{2}{x-1}+\frac{28}{x+2}\right)$

Answer Key

1. B
2. A
3. E
4. C
5. A
6. D
7. D
8. B
9. C
10. D
11. B
12. D
13. D
14. A
15. D
16. A
17. C
18. C
19. C
20. A

Name: ______________________________ Date: _____________

1. Solve the system by the method of elimination.
$$\begin{cases} 5x - y = 11 \\ -7x - y = -25 \end{cases}$$
A) $(1,18)$
B) $(3,4)$
C) $\left(-\frac{7}{6}, -\frac{157}{6}\right)$
D) $\left(-5, \frac{6}{5}\right)$
E) inconsistent

2. Find the minor M_{13} and its cofactor C_{13} of the matrix $\begin{bmatrix} 9 & -6 & 24 \\ -9 & 6 & -18 \\ 3 & -9 & 18 \end{bmatrix}$.

A) $M_{13} = \begin{vmatrix} 9 & 24 \\ 3 & 18 \end{vmatrix} = 90$
$C_{13} = -90$

B) $M_{13} = \begin{vmatrix} -6 & 24 \\ 6 & -18 \end{vmatrix} = -36$
$C_{13} = -36$

C) $M_{13} = \begin{vmatrix} -9 & 6 \\ 3 & -9 \end{vmatrix} = 63$
$C_{13} = -63$

D) $M_{13} = \begin{vmatrix} -6 & 24 \\ 6 & -18 \end{vmatrix} = -36$
$C_{13} = 36$

E) $M_{13} = \begin{vmatrix} -9 & 6 \\ 3 & -9 \end{vmatrix} = 63$
$C_{13} = 63$

3. Use a determinant to determine whether the points below are collinear.
   $(3,-1),\ (0,-3),\ (24,13)$
   A) the points are collinear
   B) the points are not collinear

4. Solve the system of linear equations $\begin{cases} -8x + 24y + 8z = 5 \\ 16x + 40y = 10 \\ 24x + 8y - 16z = -5 \end{cases}$ using an inverse matrix.

   A) $\begin{bmatrix} x \\ y \\ z \end{bmatrix} = \begin{bmatrix} -\frac{5}{8} \\ 0 \\ \frac{15}{8} \end{bmatrix}$

   B) $\begin{bmatrix} x \\ y \\ z \end{bmatrix} = \begin{bmatrix} 0 \\ \frac{5}{4} \\ -\frac{5}{8} \end{bmatrix}$

   C) $\begin{bmatrix} x \\ y \\ z \end{bmatrix} = \begin{bmatrix} \frac{5}{8} \\ -\frac{15}{8} \\ \frac{5}{4} \end{bmatrix}$

   D) $\begin{bmatrix} x \\ y \\ z \end{bmatrix} = \begin{bmatrix} \frac{5}{8} \\ 0 \\ \frac{5}{4} \end{bmatrix}$

   E) $\begin{bmatrix} x \\ y \\ z \end{bmatrix} = \begin{bmatrix} -\frac{5}{8} \\ -\frac{15}{8} \\ -\frac{5}{4} \end{bmatrix}$

5. Solve the system of linear equations

$$\begin{cases} 8x_1 - 16x_2 - 8x_3 - 16x_4 = 0 \\ 24x_1 - 40x_2 - 16x_3 - 24x_4 = -9 \\ 16x_1 - 40x_2 - 16x_3 - 40x_4 = 6 \\ -8x_1 + 32x_2 + 32x_3 + 88x_4 = 0 \end{cases}$$

using the inverse matrix $\frac{1}{8}\begin{bmatrix} -24 & 7 & 1 & -2 \\ -10 & 3 & 0 & -1 \\ -29 & 7 & 3 & -2 \\ 12 & -3 & -1 & 1 \end{bmatrix}$.

A) $\begin{bmatrix} x_1 \\ x_2 \\ x_3 \\ x_4 \end{bmatrix} = \begin{bmatrix} -\frac{57}{8} \\ -\frac{27}{8} \\ -\frac{45}{8} \\ \frac{21}{8} \end{bmatrix}$

B) $\begin{bmatrix} x_1 \\ x_2 \\ x_3 \\ x_4 \end{bmatrix} = \begin{bmatrix} -\frac{21}{8} \\ -\frac{15}{8} \\ \frac{27}{8} \\ -\frac{27}{8} \end{bmatrix}$

C) $\begin{bmatrix} x_1 \\ x_2 \\ x_3 \\ x_4 \end{bmatrix} = \begin{bmatrix} -\frac{27}{8} \\ -\frac{57}{8} \\ 0 \\ \frac{21}{8} \end{bmatrix}$

D) $\begin{bmatrix} x_1 \\ x_2 \\ x_3 \\ x_4 \end{bmatrix} = \begin{bmatrix} -\frac{9}{8} \\ \frac{3}{8} \\ 0 \\ \frac{3}{4} \end{bmatrix}$

E) $\begin{bmatrix} x_1 \\ x_2 \\ x_3 \\ x_4 \end{bmatrix} = \begin{bmatrix} 0 \\ -\frac{3}{8} \\ 0 \\ -\frac{3}{4} \end{bmatrix}$

6. Solve the system of equations algebraically.

$$\begin{cases} xy + 289 = 0 \\ 4x - 25y - 340 = 0 \end{cases}$$

A) $\left(85, -\frac{17}{5}\right)$
B) $\left(-85, \frac{17}{5}\right), \left(85, -\frac{17}{5}\right)$
C) $\left(\frac{85}{2}, -\frac{34}{5}\right)$
D) $\left(-\frac{85}{2}, \frac{34}{5}\right), \left(\frac{85}{2}, -\frac{34}{5}\right)$
E) no solution

7. An object moving vertically is at the given heights at the specified times. Find the position equation $s = \frac{1}{2}at^2 + v_o t + s_o$ for the object.

At $t = 1$ second, $s = 273$ feet
At $t = 2$ seconds, $s = 241$ feet
At $t = 3$ seconds, $s = 177$ feet

A) $s = -32t^2 + 16t + 273$
B) $s = -16t^2 + 16t - 273$
C) $s = -16t^2 + 16t + 273$
D) $s = -8t^2 + t - 273$
E) $s = -16t^2 - 16t + 273$

8. Identify the elementary row operation being performed to obtain the new row-equivalent matrix.

Original Matrix $\begin{bmatrix} 7 & 6 & 6 \\ 1 & 4 & -6 \end{bmatrix}$ *New Row-Equivalent Matrix* $\begin{bmatrix} 12 & 26 & -24 \\ 1 & 4 & -6 \end{bmatrix}$

A) Add 5 times R_1 to R_2.
B) Add –5 times R_2 to R_1.
C) Add –5 times R_1 to R_2.
D) Add 5 times R_1 to R_1.
E) Add 5 times R_2 to R_1.

9. Find the equation of the circle

$$x^2 + y^2 + Dx + Ey + F = 0$$

that passes through the points $(-1,1),(-6,-4),(4,-4)$.

A) $x^2 + y^2 + 2x + 8y - 8 = 0$
B) $x^2 + y^2 + x + 4y - 8 = 0$
C) $x^2 + y^2 + 2x + 8y + 42 = 0$
D) $x^2 + y^2 + 2x + 8y - 25 = 0$
E) $x^2 + y^2 + x + 4y - 25 = 0$

10. Use the matrix capabilities of a graphing utility to find the inverse of the matrix $\frac{1}{6}\begin{bmatrix} -4 & -5 & 3 \\ -4 & -8 & 3 \\ 1 & 2 & 0 \end{bmatrix}$ (if it exists).

A) $\begin{bmatrix} 2 & 2 & 6 \\ 0 & -2 & 0 \\ 0 & 4 & 8 \end{bmatrix}$

B) $\begin{bmatrix} -6 & 8 & 4 \\ 0 & -2 & 0 \\ 0 & 2 & -6 \end{bmatrix}$

C) $\begin{bmatrix} -4 & 0 & 6 \\ 0 & -2 & 0 \\ 0 & 2 & -4 \end{bmatrix}$

D) $\begin{bmatrix} -4 & 4 & 6 \\ 2 & -2 & 0 \\ 0 & 2 & 8 \end{bmatrix}$

E) does not exist

11. Determine which one of the ordered triples below is a solution of the given system of equations.

$$\begin{cases} 7x - 9y + 6z = 1 \\ 9x + 2y - 3z = 37 \\ 3x - 4y + 5z = 7 \end{cases}$$

A) (–8, –1, 8)
B) (4, 5, 3)
C) (5, –3, –4)
D) (–8, –1, 3)
E) (3, 5, 4)

12. Determine whether the two systems of linear equations yield the same solutions. If so, find the solutions using matrices.

$$\begin{cases} x + 5y - 5z = 42 \\ \quad\ \ y - 9z = 48 \\ \qquad\quad\ z = -5 \end{cases}$$

$$\begin{cases} x + 8y - 3z = 31 \\ \quad\ \ y + 5z = -22 \\ \qquad\quad\ z = -5 \end{cases}$$

A) $x = 2, y = 3, z = -5$
B) $x = -2, y = -3, z = -5$
C) $x = -3, y = -5, z = 2$
D) $x = 3, y = -5, z = 2$
E) The systems yield different solutions.

13. Solve the system of equations below using Gaussian elimination.

$$\begin{aligned} 5x + 4y + 3z &= 54 \\ x - 2y + 2z &= 36 \\ x - y - z &= -18 \end{aligned}$$

A) $(1, -1, 20)$
B) $(1, 0, 19)$
C) $(0, 1, 17)$
D) $(-1, 1, 16)$
E) $(0, 0, 18)$

14. Solve the system of linear equations.

$$\begin{cases} x + 4y + 2z = -2 \\ -4x + y - 4z = 9 \\ 3x - y + 2z = -3 \end{cases}$$

A) $(3, 6, 3)$
B) $(2, 1, -4)$
C) $(1, -4, 2)$
D) $(6, 3, 3)$
E) inconsistent

15. Use matrices to solve the system of equations (if possible). Use Gaussian elimination with back-substitution or Gauss-Jordan elimination.

$$\begin{cases} 7x - 3y - 9z = 22 \\ 2x + y + 3z = -3 \\ 2x - 7y - 3z = -17 \end{cases}$$

A) $x = -1, y = -4, z = -3$
B) $x = 1, y = -4, z = -3$
C) $x = 1, y = 4, z = -3$
D) $x = 4, y = -1, z = 3$
E) no solution

16. Evaluate the expression.

$$\begin{bmatrix} -7 & -3 \\ 6 & 5 \end{bmatrix} + \begin{bmatrix} -6 & 7 \\ -1 & 2 \end{bmatrix} + \begin{bmatrix} 4 & 9 \\ 9 & 4 \end{bmatrix}$$

A) $\begin{bmatrix} -17 & -5 \\ -4 & 3 \end{bmatrix}$

B) $\begin{bmatrix} -9 & 13 \\ 14 & 11 \end{bmatrix}$

C) $\begin{bmatrix} -13 & 4 \\ 5 & 7 \end{bmatrix}$

D) $\begin{bmatrix} 1 & 0 \\ 0 & 1 \end{bmatrix}$

E) not possible

17. Write the augmented matrix for the system of linear equations.

$$\begin{cases} x - 8y + 8z = 9 \\ \quad\;\; 8y - 9z = 1 \\ x \qquad\; + z = 7 \end{cases}$$

A) $\left[\begin{array}{ccc|c} 1 & -8 & 8 & 9 \\ 1 & 8 & -9 & 1 \\ 1 & 1 & 1 & 7 \end{array}\right]$

B) $\left[\begin{array}{ccc|c} 1 & -8 & 8 & 9 \\ 0 & 8 & -9 & 1 \\ 1 & 0 & 1 & 7 \end{array}\right]$

C) $\left[\begin{array}{ccc|c} 1 & -8 & 8 & 9 \\ & 8 & -9 & 1 \\ 1 & & 1 & 7 \end{array}\right]$

D) $\left[\begin{array}{ccc|c} 1 & -8 & 8 & 9 \\ 8 & -9 & 0 & 1 \\ 1 & 1 & 0 & 7 \end{array}\right]$

E) $\left[\begin{array}{ccc|c} 1 & -8 & 8 & 9 \\ 0 & 8 & -9 & 1 \\ 1 & 1 & 0 & 7 \end{array}\right]$

18. Find x and y.

$$\begin{bmatrix} x+3 & 6 & 4y \\ 5 & 2x & -1 \\ 8 & y-9 & 2 \end{bmatrix} = \begin{bmatrix} 9x+19 & 6 & -16 \\ 5 & -4 & -1 \\ 8 & -13 & 2 \end{bmatrix}$$

A) $x = -2, y = -4$
B) $x = -2, y = -5$
C) $x = -1, y = -5$
D) $x = -4, y = -1$
E) no solution

19. Solve the system by the method of elimination.

$$\begin{cases} \frac{1}{9}x - \frac{1}{8}y = 6 \\ -48x + 54y = -2591 \end{cases}$$

A) $(9,-40)$
B) no solution
C) $\left(\frac{2915}{102}, -\frac{2303}{102}\right)$
D) $\left(\frac{5183}{96}, \frac{1}{108}\right)$
E) infinitely many solutions

20. Find the inverse of the matrix $\begin{bmatrix} 8 & 16 \\ -24 & -40 \end{bmatrix}$.

A) $\frac{1}{7}\begin{bmatrix} 8 & 16 \\ -24 & -40 \end{bmatrix}$
B) $\frac{1}{8}\begin{bmatrix} -5 & -2 \\ 3 & 1 \end{bmatrix}$
C) $\begin{bmatrix} -8 & -16 \\ 40 & 24 \end{bmatrix}$
D) $\begin{bmatrix} 8 & -24 \\ 32 & -56 \end{bmatrix}$
E) $\frac{1}{13}\begin{bmatrix} -8 & -16 \\ -24 & 40 \end{bmatrix}$

Answer Key

1. B
2. E
3. A
4. D
5. A
6. C
7. C
8. E
9. A
10. D
11. B
12. E
13. E
14. B
15. C
16. B
17. B
18. A
19. B
20. B

Name: ______________________ Date: ____________

1. Solve the system by the method of substitution.

$$\begin{cases} x^2 + y^2 = 25 \\ 3x - 4y = 0 \end{cases}$$

A) (4, 3), (–4, –3)
B) (4, 3)
C) (4, 3), (–4, 3), (–4, –3), (4, –3)
D) (3, 4), (3, –4)
E) no real solution

2. Write the augmented matrix for the system of linear equations.

$$\begin{cases} x + 4y - z = -9 \\ \quad -9y + 4z = -6 \\ x \quad + 5z = -4 \end{cases}$$

A) $\left[\begin{array}{ccc|c} 1 & 4 & -1 & -9 \\ 1 & -9 & 4 & -6 \\ 1 & 1 & 5 & -4 \end{array}\right]$

B) $\left[\begin{array}{ccc|c} 1 & 4 & -1 & -9 \\ 0 & -9 & 4 & -6 \\ 1 & 0 & 5 & -4 \end{array}\right]$

C) $\left[\begin{array}{ccc|c} 1 & 4 & -1 & -9 \\ & -9 & 4 & -6 \\ 1 & & 5 & -4 \end{array}\right]$

D) $\left[\begin{array}{ccc|c} 1 & 4 & -1 & -9 \\ -9 & 4 & 0 & -6 \\ 1 & 5 & 0 & -4 \end{array}\right]$

E) $\left[\begin{array}{ccc|c} 1 & 4 & -1 & -9 \\ 0 & -9 & 4 & -6 \\ 1 & 5 & 0 & -4 \end{array}\right]$

3. Evaluate the expression.

$$\frac{1}{4}\begin{bmatrix}-9 & 1 & -1\end{bmatrix}+\begin{bmatrix}-4 & -6 & 8\end{bmatrix}$$

A) $\begin{bmatrix}-13 & -5 & 7\end{bmatrix}$

B) $\begin{bmatrix}\frac{7}{4} & \frac{25}{4} & -\frac{33}{4}\end{bmatrix}$

C) $\begin{bmatrix}-\frac{13}{4} & -\frac{5}{4} & \frac{7}{4}\end{bmatrix}$

D) $\begin{bmatrix}-\frac{25}{4} & -\frac{23}{4} & \frac{31}{4}\end{bmatrix}$

E) not possible

4. An object moving vertically is at the given heights at the specified times. Find the position equation $s=\frac{1}{2}at^2+v_o t+s_o$ for the object.

At $t = 1$ second, $s = 198$ feet
At $t = 2$ seconds, $s = 182$ feet
At $t = 3$ seconds, $s = 134$ feet

A) $s=-32t^2+32t+198$
B) $s=-16t^2+32t-182$
C) $s=-16t^2+32t+182$
D) $s=-8t^2+2t-182$
E) $s=-16t^2-32t+198$

5. Evaluate the determinant $\begin{vmatrix}7x & x\ln x \\ 7 & 2+\ln x\end{vmatrix}$ in which the entries are functions.

A) $49\ln x$
B) $14x$
C) $2x$
D) $7-\ln x$
E) $2\ln x$

6. If possible, find $5A + 4B$.

$$A=\begin{bmatrix}-8 & 9 & 2\\ -2 & 1 & -4\end{bmatrix}, B=\begin{bmatrix}8 & 2 & 1\\ -3 & 8 & 4\end{bmatrix}$$

A) $\begin{bmatrix}-72 & 37 & 6\\ 2 & -27 & -36\end{bmatrix}$

B) $\begin{bmatrix}0 & 11 & 3\\ -5 & 9 & 0\end{bmatrix}$

C) $\begin{bmatrix}1 & 0 & 0\\ 0 & 1 & 0\end{bmatrix}$

D) $\begin{bmatrix}-8 & 53 & 14\\ -22 & 37 & -4\end{bmatrix}$

E) not possible

7. Determine which one of the ordered triples below is a solution of the given system of equations.

$$\begin{cases}8x+6y-4z=74\\ x-5y+8z=-102\\ 5x-3y+2z=-46\end{cases}$$

A) (4, 7, 0)
B) (–1, 9, –7)
C) (9, 7, 1)
D) (4, 7, –7)
E) (–7, 9, –1)

8. Evaluate the expression.

$$\begin{bmatrix} 2 \\ 5 \\ -3 \end{bmatrix}\left(\begin{bmatrix} 6 & 4 \end{bmatrix}+\begin{bmatrix} -8 & 1 \end{bmatrix}\right)$$

A) $\begin{bmatrix} -4 \\ -10 \\ 6 \end{bmatrix}$

B) $\begin{bmatrix} -4 & 10 \\ -10 & 25 \\ 6 & -15 \end{bmatrix}$

C) $\begin{bmatrix} -4 & -10 & 6 \\ 10 & 25 & -15 \end{bmatrix}$

D) $\begin{bmatrix} 12 & 8 & -16 & 2 \\ 30 & 20 & -40 & 5 \\ -18 & -12 & 24 & -3 \end{bmatrix}$

E) not possible

9. Determine whether the system of linear equations is consistent or inconsistent.

$$\begin{cases} -x+9y=5 \\ -4x+36y=21 \end{cases}$$

A) inconsistent
B) consistent

10. Perform the indicated row operations on the matrix. Show the final result.

$$\begin{bmatrix} 1 & 7 & 1 \\ -7 & -48 & -8 \\ -5 & -1 & -39 \end{bmatrix}$$

Add 7 times R_1 to R_2.
Add 5 times R_1 to R_3.

A) $\begin{bmatrix} -5 & 4 & 1 \\ -7 & -48 & -8 \\ -5 & -1 & -39 \end{bmatrix}$

B) $\begin{bmatrix} 1 & 7 & 1 \\ 0 & 1 & -7 \\ 0 & 6 & -38 \end{bmatrix}$

C) $\begin{bmatrix} 1 & 7 & 1 \\ 0 & 1 & -1 \\ 0 & 34 & -34 \end{bmatrix}$

D) $\begin{bmatrix} 1 & 7 & 1 \\ -48 & -329 & -55 \\ -24 & 2 & -194 \end{bmatrix}$

E) $\begin{bmatrix} 1 & 7 & 1 \\ 0 & -13 & -3 \\ 0 & 48 & 10 \end{bmatrix}$

11. Given:

$$A = \begin{bmatrix} -7 & 3 & -2 \\ -6 & 6 & 5 \end{bmatrix},\ B = \begin{bmatrix} 3 & -6 & 5 \\ -3 & 1 & 7 \end{bmatrix} \text{ and } c = -4,$$

determine cAB^2.

A) $\begin{bmatrix} 20 & -1148 \\ 632 & -788 \end{bmatrix}$

B) $\begin{bmatrix} -1764 & -1296 & -400 \\ -1296 & -144 & -4900 \end{bmatrix}$

C) $\begin{bmatrix} 252 & -432 & 200 \\ 216 & -24 & -980 \end{bmatrix}$

D) $\begin{bmatrix} 468 & 1032 & 1876 \\ -324 & -456 & -1476 \\ -108 & 268 & -780 \end{bmatrix}$

E) not possible

12. Find the minor M_{13} and its cofactor C_{13} of the matrix $\begin{bmatrix} -3 & 2 & -8 \\ 3 & -2 & 6 \\ -1 & 3 & -6 \end{bmatrix}$.

A) $M_{13} = \begin{vmatrix} -3 & -8 \\ -1 & -6 \end{vmatrix} = 10$

$C_{13} = -10$

B) $M_{13} = \begin{vmatrix} 2 & -8 \\ -2 & 6 \end{vmatrix} = -4$

$C_{13} = -4$

C) $M_{13} = \begin{vmatrix} 3 & -2 \\ -1 & 3 \end{vmatrix} = 7$

$C_{13} = -7$

D) $M_{13} = \begin{vmatrix} 2 & -8 \\ -2 & 6 \end{vmatrix} = -4$

$C_{13} = 4$

E) $M_{13} = \begin{vmatrix} 3 & -2 \\ -1 & 3 \end{vmatrix} = 7$

$C_{13} = 7$

13. Solve the system by the method of elimination.

$$\begin{cases} 7x - 6y = -73 \\ -8x - y = 52 \end{cases}$$

A) $(2, -68)$

B) $(-7, 4)$

C) $\left(\frac{239}{15}, \frac{2812}{15}\right)$

D) $\left(-8, -\frac{121}{7}\right)$

E) inconsistent

14. Solve the system of linear equations.

$$\begin{cases} x+y+z=-11 \\ x-6y-3z=13 \\ 6y-7z=42 \end{cases}$$

A) $(-6,-1,-4)$
B) $(-1,-6,-4)$
C) $(-5,0,-6)$
D) $(0,-5,-6)$
E) $(-2,-2,-7)$

15. Use a system of equations to find the specified equation that passes through the points. Solve the system using matrices.
Parabola: $y = ax^2 + bx + c$

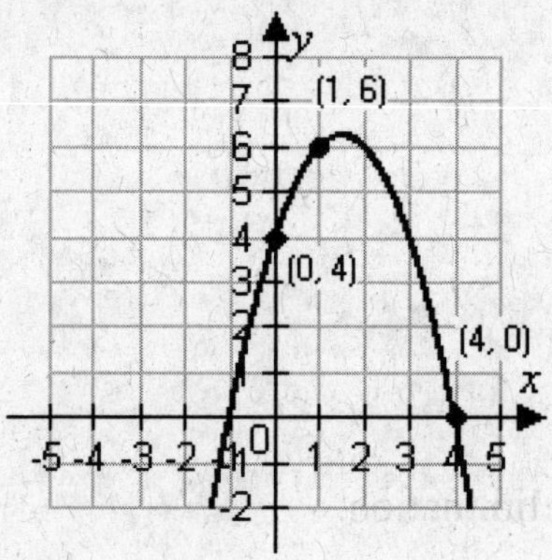

A) $y=x^2-3x+4$
B) $y=-x^2-3x+1$
C) $y=-x^2+3x+4$
D) $y=-x^2+3x+1$
E) $y=-x^2-2x+1$

16. Solve the system by the method of elimination.

$$\begin{cases} \dfrac{x+7}{5}+\dfrac{y-2}{5}=-2 \\ x-y=-14 \end{cases}$$

A) $\left(-\dfrac{29}{2},-\dfrac{1}{2}\right)$

B) $\left(-\dfrac{1}{2},-\dfrac{29}{2}\right)$

C) $\left(-\dfrac{2}{29},-\dfrac{2}{1}\right)$

D) $(-29,-15)$

E) no solution

17. Write the system of linear equations as a matrix equation $AX = B$, and use Gauss-Jordan elimination on the augmented matrix $[A \vdots B]$ to solve for the matrix X.

$$\begin{cases} x_1 + 6x_2 + 9x_3 = -44 \\ -7x_1 + 2x_2 - x_3 = -54 \\ 8x_1 + 3x_2 + 6x_3 = 26 \end{cases}$$

A) $X = \begin{bmatrix} 28 \\ 12 \\ 21 \end{bmatrix}$

B) $X = \begin{bmatrix} 12 \\ -7 \\ -4 \end{bmatrix}$

C) $X = \begin{bmatrix} 7 \\ -4 \\ -3 \end{bmatrix}$

D) $X = \begin{bmatrix} -12 \\ -28 \\ -21 \end{bmatrix}$

E) $X = \begin{bmatrix} -7 \\ 3 \\ -4 \end{bmatrix}$

18. Use the matrix capabilities of a graphing utility to reduce the augmented matrix corresponding to the system of equations, and solve the system.

$$\begin{cases} x - 7y - 3z = 64 \\ 3x + 7y + 2z = -92 \\ -2x + 14y + 6z = 64 \\ -8x + 2y + 2z = 42 \end{cases}$$

A) $x=-9, y=-7, z=-8$
B) $x=75, y=2, z=-1$
C) $x=-9, y=7, z=8$
D) $x=-9, y=-7, z=8$
E) no solution

19. Solve the system of equations below using Gaussian elimination.

$$\begin{aligned} 5x+4y+3z&=-72 \\ x-2y+2z&=-48 \\ x-y-z&=24 \end{aligned}$$

A) $(1,-1,-22)$
B) $(1,0,-23)$
C) $(0,1,-25)$
D) $(-1,1,-26)$
E) $(0,0,-24)$

20. Given:

$$A=\begin{bmatrix}4 & 2 & 5\\ 1 & 2 & 1\end{bmatrix},\ B=\begin{bmatrix}5 & 5 & -1\\ 4 & 4 & 1\end{bmatrix},\ C=\begin{bmatrix}-2 & -1\\ 3 & 1\\ -2 & -3\end{bmatrix},$$

determine $CA-CB$.

A) $\begin{bmatrix}5 & 8 & -12\\ -6 & -11 & 18\\ 11 & 12 & -12\end{bmatrix}$

B) $\begin{bmatrix}2 & -9 & -12\\ 3 & -2 & 0\end{bmatrix}$

C) $\begin{bmatrix}-19 & -20\\ 0 & 1\end{bmatrix}$

D) $\begin{bmatrix}2 & 3\\ -9 & -2\\ -12 & 0\end{bmatrix}$

E) not possible

Answer Key

1. A
2. B
3. D
4. C
5. B
6. D
7. B
8. B
9. A
10. C
11. E
12. E
13. B
14. C
15. C
16. A
17. C
18. E
19. E
20. E

Name: ______________________________ Date: _____________

1. Use the matrix capabilities of a graphing utility to find AB, if possible.

$$A=\begin{bmatrix}-7 & 5 & 2\\ 0 & 1 & 8\\ 2 & 8 & -2\end{bmatrix}, B=\begin{bmatrix}4 & 3 & -9\\ -4 & -1 & -3\\ 9 & 8 & 1\end{bmatrix}$$

A) $\begin{bmatrix}-28 & 15 & -18\\ 0 & -1 & -24\\ 18 & 64 & -2\end{bmatrix}$

B) $\begin{bmatrix}-66 & -42 & 46\\ -76 & -65 & -11\\ -6 & 14 & -40\end{bmatrix}$

C) $\begin{bmatrix}-30 & 68 & -42\\ -10 & 63 & -18\\ 50 & 5 & -44\end{bmatrix}$

D) $\begin{bmatrix}-30 & -10 & 50\\ 68 & 63 & 5\\ -42 & -18 & -44\end{bmatrix}$

E) not possible

2. Perform the indicated row operations on the matrix. Show the final result.

$$\begin{bmatrix} 1 & -1 & 9 \\ -3 & 4 & 4 \\ 6 & -8 & -8 \end{bmatrix}$$

Add 3 times R_1 to R_2.
Add –6 times R_1 to R_3.

A) $\begin{bmatrix} 6 & -14 & -224 \\ -3 & 4 & 4 \\ 6 & -8 & -8 \end{bmatrix}$

B) $\begin{bmatrix} 1 & -1 & 9 \\ 0 & 1 & 13 \\ 0 & -9 & 1 \end{bmatrix}$

C) $\begin{bmatrix} 1 & -1 & 9 \\ 0 & 1 & 31 \\ 0 & -2 & -62 \end{bmatrix}$

D) $\begin{bmatrix} 1 & -1 & 9 \\ -8 & 11 & 21 \\ -35 & 47 & 57 \end{bmatrix}$

E) $\begin{bmatrix} 1 & -1 & 9 \\ 0 & 10 & -50 \\ 0 & -11 & -11 \end{bmatrix}$

3. Solve the system by the method of substitution.

$$\begin{cases} x^2 + y^2 = 625 \\ 7x - 24y = 0 \end{cases}$$

A) (24, 7), (–24, –7)
B) (24, 7)
C) (24, 7), (–24, 7), (–24, –7), (24, –7)
D) (7, 24), (7, –24)
E) no real solution

4. Solve the system of equations.

$$\begin{cases} -\dfrac{1}{x}-\dfrac{4}{y}-\dfrac{1}{z}=-3 \\ -\dfrac{5}{x}-\dfrac{2}{y}-\dfrac{2}{z}=-5 \\ -\dfrac{1}{x}-\dfrac{5}{y}+\dfrac{2}{z}=-3 \end{cases}$$

A) $\left(\dfrac{41}{57},\dfrac{10}{19},\dfrac{10}{57}\right)$

B) $\left(\dfrac{1}{2},\dfrac{1}{5},-\dfrac{1}{19}\right)$

C) $\left(\dfrac{827}{285},\dfrac{13}{10},\dfrac{10}{57}\right)$

D) $\left(\dfrac{85}{41},\dfrac{17}{6},\dfrac{17}{2}\right)$

E) $\left(\dfrac{57}{41},\dfrac{19}{10},\dfrac{57}{10}\right)$

5. Find the minor M_{13} and its cofactor C_{13} of the matrix $\begin{bmatrix} -9 & 6 & -24 \\ 9 & -6 & 18 \\ -3 & 9 & -18 \end{bmatrix}$.

A) $M_{13} = \begin{vmatrix} -9 & -24 \\ -3 & -18 \end{vmatrix} = 90$

$C_{13} = -90$

B) $M_{13} = \begin{vmatrix} 6 & -24 \\ -6 & 18 \end{vmatrix} = -36$

$C_{13} = -36$

C) $M_{13} = \begin{vmatrix} 9 & -6 \\ -3 & 9 \end{vmatrix} = 63$

$C_{13} = -63$

D) $M_{13} = \begin{vmatrix} 6 & -24 \\ -6 & 18 \end{vmatrix} = -36$

$C_{13} = 36$

E) $M_{13} = \begin{vmatrix} 9 & -6 \\ -3 & 9 \end{vmatrix} = 63$

$C_{13} = 63$

6. Write the partial fraction decomposition of the rational expression.

$$\frac{5}{x^2 + 10x + 16}$$

A) $\frac{5}{6}\left(\frac{1}{x-2} - \frac{1}{x-8}\right)$

B) $\frac{5}{6}\left(\frac{1}{x+2} - \frac{1}{x+8}\right)$

C) $\frac{5}{6}\left(\frac{1}{x+8} - \frac{1}{x+2}\right)$

D) $\frac{5}{x^2} + \frac{1}{2x} + \frac{5}{16}$

E) $\frac{5}{x^2 + 10x + 16}$

7. An augmented matrix that represents a system of linear equations (in variables x, y, and z) has been reduced using Gauss-Jordan elimination. Write the solution represented by the augmented matrix.

$$\left[\begin{array}{ccc|c} 1 & 0 & 0 & -1 \\ 0 & 1 & 0 & 3 \\ 0 & 0 & 1 & 8 \end{array}\right]$$

A) $x = -1, y = 3, z = 8$
B) $x = 1, y = -3, z = -8$
C) $x = 0, y = 0, z = 0$
D) $x = -x, y = 8y, z = 3z$
E) $x = -1, y = 0, z = 0$

8. Solve using any method.

$$\begin{cases} 7x - 9y = 3 \\ y = x - 4 \end{cases}$$

A) $(-4,-8)$
B) $\left(7, \frac{66}{7}\right)$
C) $\left(-\frac{39}{2}, \frac{31}{2}\right)$
D) $\left(\frac{33}{2}, \frac{25}{2}\right)$
E) inconsistent

9. Find a solution of the system of equations and verify your result using matrix multiplication.

$$\begin{cases} -8x + 3y = 20 \\ 3x - 5y = -9 \end{cases}$$

A) $\begin{bmatrix} -\frac{3}{8} \\ \frac{17}{3} \end{bmatrix}$

B) $\begin{bmatrix} -\frac{65}{38} \\ \frac{40}{19} \end{bmatrix}$

C) $\begin{bmatrix} -\frac{73}{49} \\ \frac{132}{49} \end{bmatrix}$

D) $\begin{bmatrix} -\frac{73}{31} \\ \frac{12}{31} \end{bmatrix}$

E) $\begin{bmatrix} -\frac{31}{73} \\ \frac{31}{12} \end{bmatrix}$

10. Determine which ordered pair is a solution of the system.

$$\begin{cases} 3x - 6y = -8 \\ 9x + 8y = 5 \end{cases}$$

A) $\left(-\frac{17}{39}, -\frac{19}{26}\right)$

B) $\left(-\frac{47}{39}, -\frac{19}{26}\right)$

C) $(-5, 7)$

D) $\left(-\frac{17}{39}, \frac{29}{26}\right)$

E) $(-5, 50)$

11. Find $|AB|$, if

$$A=\begin{bmatrix}-3 & 4 & 5\\ -3 & -5 & 1\\ 5 & -5 & 1\end{bmatrix},\ B=\begin{bmatrix}-1 & -2 & 5\\ 4 & 1 & -3\\ -1 & -1 & 4\end{bmatrix}$$

A) 2320
B) 2736
C) 202
D) 0
E) –2644

12. Solve system of equations by the method of substitution.

$$\begin{cases}12x^3-9y=0\\ 12x-y=0\end{cases}$$

A) $(0,0)$, $(-3,-36)$, $(3,36)$
B) $(0,0)$, $(-3,36)$, $(3,36)$
C) $(0,0)$, $(3,-36)$, $(3,36)$
D) $(0,0)$, $(-3,37)$, $(3,37)$
E) $(0,0)$, $(-3,-35)$, $(3,35)$

13. Use the matrix capabilities of a graphing utility to reduce the augmented matrix corresponding to the system of equations, and solve the system.

$$\begin{cases}x+4y+8z=5\\ 7x-3y-4z=-64\\ -x-4y-8z=-5\\ 6x+5y+7z=-18\end{cases}$$

A) $x=-7, y=9, z=-3$
B) $x=5, y=-2, z=1$
C) $x=-7, y=-9, z=3$
D) $x=-7, y=9, z=3$
E) no solution

14. Use a system of equations to find the specified equation that passes through the points. Solve the system using matrices.
    Parabola: $y = ax^2 + bx + c$

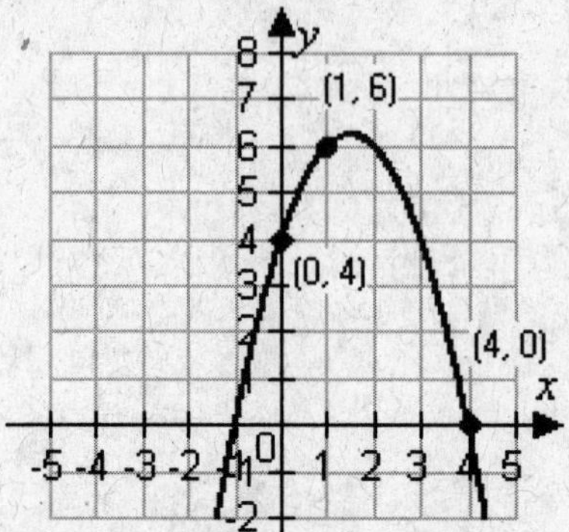

A) $y = x^2 - 3x + 4$
B) $y = -x^2 - 3x + 1$
C) $y = -x^2 + 3x + 4$
D) $y = -x^2 + 3x + 1$
E) $y = -x^2 - 2x + 1$

15. Find the equation of the parabola $y = ax^2 + bx + c$ that passes through the points.
    $(-3,9),(-2,7),(-1,7)$
A) $y = x^2 - 3x - 9$
B) $y = x^2 + 3x - 9$
C) $y = 2x^2 + 3x + 0$
D) $y = 2x^2 + 3x + 5$
E) $y = x^2 + 3x + 9$

16. Write the form of the partial fraction decomposition of the rational expression. Do not solve for the constants.

$$\frac{x+9}{x\left(x^2+3\right)^2}$$

A) $\dfrac{Ax+B}{x^2+3}+\dfrac{Cx+D}{\left(x^2+3\right)^2}$

B) $\dfrac{A}{x}+\dfrac{B}{x^2+3}+\dfrac{C}{\left(x^2+3\right)^2}$

C) $\dfrac{A}{x}+\dfrac{Bx+C}{\left(x^2+3\right)^2}$

D) $\dfrac{Ax+B}{x}+\dfrac{Cx+D}{x^2+3}+\dfrac{Ex+F}{\left(x^2+3\right)^2}$

E) $\dfrac{A}{x}+\dfrac{Bx+C}{x^2+3}+\dfrac{Dx+E}{\left(x^2+3\right)^2}$

17. Solve the system of linear equations.

$$\begin{cases} x+5y+5z=9 \\ -2x \qquad\quad +z=-6 \\ 2x+4y-3z=-2 \end{cases}$$

A) $(-4,0,-5)$

B) $(4,-1,2)$

C) $(-1,2,4)$

D) $(0,-5,-4)$

E) $(4,-1,1)$

18. Find a system of linear equations that has the following solution.

$\left(-\frac{4}{7}, 10\right)$

A) $\begin{cases} 28x+15y=\frac{1900}{7} \\ 56x+10y=\frac{3880}{7} \end{cases}$

B) $\begin{cases} 28y+15x=134 \\ 56y+10x=68 \end{cases}$

C) $\begin{cases} 56x+10y=-166 \\ 28x+15y=-132 \end{cases}$

D) $\begin{cases} 28y+15x=\frac{1900}{7} \\ 56y+10x=\frac{3880}{7} \end{cases}$

E) $\begin{cases} 28x+15y=134 \\ 56x+10y=68 \end{cases}$

19. Solve the system of equations below using Gaussian elimination.

$$\begin{aligned} 5x+4y+3z&=90 \\ x-2y+2z&=60 \\ x-y-z&=-30 \end{aligned}$$

A) $(1,-1,32)$
B) $(1,0,31)$
C) $(0,1,29)$
D) $(-1,1,28)$
E) $(0,0,30)$

20. Given:

$$A=\begin{bmatrix}4 & 2 & -3\\ 1 & -2 & 2\end{bmatrix},\ B=\begin{bmatrix}0 & -2 & 5\\ 5 & 4 & 2\end{bmatrix},\ C=\begin{bmatrix}-4 & 1\\ -5 & -1\\ -3 & -3\end{bmatrix},$$

determine $CA-CB$.

A) $\begin{bmatrix}-20 & -22 & 32\\ -16 & -14 & 40\\ 0 & 6 & 24\end{bmatrix}$

B) $\begin{bmatrix}-16 & -20 & 24\\ -4 & 6 & 0\end{bmatrix}$

C) $\begin{bmatrix}-12 & 24\\ 46 & 2\end{bmatrix}$

D) $\begin{bmatrix}-16 & -4\\ -20 & 6\\ 24 & 0\end{bmatrix}$

E) not possible

Answer Key

1. D
2. C
3. A
4. E
5. E
6. B
7. A
8. D
9. D
10. D
11. A
12. A
13. A
14. C
15. E
16. E
17. B
18. E
19. E
20. E

Name: ______________________________ Date: ______________

1. Use mathematical induction to prove the following for every positive integer n.

$$\sum_{i=1}^{n} 36i^5 = 3n^2(n+1)^2(2n^2+2n-1)$$

2. Write the first five terms of the sequence. (Assume that n begins with 0.)

$$a_n = \frac{3^n}{(n+1)!}$$

A) $1, \frac{3}{2}, \frac{3}{2}, \frac{9}{8}, \frac{27}{40}$
B) $\frac{3}{2}, \frac{9}{8}, \frac{27}{40}, \frac{27}{80}, \frac{81}{560}$
C) $0, \frac{3}{2}, \frac{9}{8}, \frac{27}{40}, \frac{27}{80}$
D) $3, \frac{3}{8}, \frac{9}{40}, \frac{9}{80}, \frac{27}{560}$
E) $3, \frac{3}{2}, \frac{3}{8}, \frac{3}{40}, \frac{1}{80}$

3. Use mathematical induction to prove the following for every positive integer n.

$$\sum_{i=1}^{n} \frac{8}{(2i-1)(2i+1)} = \frac{8n}{2n+1}$$

4. Evaluate: ${}_8P_5$
A) 56
B) 336
C) 6720
D) 40
E) undefined

5. Find the sum.

$$\sum_{i=1}^{4}(-i-4)$$

A) –18
B) –30
C) –26
D) –8
E) –10

6. Find the sum of the integers from –1 to 27.

A) 378
B) 28
C) 754
D) 377
E) 756

7. Determine the sample space for the experiment.

Four coins are flipped and the number of heads observed is recorded.

A) $S = \{1, 2, 3, 4\}$
B) $S = \{4\}$
C) $S = \{0, 1, 2, 3, 4\}$
D) $S = \{HT, TH\}$
E) $S = \{HHHH\}$

8. Find a formula for a_n for the arithmetic sequence.

$a_3 = 19, a_{13} = 99$

A) $a_n = 3 + 8n$
B) $a_n = 5 + 3n$
C) $a_n = 8 + 3n$
D) $a_n = 3(8)^n$
E) $a_n = -5 + 8n$

9. Use the Binomial Theorem to expand and simplify the expression.

   $(w-2)^5$

   A) $w^5-10w^4+40w^3-80w^2+80w$
   B) $w^4-8w^3+24w^2-32w+16$
   C) $w^5-10w^4+60w^3-120w^2+80w-32$
   D) $w^5-8w^4+36w^3-72w^2+64w-32$
   E) $w^5-10w^4+40w^3-80w^2+80w-32$

10. Find the rational number representation of the repeating decimal.

    $0.\overline{167}$

    A) $\dfrac{167}{9999}$
    B) $\dfrac{16.7}{999}$
    C) $\dfrac{167}{9}$
    D) $\dfrac{167}{999}$
    E) $\dfrac{167}{99}$

11. Solve for n.

    $42 \cdot {}_{n-1}P_5 = {}_{n+1}P_6$

    A) $n=35$
    B) $n=7, 34$
    C) $n=6, 35$
    D) $n=6$
    E) no solution

12. Determine whether the sequence is geometric. If so, find the common ratio.

    $-1, 2, 5, 8, \ldots$

    A) 3
    B) –1
    C) $\dfrac{1}{3}$
    D) –3
    E) not geometric

13. Use mathematical induction to prove that 80 is a factor of $2^{8n+2}+16$ for all positive n.

14. Find the sum of the following infinite geometric series.

$$-14+13-\frac{169}{14}+\frac{2197}{196}-\ ...$$

A) $-\frac{27}{196}$
B) $-\frac{1}{196}$
C) $\frac{27}{196}$
D) -14
E) $-\frac{196}{27}$

15. Determine whether the sequence is geometric. If so, find the common ratio.

$-2, 6, -18, 54, ...$

A) -3
B) -2
C) $-\frac{1}{3}$
D) 3
E) not geometric

16. Find the sum using the formulas for the sums of powers of integers.

$$\sum_{n=1}^{12} n^3$$

A) 4356
B) 12,168
C) 1728
D) 650
E) 6084

17. Use mathematical induction to prove the property for all positive integers n.

$$\left[a^n\right]^4 = a^{4n}$$

18. Find the specified nth term in the expansion of the binomial. (Write the expansion in descending powers of x.)

$(2x-3y)^8, n=5$

A) $448x^3y^5$

B) $90,720x^4y^4$

C) $56x^3y^5$

D) $6561y^8$

E) $6720x^3y^5$

19. Evaluate using a graphing utility: ${}_{20}P_5$

A) 15,504
B) 116,280
C) 1,860,480
D) 100
E) undefined

20. Use mathematical induction to prove the following for every positive integer n.

$$\sum_{i=1}^{n} \frac{17}{i(i+1)(i+2)} = \frac{17n(n+3)}{4(n+1)(n+2)}$$

Answer Key

1. i. Show true for $n=1$

$$36\cdot 1^5 = 3\cdot 1^2\cdot(1+1)^2(2\cdot 1^2+2\cdot 1-1)$$
$$36 = 3\cdot 4\cdot 3$$
$$36 = 36$$
$$1 = 1$$

ii. Assume true for $n=k$

$$\sum_{i=1}^{k} 36i^5 = 3k^2(k+1)^2(2k^2+2k-1)$$

iii. Prove true for $n=k+1$

$$\begin{aligned}
\sum_{i=1}^{k+1} 36i^5 &= \sum_{i=1}^{k} 36i^5 + 36(k+1)^5 \\
&= 3k^2(k+1)^2(2k^2+2k-1)+36(k+1)^5 \\
&= 3(k+1)^2\left[k^2(2k^2+2k-1)+12(k+1)^3\right] \\
&= 3(k+1)^2\left[2k^4+14k^3+35k^2+36k+12\right] \\
&= 3(k+1)^2\left[2k^4+8k^3+8k^2+6k^3+24k^2+24k+3k^2+12k+12\right] \\
&= 3(k+1)^2\left[2k^2\left(k^2+4k+4\right)+6k\left(k^2+4k+4\right)+3\left(k^2+4k+4\right)\right] \\
&= 3(k+1)^2\left[2k^2(k+2)^2+6k(k+2)^2+3(k+2)^2\right] \\
&= 3(k+1)^2(k+2)^2\left[2k^2+6k+3\right] \\
&= 3(k+1)^2\left[(k+1)+1\right]^2\left[2k^2+4k+2+2k+2-1\right] \\
&= 3(k+1)^2\left[(k+1)+1\right]^2\left[2\left(k^2+2k+1\right)+2\left(k+1\right)-1\right] \\
&= 3(k+1)^2\left[(k+1)+1\right]^2\left[2(k+1)^2+2\left(k+1\right)-1\right]
\end{aligned}$$

2. A

3. i. Show true for $n = 1$

$$\frac{8}{(2\cdot1-1)(2\cdot1+1)} = \frac{8\cdot1}{2\cdot1+1}$$
$$\frac{8}{3} = \frac{8}{3}$$
$$1 = 1$$

ii. Assume true for $n = k$

$$\sum_{i=1}^{k}\frac{8}{(2i-1)(2i+1)} = \frac{8k}{2k+1}$$

iii. Prove true for $n = k+1$

$$\begin{aligned}
\sum_{i=1}^{k+1}\frac{8}{(2i-1)(2i+1)} &= \sum_{i=1}^{k}\frac{8}{(2i-1)(2i+1)} + \frac{8}{[2(k+1)-1][(2(k+1)+1)]} \\
&= \frac{8k}{2k+1} + \frac{8}{(2k+1)(2k+3)} \\
&= \frac{8k(2k+3)+8}{(2k+1)(2k+3)} \\
&= \frac{16k^2+24k+8}{(2k+1)(2k+3)} \\
&= \frac{16k^2+16k+8k+8}{(2k+1)(2k+3)} \\
&= \frac{16k(k+1)+8(k+1)}{(2k+1)(2k+3)} \\
&= \frac{8(k+1)(2k+1)}{(2k+1)(2k+3)} \\
&= \frac{8(k+1)}{2(k+1)+1}
\end{aligned}$$

4. C
5. C
6. D
7. C
8. E
9. E
10. D
11. C
12. E

13. i. Show true for $n=1$

$$\left(2^{8\cdot1+2}+16\right)=1{,}040$$
$$=13\cdot80$$

ii. Assume true for $n=k$

80 is a factor of $\left(2^{8k+2}+16\right)$

iii. Prove true for $n=k+1$

$$\left(2^{8(k+1)+2}+16\right)=\left(2^{(8k+2)+16}+16\right)$$
$$=65{,}536\cdot2^{(8k+2)}+16$$
$$=65{,}536\cdot2^{(8k+2)}+16\cdot65{,}536-1{,}048{,}560$$
$$=65{,}536\cdot\left(2^{(8k+2)}+16\right)-13{,}107\cdot80$$

14. E
15. A
16. E
17. 1) $n=1$:

$$\left[a^1\right]^4\overset{?}{=}a^{4\cdot1}$$
$$\left[a\right]^4\overset{?}{=}a^4$$
$$a^4=a^4$$

The statement is true for $n=1$.

2) Assume $\left[a^k\right]^4=a^{4k}$. Then,

$$(a)^4\left[a^k\right]^4=(a)^4\left(a^{4k}\right)$$
$$\left[a\cdot a^k\right]^4=a^{4+4k}$$
$$\left[a^{k+1}\right]^4=a^{4(k+1)}$$

By mathematical induction, the property is true for all positive values of n.

18. B
19. C

20. i. Show true for $n=1$

$$\frac{17}{1(1+1)(1+2)}=\frac{17\cdot1\cdot(1+3)}{4(1+1)(1+2)}$$

$$\frac{17}{2\cdot3}=\frac{17\cdot4}{4\cdot2\cdot3}$$

$$\frac{17}{6}=\frac{17}{6}$$

$$1=1$$

ii. Assume true for $n=k$

$$\sum_{i=1}^{k}\frac{17}{i(i+1)(i+2)}=\frac{17k(k+3)}{4(k+1)(k+2)}$$

iii. Prove true for $n = k+1$

$$\sum_{i=1}^{k+1}\frac{17}{i(i+1)(i+2)} = \sum_{i=1}^{k}\frac{17}{i(i+1)(i+2)} + \frac{17}{(k+1)\left[(k+1)+1\right]\left[(k+1)+2\right]}$$

$$= \frac{17k(k+3)}{4(k+1)(k+2)} + \frac{17}{(k+1)(k+2)(k+3)}$$

$$= \frac{17k(k+3)^2+68}{4(k+1)(k+2)(k+3)}$$

$$= \frac{17(k^3+6k^2+9k+4)}{4(k+1)(k+2)(k+3)}$$

$$= \frac{17(k^3+k^2+5k^2+5k+4k+4)}{4(k+1)(k+2)(k+3)}$$

$$= \frac{17\left[k^2(k+1)+5k(k+1)+4(k+1)\right]}{4(k+1)(k+2)(k+3)}$$

$$= \frac{17(k+1)(k^2+5k+4)}{4(k+1)(k+2)(k+3)}$$

$$= \frac{17(k^2+k+4k+4)}{4(k+2)(k+3)}$$

$$= \frac{17\left[k(k+1)+4(k+1)\right]}{4(k+2)(k+3)}$$

$$= \frac{17(k+1)(k+4)}{4(k+2)(k+3)}$$

$$= \frac{17(k+1)\left[(k+1)+3\right]}{4\left[(k+1)+1\right]\left[(k+1)+2\right]}$$

Name: ______________________ Date: ___________

1. Determine whether the sequence is geometric. If so, find the common ratio.
   $-1, -2, -4, -8, \ldots$
   A) 2
   B) -1
   C) $\frac{1}{2}$
   D) -2
   E) not geometric

2. Expand the binomial by using Pascal's triangle to determine the coefficients.
   $(3x+2y)^6$
   A) $729x^6 + 2916x^5y + 4860x^4y^2 + 4320x^3y^3 + 4860x^2y^4 + 576xy^5 + 64y^6$
   B) $729x^6 + 2916x^5y + 3240x^4y^2 + 4320x^3y^3 + 2160x^2y^4 + 576xy^5 + 64y^6$
   C) $729x^6 + 2916x^5y + 4860x^4y^2 + 4320x^3y^3 + 2880x^2y^4 + 576xy^5 + 64y^6$
   D) $729x^6 + 2916x^5y + 4860x^4y^2 + 4320x^3y^3 + 2160x^2y^4 + 576xy^5 + 64y^6$
   E) $729x^6 + 2916x^5y + 2160x^4y^2 + 4320x^3y^3 + 2160x^2y^4 + 576xy^5 + 64y^6$

3. Write an expression for the apparent *n*th term of the sequence. (Assume that *n* begins with 1.)
   $-5, -2, 1, 4, 7$
   A) $a_n = -8n + 3$
   B) $a_n = (-1)^n(3n-8)$
   C) $a_n = 3n$
   D) $a_n = 3^n - 8$
   E) $a_n = 3n - 8$

4. Find a formula for a_n for the arithmetic sequence.

$$a_4 = -13, a_{13} = -31$$

A) $a_n = -7 - 2n$
B) $a_n = 5 - 7n$
C) $a_n = -2 - 7n$
D) $a_n = -7(-2)^n$
E) $a_n = -5 - 2n$

5. Find the sum.

$$\sum_{k=1}^{3} \frac{1}{k^2 + 5}$$

A) $\frac{22}{63}$
B) $\frac{1}{14}$
C) 1
D) $\frac{73}{168}$
E) $\frac{12}{11}$

6. Use the Binomial Theorem to expand the following complex number. Write your answer in standard form.

$$\left(-\frac{3}{7}-\frac{\sqrt{7}}{7}i\right)^3$$

A) $\frac{36}{343}-\frac{20\sqrt{7}}{343}i$

B) $\frac{36}{343}+\frac{20\sqrt{7}}{49}i$

C) $-\frac{36}{49}-\frac{20\sqrt{7}}{343}i$

D) $-\frac{27}{343}+\frac{\sqrt{7}}{49}i$

E) $-\frac{27}{343}-\frac{\sqrt{7}}{49}i$

7. Find the sum of the finite geometric sequence.

$$\sum_{n=1}^{6} 2\left(-\frac{2}{3}\right)^{n-1}$$

A) $-\frac{463}{243}$

B) $-\frac{1024}{3}$

C) 4256

D) $\frac{55}{81}$

E) $\frac{266}{243}$

8. Find the indicated *n*th term of the geometric sequence.

   5th term: $a_3 = -\frac{3}{16}, a_9 = -\frac{3}{65,536}$

   A) $\frac{3}{1024}$
   B) $-\frac{4}{81}$
   C) $-\frac{3}{4096}$
   D) $-\frac{3}{256}$
   E) $\frac{3}{64}$

9. Determine whether the sequence is arithmetic. If so, find the common difference. (Assume that *n* begins with 1.)

   $a_n = -6 + 4n$

   A) –4
   B) 6
   C) –6
   D) 4
   E) not arithmetic

10. Determine the number of ways a computer can randomly generate a prime integer between 10 and 20.

   A) 4
   B) 3
   C) 2
   D) 5
   E) 11

11. Find the sum of the finite geometric sequence. Round to the nearest thousandth.

$$\sum_{i=0}^{6} 200(1.07)^i$$

A) 1730.804
B) 1530.804
C) 1153.308
D) 1430.658
E) 1851.961

12. Use mathematical induction to prove the following inequality for all $n \geq 2$.

$$\frac{1}{\sqrt{23}}+\frac{1}{\sqrt{46}}+\frac{1}{\sqrt{69}}+\ldots+\frac{1}{\sqrt{23n}}>\frac{\sqrt{n}}{\sqrt{23}}$$

13. Expand the following binomial by using Pascal's Triangle.

$(2x-4)^5$

A) $32x^5-1024$
B) $-1024x^5+1280x^3+32$
C) $32x^5-320x^4+1280x^3-2560x^2+2560x-1024$
D) $32x^5+2560x^4-2560x^3+1280x^2-320x-1024$
E) $32x^5-64x^4+128x^3-256x^2+512x-1024$

14. Given the sequence $4+\frac{12}{13}, 4+\frac{19}{20}, 4+\frac{26}{27}, 4+\frac{33}{34}, 4+\frac{40}{41}, \ldots$, write an expression for the *apparent* nth term assuming n begins with 1.

A) $\frac{36n+29}{7n+6}$
B) $\frac{35n+30}{7n+6}$
C) $\frac{35n+29}{7n+6}$
D) $\frac{35n+29}{7n+7}$
E) $\frac{36n+30}{7n+7}$

15. Use mathematical induction to prove the following for every positive integer *n*.

$$\sum_{i=1}^{n} 96i^5 = 8n^2(n+1)^2(2n^2+2n-1)$$

16. Use mathematical induction to prove the following for every positive integer *n*.

$$\sum_{i=1}^{n} \frac{15}{(2i-1)(2i+1)} = \frac{15n}{2n+1}$$

17. Find the probability for the experiment of drawing two marbles (without replacement) from a bag containing four green, six yellow, and five red marbles such that both marbles are yellow.

A) $\frac{1}{7}$
B) $\frac{4}{25}$
C) $\frac{6}{35}$
D) $\frac{1}{3}$
E) $\frac{2}{3}$

18. Determine whether the sequence is arithmetic. If so, find the common difference.
7, 8, 9, 10, 11

A) 6
B) 1
C) 7
D) –1
E) not arithmetic

19. Find the number of distinguishable permutations of the group of letters.
    G, A, U, S, S
    A) 60
    B) 120
    C) 30
    D) 5
    E) 15

20. Write the first five terms of the arithmetic sequence.
    $a_1 = 5, d = 7$
    A) 5, 12, 19, 26, 33
    B) 5, –2, –9, –16, –23
    C) 12, 19, 26, 33, 40
    D) 5, 35, 245, 1715, 12005
    E) 5, 12, 17, 22, 27

Answer Key

1. A
2. D
3. E
4. E
5. A
6. A
7. E
8. D
9. D
10. A
11. A
12. i. Show true for $n=2$

$$\frac{1}{\sqrt{23}}+\frac{1}{\sqrt{46}}>\frac{\sqrt{2}}{\sqrt{23}}$$

$$\sqrt{23}\left(\frac{1}{\sqrt{23}}+\frac{1}{\sqrt{46}}\right)>\sqrt{23}\cdot\frac{\sqrt{2}}{\sqrt{23}}$$

$$1+\frac{1}{\sqrt{2}}>\sqrt{2}$$

$$\sqrt{2}+1>2$$

$$\sqrt{2}>1$$

ii. Assume true for $n=k$

$$\frac{1}{\sqrt{23}}+\frac{1}{\sqrt{46}}+\frac{1}{\sqrt{69}}+...+\frac{1}{\sqrt{23k}}>\frac{\sqrt{k}}{\sqrt{23}}$$

iii. Prove true for $n=k+1$

$$\frac{1}{\sqrt{23}}+\frac{1}{\sqrt{46}}+\frac{1}{\sqrt{69}}+...+\frac{1}{\sqrt{23k}}+\frac{1}{\sqrt{23(k+1)}}>\frac{\sqrt{k}}{\sqrt{23}}+\frac{1}{\sqrt{23(k+1)}}$$

$$>\frac{\sqrt{k(k+1)}+1}{\sqrt{23(k+1)}}$$

Want to show

$$\frac{\sqrt{k(k+1)}+1}{\sqrt{23(k+1)}} > \frac{\sqrt{(k+1)}}{\sqrt{23}}$$

$$\sqrt{23}\cdot\frac{\sqrt{k(k+1)}+1}{\sqrt{23(k+1)}} > \sqrt{23}\cdot\frac{\sqrt{(k+1)}}{\sqrt{23}}$$

$$\frac{\sqrt{k(k+1)}+1}{\sqrt{(k+1)}} > \sqrt{(k+1)}$$

$$\sqrt{k(k+1)}+1 > k+1$$

$$\sqrt{k(k+1)} > k$$

$$k^2+k > k^2$$

$$k > 0$$

Since these manipulations are reversible for positive integers larger than 1, we may conclude that the inequality holds for all integers larger than 1.

13. C
14. C
15. i. Show true for $n=1$

$$96\cdot 1^5 = 8\cdot 1^2\cdot(1+1)^2(2\cdot 1^2+2\cdot 1-1)$$

$$96 = 8\cdot 4\cdot 3$$

$$96 = 96$$

$$1 = 1$$

ii. Assume true for $n=k$

$$\sum_{i=1}^{k} 96i^5 = 8k^2(k+1)^2(2k^2+2k-1)$$

iii. Prove true for $n = k+1$

$$\begin{aligned}
\sum_{i=1}^{k+1} 96i^5 &= \sum_{i=1}^{k} 96i^5 + 96(k+1)^5 \\
&= 8k^2(k+1)^2(2k^2+2k-1)+96(k+1)^5 \\
&= 8(k+1)^2\left[k^2(2k^2+2k-1)+12(k+1)^3\right] \\
&= 8(k+1)^2\left[2k^4+14k^3+35k^2+36k+12\right] \\
&= 8(k+1)^2\left[2k^4+8k^3+8k^2+6k^3+24k^2+24k+3k^2+12k+12\right] \\
&= 8(k+1)^2\left[2k^2\left(k^2+4k+4\right)+6k\left(k^2+4k+4\right)+3\left(k^2+4k+4\right)\right] \\
&= 8(k+1)^2\left[2k^2(k+2)^2+6k(k+2)^2+3(k+2)^2\right] \\
&= 8(k+1)^2(k+2)^2\left[2k^2+6k+3\right] \\
&= 8(k+1)^2\left[(k+1)+1\right]^2\left[2k^2+4k+2+2k+2-1\right] \\
&= 8(k+1)^2\left[(k+1)+1\right]^2\left[2\left(k^2+2k+1\right)+2\left(k+1\right)-1\right] \\
&= 8(k+1)^2\left[(k+1)+1\right]^2\left[2(k+1)^2+2\left(k+1\right)-1\right]
\end{aligned}$$

16. i. Show true for $n = 1$

$$\frac{15}{(2\cdot1-1)(2\cdot1+1)} = \frac{15\cdot1}{2\cdot1+1}$$

$$\frac{15}{3} = \frac{15}{3}$$

$$1 = 1$$

ii. Assume true for $n = k$

$$\sum_{i=1}^{k} \frac{15}{(2i-1)(2i+1)} = \frac{15k}{2k+1}$$

iii. Prove true for $n = k+1$

$$\sum_{i=1}^{k+1} \frac{15}{(2i-1)(2i+1)} = \sum_{i=1}^{k} \frac{15}{(2i-1)(2i+1)} + \frac{15}{[2(k+1)-1][(2(k+1)+1)]}$$

$$= \frac{15k}{2k+1} + \frac{15}{(2k+1)(2k+3)}$$

$$= \frac{15k(2k+3)+15}{(2k+1)(2k+3)}$$

$$= \frac{30k^2+45k+15}{(2k+1)(2k+3)}$$

$$= \frac{30k^2+30k+15k+15}{(2k+1)(2k+3)}$$

$$= \frac{30k(k+1)+15(k+1)}{(2k+1)(2k+3)}$$

$$= \frac{15(k+1)(2k+1)}{(2k+1)(2k+3)}$$

$$= \frac{15(k+1)}{2(k+1)+1}$$

17. A
18. B
19. A
20. A

Name: ______________________________ Date: _____________

1. Prove the inequality for the indicated integer values of n.

$$\left(\frac{9}{8}\right)^n > n, n \geq 29$$

2. Use mathematical induction to prove the following inequality for all $n \geq 2$.

$$(3)^{n+2} > \frac{9}{32}n(2)^{n+5}$$

3. Find the coefficient a of the term in the expansion of the binomial.

Binomial	*Term*
$(4x+3y)^6$	ax^2y^4

A) $a = 10$
B) $a = 19,440$
C) $a = 12$
D) $a = 729$
E) $a = 360$

4. Find the indicated nth term of the geometric sequence.

6th term: $a_5 = \frac{4}{81}, a_8 = \frac{4}{2187}$

A) $\frac{4}{729}$
B) $\frac{3}{1024}$
C) $\frac{4}{2187}$
D) $\frac{4}{243}$
E) $\frac{4}{81}$

5. Find the number of distinguishable permutations of the group of letters.
   G, A, U, S, S

   A) 60
   B) 120
   C) 30
   D) 5
   E) 15

6. Find the indicated partial sum of the series.

   $$\sum_{i=1}^{\infty} 5\left(-\frac{1}{3}\right)^{i}$$

   fourth partial sum

   A) $\frac{5}{81}$
   B) $\frac{305}{81}$
   C) $-\frac{100}{81}$
   D) $\frac{35}{81}$
   E) $-\frac{205}{162}$

7. Find the sum of the infinite series.

$$\sum_{i=1}^{\infty} 2\left(\frac{1}{4}\right)^i$$

A) undefined
B) $\frac{2}{5}$
C) 2
D) $\frac{4}{3}$
E) $\frac{2}{3}$

8. Simplify the factorial expression.

$$\frac{11!}{9!}$$

A) 990
B) 110
C) $\frac{11}{9}$
D) 11
E) 1320

9. Find a quadratic model for the sequence with the indicated terms.

$a_0 = 7, a_2 = 3, a_5 = 12$

A) $a_n = n^2 + 7$
B) $a_n = n^2 - 4n + 7$
C) $a_n = n^2 - 7n + 7$
D) $a_n = n^2 - 7n + 4$
E) $a_n = n^2 - n + 7$

10. Evaluate using a graphing utility: ${}_{15}P_6$
    A) 5005
    B) 360,360
    C) 3,603,600
    D) 90
    E) undefined

11. Use mathematical induction to prove the following equality.
    $\ln\left(3^n x_1 x_2 \ldots x_n\right) = \ln\left(3x_1\right) + \ln\left(3x_2\right) + \ldots + \ln\left(3x_n\right)$, where
    $x_1 > 0, x_2 > 0, \ldots, x_n > 0$

12. Find the sum of the infinite geometric series.
    $$\sum_{n=0}^{\infty} 4\left(-\frac{1}{5}\right)^n$$
    A) $-\frac{10}{3}$
    B) $\frac{10}{3}$
    C) $\frac{2}{3}$
    D) $-\frac{2}{3}$
    E) undefined

13. How many 3-digit numbers can be formed if the leading digit cannot be zero and repeats are not allowed?
    A) 997
    B) 504
    C) 810
    D) 648
    E) 900

14. Find a formula for the *nth* term of the following geometric sequence, then find the $4th$ term of the sequence.

    9, 36, 144,...

    A) $a_n = 9(4)^n$; $a_4 = 576$
    B) $a_n = 9(4)^n$; $a_4 = 2304$
    C) $a_n = 9(4)^{n-1}$; $a_4 = 2304$
    D) $a_n = 9(4)^{n-1}$; $a_4 = 576$
    E) $a_n = 9(4)^{n-1}$; $a_4 = 144$

15. Expand the following expression in the difference quotient and simplify.

    $\dfrac{f(x+h)-f(x)}{h}$, $h \neq 0$. $f(x) = \dfrac{5}{2x}$

    A) $\dfrac{5}{2(x^2+h)}$
    B) $-\dfrac{5}{2(x+h)^2}$
    C) $-\dfrac{5}{2x(x+h)}$
    D) $\dfrac{5}{2(x+h)}$
    E) $-\dfrac{h}{x(x+h)}$

16. Write the first five terms of the geometric sequence.

    $a_1 = 4, r = -\dfrac{1}{5}$

    A) $4, -1, -6, -11, -16$
    B) $4, -20, 100, -500, 2500$
    C) $-\dfrac{4}{5}, \dfrac{4}{25}, -\dfrac{4}{125}, \dfrac{4}{625}, -\dfrac{4}{3125}$
    D) $1, 4, -\dfrac{4}{5}, \dfrac{4}{25}, -\dfrac{4}{125}$
    E) $4, -\dfrac{4}{5}, \dfrac{4}{25}, -\dfrac{4}{125}, \dfrac{4}{625}$

17. Use the Binomial Theorem to expand the complex number. Simplify your result.

$$(-2+5i)^4$$

A) $41 + 840i$
B) $41 - 840i$
C) $-41 + 840i$
D) $-41 - 840i$
E) 16

18. Write the first five terms of the sequence defined recursively. Use the pattern to write the *n*th term of the sequence as a function of *n*. (Assume that *n* begins with 1.)

$$a_1 = 21, a_{k+1} = a_k - 5$$

A) $a_n = 26 - 5n$
B) $a_n = 21 - 5n$
C) $a_n = 31 - 5n$
D) $a_n = 26$
E) $a_n = 21(n-1)$

19. Find the sum of the following infinite geometric series.

$$-10 + 9 - \frac{81}{10} + \frac{729}{100} - \ ...$$

A) $-\frac{19}{100}$
B) $-\frac{1}{100}$
C) $\frac{19}{100}$
D) -10
E) $-\frac{100}{19}$

20. Find the sum of the integers from –34 to –12.
A) 66
B) 22
C) –1058
D) –529
E) 132

Answer Key

1.

1) $\left(\frac{9}{8}\right)^{29} \approx 30.4 > 29$. The statement is true for $n = 29$.

2) Assuming that $\left(\frac{9}{8}\right)^{k} > k$ for $k > 29$, show that $\left(\frac{9}{8}\right)^{k+1} > k+1$.

$\left(\frac{9}{8}\right)^{k} > k$, by assumption

$\left(\frac{9}{8}\right)\left(\frac{9}{8}\right)^{k} > \frac{9}{8}k$

$\left(\frac{9}{8}\right)^{k+1} > k + \frac{1}{8}k$

Since $k > 29 > \frac{8}{1}$, then $k > \frac{8}{1}$ or $\frac{1}{8}k > 1$ or $k + \frac{1}{8}k > k+1$. Therefore,

$\left(\frac{9}{8}\right)^{k+1} > k + \frac{1}{8}k > k+1$ or $\left(\frac{9}{8}\right)^{k+1} > k+1$.

By mathematical induction, the relation is true for all $n \geq 29$.

2. i. Show true for $n = 2$

$(3)^{2+2} > \frac{9}{32}(2)(2)^{2+5}$

$(3)^{4} > \frac{9}{32}\cdot 2\cdot (2)^{7}$

$32\cdot 81 > 9\cdot 256$

$2592 > 2304$

ii. Assume true for $n = k$

$(3)^{k+2} > \frac{9}{32}k(2)^{k+5}$

iii. Prove true for $n = k+1$

$$(3)^{(k+1)+2} > \frac{9}{32}(k+1)(2)^{(k+1)+5}$$

$$3(3)^{k+2} > \frac{9}{32}(k+1)2(2)^{k+5}$$

$$2(3)^{k+2} + (3)^{k+2} > \frac{9}{32}2k(2)^{k+5} + \frac{9}{32}2(2)^{k+5}$$

$$2(3)^{k+2} - \frac{9}{32}2k(2)^{k+5} + (3)^{k+2} > \frac{9}{32}2(2)^{k+5}$$

$$2\left[(3)^{k+2} - \frac{9}{32}k(2)^{k+5}\right] + (3)^{k+2} > \frac{9}{32}2(2)^{k+5}$$

$$(3)^{k+2} > \frac{9}{32}2(2)^{k+5}$$

$$9(3)^{k} > \frac{9}{32}2\cdot 32\cdot(2)^{k}$$

$$(3)^{k} > 2(2)^{k}$$

$$\left(\frac{3}{2}\right)^{k} > 2$$

Since these manipulations are reversible, we may conclude that the inequality is true for $n = k+1$.

3. B
4. D
5. A
6. C
7. E
8. B
9. B
10. C
11. i. Show true for $n = 1$

$$\ln\left(3^{1}x_1\right) = \ln\left(3x_1\right)$$

$$\ln\left(3x_1\right) = \ln\left(3x_1\right)$$

ii. Assume true for $n = k$

$$\ln\left(3^{k}x_1x_2\ldots x_k\right) = \ln\left(3x_1\right) + \ln\left(3x_2\right) + \ldots + \ln\left(3x_k\right)$$

iii. Prove true for $n = k+1$

$$\begin{aligned}\ln\left(3^{k+1}x_1x_2...x_kx_{k+1}\right) &= \ln\left(3\cdot 3^k x_1x_2...x_kx_{k+1}\right)\\ &= \ln\left(\left[3^k x_1x_2...x_k\right]3x_{k+1}\right)\\ &= \ln\left(3^k x_1x_2...x_k\right)+\ln\left(3x_{k+1}\right)\\ &= \ln\left(3x_1\right)+\ln\left(3x_2\right)+...+\ln\left(3x_k\right)+\ln\left(3x_{k+1}\right)\end{aligned}$$

12. B
13. D
14. D
15. C
16. E
17. A
18. A
19. E
20. D

Name: ______________________ Date: ___________

1. Find P_{k+1} for the given P_k.

   $P_k = 7+13+19+...+[6(k-1)+1]+[6k+1]$

   A) $P_{k+1} = 7+13+19+...+[6k+7]+[6(k+1)+1]$
   B) $P_{k+1} = 7+13+19+...+[6k+1]+[6(k+1)+7]$
   C) $P_{k+1} = 7+13+19+...+[6k+7]+[6(k+1)+7]$
   D) $P_{k+1} = 7+13+19+...+[6k+1]+[6(k+1)+1]$
   E) $P_{k+1} = 7+13+19+...+[6(k-1)+1]+[6k+1]$

2. Find the sum using the formulas for the sums of powers of integers.

   $$\sum_{n=1}^{9} n^3$$

   A) 1296
   B) 4050
   C) 729
   D) 285
   E) 2025

3. Use mathematical induction to prove the following for every positive integer *n*.

   $$1+6+36+216+\ldots+6^{n-1} = \frac{6^n-1}{5}$$

4. Write an expression for the apparent *n*th term of the sequence. (Assume that *n* begins with 1.)

   –4, –2, 0, 2, 4

   A) $a_n = -6n+2$
   B) $a_n = (-1)^n(2n-6)$
   C) $a_n = 2n$
   D) $a_n = 2^n - 6$
   E) $a_n = 2n-6$

5. Use mathematical induction to prove the following inequality for all $n \geq 2$.

$$(3)^{n+3} > \frac{27}{2}n(2)^{n+1}$$

6. Find the probability for the experiment of tossing a coin four times and getting at least two heads. Use the sample space

$$S = \begin{matrix} HHHH & HHHT & HHTH & HTHH \\ THHH & HHTT & HTHT & THHT \\ THTH & HTTH & TTHH & TTTH \\ TTHT & THTT & HTTT & TTTT \end{matrix}$$

A) $\frac{11}{16}$

B) $\frac{1}{2}$

C) $\frac{13}{16}$

D) $\frac{7}{8}$

E) $\frac{5}{8}$

7. Find the probability for the experiment of selecting one card from a standard deck of 52 playing cards such that the card is *not* a red face card.

A) $\frac{3}{26}$

B) $\frac{10}{13}$

C) $\frac{11}{13}$

D) $\frac{23}{26}$

E) $\frac{49}{52}$

8. Find the specified *n*th term in the expansion of the binomial. (Write the expansion in descending powers of *x*.)

$(2x+3y)^6, n=3$

A) $160x^3y^3$
B) $2160x^4y^2$
C) $20x^3y^3$
D) $729y^6$
E) $120x^3y^3$

9. Calculate the binomial coefficient: $\binom{11}{6}$

A) 332,640
B) 66
C) 462
D) 1
E) 0

10. Find the indicated partial sum of the series.

$$\sum_{i=1}^{\infty} 5\left(\frac{1}{3}\right)^{i}$$

fourth partial sum

A) $\frac{5}{81}$
B) $\frac{605}{81}$
C) $\frac{200}{81}$
D) $\frac{65}{81}$
E) $\frac{205}{81}$

11. Use mathematical induction to prove the property for all positive integers n.

$$\left[a^n\right]^3 = a^{3n}$$

12. Use mathematical induction to prove the following inequality for all $n \geq 2$.

$$\frac{1}{\sqrt{19}} + \frac{1}{\sqrt{38}} + \frac{1}{\sqrt{57}} + \ldots + \frac{1}{\sqrt{19n}} > \frac{\sqrt{n}}{\sqrt{19}}$$

13. Determine whether the sequence is geometric. If so, find the common ratio.

$-2, -6, -18, -54, \ldots$

A) 3
B) -2
C) $\frac{1}{3}$
D) -3
E) not geometric

14. Determine whether the sequence is arithmetic. If so, find the common difference. (Assume that n begins with 1.)

$$a_n = -\left(\frac{1}{6}\right)^n$$

A) -1
B) 6
C) $\frac{1}{6}$
D) 1
E) not arithmetic

15. You are given the probability that an event *will* happen. Find the probability that the event *will not* happen.

$P(E) = 0.33$

A) 0.33
B) 0.67
C) 0.335
D) 0
E) 1

16. Find the indicated term of the sequence.

$$a_n = (-1)^n (5n-1)$$

$$a_{29} = \boxed{}$$

A) 139
B) -144
C) -146
D) 4
E) -140

17. Write the first five terms of the sequence. (Assume that n begins with 1.)

$$a_n = (-1)^n (n-1)(n-2)$$

A) 4, 0, 0, –2, 6
B) 0, 0, –2, 6, –12
C) 0, 0, –2, –6, –12
D) 0, 2, –6, 12, –20
E) 0, –1, 2, –3, 4

18. Determine whether the sequence is geometric. If so, find the common ratio.
    $-1, 3, 7, 11, ...$
    A) 4
    B) –1
    C) $\frac{1}{4}$
    D) –4
    E) not geometric

19. Determine the sample space for the experiment.
    Two marbles are selected from marbles labeled A through D where the marbles are not replaced and the order of selection does not matter.
    A) S = {AB, AC, AD, BC, BD}
    B) S = {AB, AC, AD, BC, BD, CD}
    C) S = {AB, AC, AD, BC, CD}
    D) S = {AB, AC, AD, BC, BD, CD, DA}
    E) S = {AB, AC, AD, BC, BD, CA, CB, CD}

20. Find the indicated nth partial sum of the arithmetic sequence.
    $3.6, 5.7, 7.8, 9.9, ..., n = 60$
    A) 6498
    B) 4059
    C) 3933
    D) 3931.5
    E) 3934.5

Answer Key

1. D
2. E
3. i. Show true for $n=1$

$$6^{1-1}=\frac{6^1-1}{5}$$
$$6^0=\frac{6-1}{5}$$
$$1=\frac{5}{5}$$
$$1=1$$

ii. Assume true for $n=k$

$$1+6+36+216+\ldots+6^{k-1}=\frac{6^k-1}{5}$$

iii. Prove true for $n=k+1$

$$\begin{aligned}1+6+36+216+\ldots+6^{k-1}+6^{(k+1)-1}&=\frac{6^k-1}{5}+6^{(k+1)-1}\\&=\frac{6^k-1}{5}+6^k\\&=\frac{6^k-1+5\cdot 6^k}{5}\\&=\frac{6\cdot 6^k-1}{5}\\&=\frac{6^{k+1}-1}{5}\end{aligned}$$

4. E
5. i. Show true for $n=2$

$$(3)^{2+3}>\frac{27}{2}(2)(2)^{2+1}$$
$$(3)^5>\frac{27}{2}\cdot 2\cdot(2)^3$$
$$2\cdot 243>27\cdot 16$$
$$486>432$$

ii. Assume true for $n=k$

$$(3)^{k+3}>\frac{27}{2}k(2)^{k+1}$$

iii. Prove true for $n = k + 1$

$$(3)^{(k+1)+3} > \frac{27}{2}(k+1)(2)^{(k+1)+1}$$

$$3(3)^{k+3} > \frac{27}{2}(k+1)2(2)^{k+1}$$

$$2(3)^{k+3} + (3)^{k+3} > \frac{27}{2}2k(2)^{k+1} + \frac{27}{2}2(2)^{k+1}$$

$$2(3)^{k+3} - \frac{27}{2}2k(2)^{k+1} + (3)^{k+3} > \frac{27}{2}2(2)^{k+1}$$

$$2\left[(3)^{k+3} - \frac{27}{2}k(2)^{k+1}\right] + (3)^{k+3} > \frac{27}{2}2(2)^{k+1}$$

$$(3)^{k+3} > \frac{27}{2}2(2)^{k+1}$$

$$27(3)^{k} > \frac{27}{2}2 \cdot 2 \cdot (2)^{k}$$

$$(3)^{k} > 2(2)^{k}$$

$$\left(\frac{3}{2}\right)^{k} > 2$$

Since these manipulations are reversible, we may conclude that the inequality is true for $n = k + 1$.

6. A
7. D
8. B
9. C
10. C
11. 1) $n = 1$:

$$\left[a^{1}\right]^{3} \stackrel{?}{=} a^{3 \cdot 1}$$

$$[a]^{3} \stackrel{?}{=} a^{3}$$

$$a^{3} = a^{3}$$

The statement is true for $n = 1$.

2) Assume $\left[a^{k}\right]^{3} = a^{3k}$. Then,

$$(a)^{3}\left[a^{k}\right]^{3} = (a)^{3}\left(a^{3k}\right)$$

$$\left[a \cdot a^{k}\right]^{3} = a^{3+3k}$$

$$\left[a^{k+1}\right]^{3} = a^{3(k+1)}$$

By mathematical induction, the property is true for all positive values of n.

12. i. Show true for $n = 2$

$$\frac{1}{\sqrt{19}}+\frac{1}{\sqrt{38}}>\frac{\sqrt{2}}{\sqrt{19}}$$

$$\sqrt{19}\left(\frac{1}{\sqrt{19}}+\frac{1}{\sqrt{38}}\right)>\sqrt{19}\cdot\frac{\sqrt{2}}{\sqrt{19}}$$

$$1+\frac{1}{\sqrt{2}}>\sqrt{2}$$

$$\sqrt{2}+1>2$$

$$\sqrt{2}>1$$

ii. Assume true for $n = k$

$$\frac{1}{\sqrt{19}}+\frac{1}{\sqrt{38}}+\frac{1}{\sqrt{57}}+\ldots+\frac{1}{\sqrt{19k}}>\frac{\sqrt{k}}{\sqrt{19}}$$

iii. Prove true for $n = k+1$

$$\frac{1}{\sqrt{19}}+\frac{1}{\sqrt{38}}+\frac{1}{\sqrt{57}}+\ldots+\frac{1}{\sqrt{19k}}+\frac{1}{\sqrt{19(k+1)}}>\frac{\sqrt{k}}{\sqrt{19}}+\frac{1}{\sqrt{19(k+1)}}$$

$$>\frac{\sqrt{k(k+1)}+1}{\sqrt{19(k+1)}}$$

Want to show

$$\frac{\sqrt{k(k+1)}+1}{\sqrt{19(k+1)}}>\frac{\sqrt{(k+1)}}{\sqrt{19}}$$

$$\sqrt{19}\cdot\frac{\sqrt{k(k+1)}+1}{\sqrt{19(k+1)}}>\sqrt{19}\cdot\frac{\sqrt{(k+1)}}{\sqrt{19}}$$

$$\frac{\sqrt{k(k+1)}+1}{\sqrt{(k+1)}}>\sqrt{(k+1)}$$

$$\sqrt{k(k+1)}+1>k+1$$

$$\sqrt{k(k+1)}>k$$

$$k^2+k>k^2$$

$$k>0$$

Since these manipulations are reversible for positive integers larger than 1, we may conclude that the inequality holds for all integers larger than 1.

13. A
14. E
15. B
16. B
17. B
18. E
19. B
20. C

Name: ____________________________ Date: ______________

1. Find the indicated sum.

$$\sum_{n=1}^{8} n^4$$

A) 4096
B) 1,679,616
C) 36
D) 87,380
E) 8772

2. Use mathematical induction to prove the following equality.

$$\ln\left(2^n x_1 x_2 \ldots x_n\right) = \ln\left(2x_1\right) + \ln\left(2x_2\right) + \ldots + \ln\left(2x_n\right),\ \text{where}$$

$x_1 > 0, x_2 > 0, \ldots, x_n > 0$

3. Write the first five terms of the arithmetic sequence.

$a_5 = -29, a_{10} = -59$

A) –5, 1, 7, 13, 19
B) –11, –17, –23, –29, –35
C) –5, –11, –17, –23, –29
D) –5, 30, –180, 1080, –6480
E) –5, –11, –16, –21, –26

4. Find the sum of the infinite series.

$$\sum_{i=1}^{\infty} 3\left(\frac{1}{3}\right)^i$$

A) undefined
B) $\frac{3}{4}$
C) 3
D) $\frac{9}{4}$
E) $\frac{3}{2}$

5. Determine the sample space for the experiment.
   Two marbles are selected from marbles labeled A through E where the marbles are not replaced and the order of selection does not matter.
   A) $S = \{AB, AC, AD, AE, BC, BD, BE, CD\}$
   B) $S = \{AB, AC, AD, AE, BC, BD, BE, CD, CE, DE\}$
   C) $S = \{AB, AC, AD, AE, BC, BD, BE, DE\}$
   D) $S = \{AB, AC, AD, AE, BC, BD, BE, CD, CE, DE, DA, DB\}$
   E) $S = \{AB, AC, AD, AE, BC, BD, BE, CA, CB, CD, CE, DE\}$

6. Expand the binomial by using Pascal's triangle to determine the coefficients.
   $(5x-4y)^6$
   A) $15625x^6 - 75000x^5y + 150000x^4y^2 - 160000x^3y^3 + 150000x^2y^4 - 30720xy^5 + 4096y^6$
   B) $15625x^6 - 75000x^5y + 100000x^4y^2 - 160000x^3y^3 + 96000x^2y^4 - 30720xy^5 + 4096y^6$
   C) $15625x^6 - 75000x^5y + 150000x^4y^2 - 160000x^3y^3 + 128000x^2y^4 - 30720xy^5 + 4096y^6$
   D) $15625x^6 - 75000x^5y + 150000x^4y^2 - 160000x^3y^3 + 96000x^2y^4 - 30720xy^5 + 4096y^6$
   E) $15625x^6 - 75000x^5y + 96000x^4y^2 - 160000x^3y^3 + 96000x^2y^4 - 30720xy^5 + 4096y^6$

7. Find the sum of the finite geometric sequence.
   $$\sum_{n=1}^{5} -\left(-\frac{1}{6}\right)^{n-1}$$
   A) $\frac{6665}{1296}$
   B) $-\frac{1}{6}$
   C) 1111
   D) $\frac{185}{216}$
   E) $-\frac{1111}{1296}$

8. Use mathematical induction to prove the following inequality for all $n \geq 2$.
   $$\frac{1}{\sqrt{2}} + \frac{1}{\sqrt{4}} + \frac{1}{\sqrt{6}} + \ldots + \frac{1}{\sqrt{2n}} > \frac{\sqrt{n}}{\sqrt{2}}$$

9. Find a formula for a_n for the arithmetic sequence.

   $a_4 = 10, a_8 = 34$

   A) $a_n = -8 + 6n$
   B) $a_n = 14 - 8n$
   C) $a_n = 6 - 8n$
   D) $a_n = -8(6)^n$
   E) $a_n = -14 + 6n$

10. Determine whether the sequence is arithmetic. If so, find the common difference. (Assume that n begins with 1.)

    $a_n = 6 + 4n$

    A) –4
    B) –6
    C) 6
    D) 4
    E) not arithmetic

11. Find the number of distinguishable permutations of the group of letters.

    E, S, T, I, M, A, T, E

    A) 10,080
    B) 8
    C) 20,160
    D) 40,320
    E) 3360

12. Find a formula for the *nth* term of the following geometric sequence, then find the *4th* term of the sequence.

    7, 28, 112,...

    A) $a_n = 7(4)^n$; $a_4 = 448$

    B) $a_n = 7(4)^n$; $a_4 = 1792$

    C) $a_n = 7(4)^{n-1}$; $a_4 = 1792$

    D) $a_n = 7(4)^{n-1}$; $a_4 = 448$

    E) $a_n = 7(4)^{n-1}$; $a_4 = 112$

13. Determine whether the sequence is geometric. If so, find the common ratio.

    2, –2, –6, –10, ...

    A) –4
    B) 2
    C) $-\frac{1}{4}$
    D) 4
    E) not geometric

14. Determine whether the sequence is arithmetic. If so, find the common difference.

    –3, –7, –11, –15, –19

    A) 1
    B) –4
    C) –3
    D) 4
    E) not arithmetic

15. Use mathematical induction to prove the formula for every positive integer *n*. Show all your work.

$$7 + 10 + 13 + 16 + \ldots + (3n + 4) = \frac{n}{2}(3n + 11)$$

16. Find the coefficient a of the term in the expansion of the binomial.

Binomial	*Term*
$(2x-5y)^7$	ax^2y^5

A) $a = 12$
B) $a = -262,500$
C) $a = 14$
D) $a = -78,125$
E) $a = 2520$

17. Calculate the binomial coefficient: $\binom{8}{7}$

A) 40,320
B) 56
C) 8
D) 1
E) 0

18. Use mathematical induction to prove the following for every positive integer n.

$$\sum_{i=1}^{n} \frac{11}{i(i+1)(i+2)} = \frac{11n(n+3)}{4(n+1)(n+2)}$$

19. Use the Binomial Theorem to expand and simplify the expression.

$$\left(x^{3/4}-5\right)^4$$

A) x^3+625
B) $x^{12}-20x^9+150x^6-500x^3+625$
C) $x^3-20x^{9/4}+150x^{3/2}-500x^{3/4}+625$
D) $x^3+20x^{9/4}+150x^{3/2}+500x^{3/4}+625$
E) $x^{12}+20x^9+150x^6+500x^3+625$

20. Find P_{k+1} for the given P_k.

$$P_k = \frac{4}{k(k+1)}$$

A) $P_{k+1} = \dfrac{4}{k(k+1)} + 1$

B) $P_{k+1} = \dfrac{4}{k(k+1)} + \dfrac{4}{(k+1)(k+2)}$

C) $P_{k+1} = \dfrac{4}{(k+1)(k+2)}$

D) $P_{k+1} = \dfrac{4}{k(k+2)}$

E) $P_{k+1} = \dfrac{16}{(k+1)(k+2)}$

Answer Key

1. E
2. i. Show true for $n=1$

$$\ln\left(2^1 x_1\right)=\ln\left(2x_1\right)$$
$$\ln\left(2x_1\right)=\ln\left(2x_1\right)$$

ii. Assume true for $n=k$

$$\ln\left(2^k x_1x_2...x_k\right)=\ln\left(2x_1\right)+\ln\left(2x_2\right)+...+\ln\left(2x_k\right)$$

iii. Prove true for $n=k+1$

$$\begin{aligned}\ln\left(2^{k+1}x_1x_2...x_kx_{k+1}\right)&=\ln\left(2\cdot 2^k x_1x_2...x_kx_{k+1}\right)\\&=\ln\left(\left[2^k x_1x_2...x_k\right]2x_{k+1}\right)\\&=\ln\left(2^k x_1x_2...x_k\right)+\ln\left(2x_{k+1}\right)\\&=\ln\left(2x_1\right)+\ln\left(2x_2\right)+...+\ln\left(2x_k\right)+\ln\left(2x_{k+1}\right)\end{aligned}$$

3. C
4. E
5. B
6. D
7. E
8. i. Show true for $n=2$

$$\frac{1}{\sqrt{2}}+\frac{1}{\sqrt{4}}>\frac{\sqrt{2}}{\sqrt{2}}$$
$$\sqrt{2}\left(\frac{1}{\sqrt{2}}+\frac{1}{\sqrt{4}}\right)>\sqrt{2}\cdot\frac{\sqrt{2}}{\sqrt{2}}$$
$$1+\frac{1}{\sqrt{2}}>\sqrt{2}$$
$$\sqrt{2}+1>2$$
$$\sqrt{2}>1$$

ii. Assume true for $n=k$

$$\frac{1}{\sqrt{2}}+\frac{1}{\sqrt{4}}+\frac{1}{\sqrt{6}}+...+\frac{1}{\sqrt{2k}}>\frac{\sqrt{k}}{\sqrt{2}}$$

iii. Prove true for $n = k+1$

$$\frac{1}{\sqrt{2}}+\frac{1}{\sqrt{4}}+\frac{1}{\sqrt{6}}+\ldots+\frac{1}{\sqrt{2k}}+\frac{1}{\sqrt{2(k+1)}} > \frac{\sqrt{k}}{\sqrt{2}}+\frac{1}{\sqrt{2(k+1)}}$$

$$> \frac{\sqrt{k(k+1)}+1}{\sqrt{2(k+1)}}$$

Want to show

$$\frac{\sqrt{k(k+1)}+1}{\sqrt{2(k+1)}} > \frac{\sqrt{(k+1)}}{\sqrt{2}}$$

$$\sqrt{2}\cdot\frac{\sqrt{k(k+1)}+1}{\sqrt{2(k+1)}} > \sqrt{2}\cdot\frac{\sqrt{(k+1)}}{\sqrt{2}}$$

$$\frac{\sqrt{k(k+1)}+1}{\sqrt{(k+1)}} > \sqrt{(k+1)}$$

$$\sqrt{k(k+1)}+1 > k+1$$

$$\sqrt{k(k+1)} > k$$

$$k^2+k > k^2$$

$$k > 0$$

Since these manipulations are reversible for positive integers larger than 1, we may conclude that the inequality holds for all integers larger than 1.

9. E
10. D
11. A
12. D
13. E
14. B
15.

1) When $n = 1$, $S_1 = 7 = \frac{1}{2}(3+11)$. The formula is valid for $n = 1$.

2) Assume that $7+10+13+16+\ldots+(3k+4) = \frac{k}{2}(3k+11)$ is true.

$$S_{k+1} = 7+10+13+16+\ldots+(3(k+1)+4) = \frac{k}{2}(3k+11)+3k+7$$

$$= \frac{1}{2}\left(3k^2+11k\right)+3k+7$$

$$= \frac{1}{2}\left(3k^2+17k+14\right)$$

$$= \frac{k+1}{2}\left(3(k+1)+11\right)$$

By mathematical induction, the formula is true for all positive integers n.

16. B
17. C
18. i. Show true for $n=1$

$$\frac{11}{1(1+1)(1+2)}=\frac{11\cdot 1\cdot(1+3)}{4(1+1)(1+2)}$$
$$\frac{11}{2\cdot 3}=\frac{11\cdot 4}{4\cdot 2\cdot 3}$$
$$\frac{11}{6}=\frac{11}{6}$$
$$1=1$$

ii. Assume true for $n=k$

$$\sum_{i=1}^{k}\frac{11}{i(i+1)(i+2)}=\frac{11k(k+3)}{4(k+1)(k+2)}$$

iii. Prove true for $n = k+1$

$$\sum_{i=1}^{k+1} \frac{11}{i(i+1)(i+2)} = \sum_{i=1}^{k} \frac{11}{i(i+1)(i+2)} + \frac{11}{(k+1)\left[(k+1)+1\right]\left[(k+1)+2\right]}$$

$$= \frac{11k(k+3)}{4(k+1)(k+2)} + \frac{11}{(k+1)(k+2)(k+3)}$$

$$= \frac{11k(k+3)^2 + 44}{4(k+1)(k+2)(k+3)}$$

$$= \frac{11(k^3 + 6k^2 + 9k + 4)}{4(k+1)(k+2)(k+3)}$$

$$= \frac{11(k^3 + k^2 + 5k^2 + 5k + 4k + 4)}{4(k+1)(k+2)(k+3)}$$

$$= \frac{11\left[k^2(k+1) + 5k(k+1) + 4(k+1)\right]}{4(k+1)(k+2)(k+3)}$$

$$= \frac{11(k+1)(k^2 + 5k + 4)}{4(k+1)(k+2)(k+3)}$$

$$= \frac{11(k^2 + k + 4k + 4)}{4(k+2)(k+3)}$$

$$= \frac{11\left[k(k+1) + 4(k+1)\right]}{4(k+2)(k+3)}$$

$$= \frac{11(k+1)(k+4)}{4(k+2)(k+3)}$$

$$= \frac{11(k+1)\left[(k+1)+3\right]}{4\left[(k+1)+1\right]\left[(k+1)+2\right]}$$

19. C
20. C

Name: ______________________________ Date: ______________

1. Determine whether the sequence is geometric. If so, find the common ratio.
   3, 6, 12, 24, ...
   A) 2
   B) 3
   C) $\frac{1}{2}$
   D) –2
   E) not geometric

2. Use mathematical induction to prove the following for every positive integer n.
   $$\sum_{i=1}^{n} \frac{5}{(2i-1)(2i+1)} = \frac{5n}{2n+1}$$

3. Write the nth term of the geometric sequence as a function of n.
   $a_1 = 4, a_{k+1} = 2a_k$
   A) $a_n = 4(2)^{n-1}$
   B) $a_n = 2(4)^{n-1}$
   C) $a_n = 4\left(\frac{1}{2}\right)^{n-1}$
   D) $a_n = 2 + 2n$
   E) $a_n = 4(2)^n$

4. Find the indicated nth term of the geometric sequence.
   4th term: 4, –12, 36,...
   A) –5
   B) –108
   C) 324
   D) –192
   E) –768

5. Use mathematical induction to prove that 160 is a factor of $2^{4n+3} + 32$ for all positive n.

6. Use summation notation to write the sum.

$2-6+18-\ldots-486$

A) $\sum_{n=0}^{5} 2(-3)^{n-1}$

B) $\sum_{n=1}^{4} 2(-3)^{n}$

C) $\sum_{n=1}^{6} 2(-3)^{n-1}$

D) $\sum_{n=1}^{4} 2(-3)^{n-1}$

E) $\sum_{n=1}^{5} 2(-3)^{n+1}$

7. Use sigma notation to write the sum.

$\frac{1}{3\cdot 2}+\frac{1}{4\cdot 3}+\cdots+\frac{1}{9\cdot 8}$

A) $\sum_{n=1}^{7} \frac{1}{(n+1)(n+2)}$

B) $\sum_{n=1}^{7} \frac{1}{n(n+1)}$

C) $\sum_{n=1}^{7} \frac{n}{(n+2)!}$

D) $\sum_{n=1}^{5} \frac{1}{(n+1)(n+2)}$

E) $\sum_{n=0}^{6} \frac{1}{(n+1)(n+2)}$

8. Use mathematical induction to prove the following for every positive integer n.

$$\sum_{i=1}^{n} \frac{9}{i(i+1)(i+2)} = \frac{9n(n+3)}{4(n+1)(n+2)}$$

9. Find the rational number representation of the repeating decimal.

   $0.\overline{157}$

   A) $\dfrac{157}{9999}$

   B) $\dfrac{15.7}{999}$

   C) $\dfrac{157}{9}$

   D) $\dfrac{157}{999}$

   E) $\dfrac{157}{99}$

10. The first two terms of the arithmetic sequence are given. Find the indicated term.

    $a_1 = 3, a_2 = 7, a_{11} = \boxed{}$

    A) 47
    B) 34
    C) 37
    D) 51
    E) 43

11. Use mathematical induction to prove the following inequality for all $n \geq 1$.

    $15^n \geq 14n$

12. Given the sequence $4+\dfrac{7}{8},\ 4+\dfrac{8}{9},\ 4+\dfrac{9}{10},\ 4+\dfrac{10}{11},\ 4+\dfrac{11}{12},\ldots$, write an expression for the *apparent* nth term assuming n begins with 1.

    A) $\dfrac{6n+34}{n+7}$

    B) $\dfrac{5n+35}{n+7}$

    C) $\dfrac{5n+34}{n+7}$

    D) $\dfrac{5n+34}{n+8}$

    E) $\dfrac{6n+35}{n+8}$

13. Write the first five terms of the geometric sequence.

$$a_1 = -5, r = -\frac{1}{9}$$

A) $-5, -14, -23, -32, -41$
B) $-5, 45, -405, 3645, -32{,}805$
C) $\frac{5}{9}, -\frac{5}{81}, \frac{5}{729}, -\frac{5}{6561}, \frac{5}{59{,}049}$
D) $1, -5, \frac{5}{9}, -\frac{5}{81}, \frac{5}{729}$
E) $-5, \frac{5}{9}, -\frac{5}{81}, \frac{5}{729}, -\frac{5}{6561}$

14. Use mathematical induction to prove the following for every positive integer n.

$$1+3+9+27+\ldots+3^{n-1} = \frac{3^n - 1}{2}$$

15. Use the Binomial Theorem to expand the complex number. Simplify your result.

$$(5+4i)^4$$

A) $-1519 + 720i$
B) $-1519 - 720i$
C) $1519 + 720i$
D) $1519 - 720i$
E) 625

16. Use the first and second differences of the first five terms of the given sequence to determine whether the sequence has a linear model, a quadratic model, or neither.

$$a_1 = 0$$
$$a_n = a_{n-1} + 8n$$

A) first differences: 1, 2, 3, 4
second differences: 1, 1, 1
linear model
B) first differences: 8, 24, 56, 120
second differences: 16, 32, 64
linear model
C) first differences: 8, 16, 32, 64
second differences: 8, 16, 32
quadratic model
D) first differences: 8, 24, 56, 120
second differences: 16, 32, 64
quadratic model
E) first differences: 8, 16, 32, 64
second differences: 8, 16, 32
neither linear nor quadratic

17. You are given the probability that an event *will not* happen. Find the probability that the event *will* happen.

$$P(E') = \frac{5}{9}$$

A) $\frac{5}{9}$
B) 0
C) 1
D) $\frac{4}{9}$
E) $\frac{2}{9}$

18. Find the coefficient a of the term in the expansion of the binomial.

Binomial	*Term*
$(x-3y)^6$	ax^3y^3

A) $a=9$
B) $a=-540$
C) $a=18$
D) $a=729$
E) $a=120$

19. Find the sum of the following infinite geometric series.

$$-15+14-\frac{196}{15}+\frac{2744}{225}-\ldots$$

A) $-\frac{29}{225}$
B) $-\frac{1}{225}$
C) $\frac{29}{225}$
D) -15
E) $-\frac{225}{29}$

20. Determine the sample space for the experiment.
Three marbles are selected from marbles labeled A through D where the marbles are not replaced and the order of selection does not matter.

A) $S=$ {ABC, ABD, BCD}
B) $S=$ {ABC, ABD, ACD, BCD}
C) $S=$ {ABC, BCD, CDA}
D) $S=$ {ABC, ABD, BCD, CBA, CDA}
E) $S=$ {ABC, ABD, BCD, CDB}

Answer Key

1. A
2. i. Show true for $n=1$

$$\frac{5}{(2\cdot1-1)(2\cdot1+1)}=\frac{5\cdot1}{2\cdot1+1}$$
$$\frac{5}{3}=\frac{5}{3}$$
$$1=1$$

ii. Assume true for $n=k$

$$\sum_{i=1}^{k}\frac{5}{(2i-1)(2i+1)}=\frac{5k}{2k+1}$$

iii. Prove true for $n=k+1$

$$\begin{aligned}\sum_{i=1}^{k+1}\frac{5}{(2i-1)(2i+1)} &= \sum_{i=1}^{k}\frac{5}{(2i-1)(2i+1)}+\frac{5}{[2(k+1)-1][(2(k+1)+1)]}\\ &= \frac{5k}{2k+1}+\frac{5}{(2k+1)(2k+3)}\\ &= \frac{5k(2k+3)+5}{(2k+1)(2k+3)}\\ &= \frac{10k^2+15k+5}{(2k+1)(2k+3)}\\ &= \frac{10k^2+10k+5k+5}{(2k+1)(2k+3)}\\ &= \frac{10k(k+1)+5(k+1)}{(2k+1)(2k+3)}\\ &= \frac{5(k+1)(2k+1)}{(2k+1)(2k+3)}\\ &= \frac{5(k+1)}{2(k+1)+1}\end{aligned}$$

3. A
4. B
5. i. Show true for $n=1$

$$\left(2^{4\cdot1+3}+32\right)=160$$
$$=1\cdot160$$

ii. Assume true for $n=k$

160 is a factor of $\left(2^{4k+3}+32\right)$

iii. Prove true for $n = k+1$

$$\left(2^{4(k+1)+3}+32\right)=\left(2^{(4k+3)+12}+32\right)$$
$$=4096\cdot 2^{(4k+3)}+32$$
$$=4096\cdot 2^{(4k+3)}+32\cdot 4096-131{,}040$$
$$=4096\cdot\left(2^{(4k+3)}+32\right)-819\cdot 160$$

6. C
7. A
8. i. Show true for $n=1$

$$\frac{9}{1(1+1)(1+2)}=\frac{9\cdot 1\cdot(1+3)}{4(1+1)(1+2)}$$
$$\frac{9}{2\cdot 3}=\frac{9\cdot 4}{4\cdot 2\cdot 3}$$
$$\frac{9}{6}=\frac{9}{6}$$
$$1=1$$

ii. Assume true for $n=k$

$$\sum_{i=1}^{k}\frac{9}{i(i+1)(i+2)}=\frac{9k(k+3)}{4(k+1)(k+2)}$$

iii. Prove true for $n = k + 1$

$$\sum_{i=1}^{k+1} \frac{9}{i(i+1)(i+2)} = \sum_{i=1}^{k} \frac{9}{i(i+1)(i+2)} + \frac{9}{(k+1)\left[(k+1)+1\right]\left[(k+1)+2\right]}$$

$$= \frac{9k(k+3)}{4(k+1)(k+2)} + \frac{9}{(k+1)(k+2)(k+3)}$$

$$= \frac{9k(k+3)^2 + 36}{4(k+1)(k+2)(k+3)}$$

$$= \frac{9(k^3 + 6k^2 + 9k + 4)}{4(k+1)(k+2)(k+3)}$$

$$= \frac{9(k^3 + k^2 + 5k^2 + 5k + 4k + 4)}{4(k+1)(k+2)(k+3)}$$

$$= \frac{9\left[k^2(k+1) + 5k(k+1) + 4(k+1)\right]}{4(k+1)(k+2)(k+3)}$$

$$= \frac{9(k+1)(k^2 + 5k + 4)}{4(k+1)(k+2)(k+3)}$$

$$= \frac{9(k^2 + k + 4k + 4)}{4(k+2)(k+3)}$$

$$= \frac{9\left[k(k+1) + 4(k+1)\right]}{4(k+2)(k+3)}$$

$$= \frac{9(k+1)(k+4)}{4(k+2)(k+3)}$$

$$= \frac{9(k+1)\left[(k+1)+3\right]}{4\left[(k+1)+1\right]\left[(k+1)+2\right]}$$

9. D
10. E

11. i. Show true for $n=1$

$$15^1 \geq 14 \cdot 1$$
$$15 \geq 14$$

ii. Assume true for $n=k$

$$15^k \geq 14k$$

iii. Prove true for $n=k+1$

$$\begin{aligned}(15)^{k+1} &= 15(15)^k \\ &\geq 15 \cdot 14k \\ &= 210k \\ &= 14k+14+196k-14 \\ &= 14(k+1)+196k-14 \\ &\geq 14(k+1)\end{aligned}$$

12. C
13. E
14. i. Show true for $n=1$

$$3^{1-1} = \frac{3^1-1}{2}$$
$$3^0 = \frac{3-1}{2}$$
$$1 = \frac{2}{2}$$
$$1 = 1$$

ii. Assume true for $n=k$

$$1+3+9+27+\ldots+3^{k-1} = \frac{3^k-1}{2}$$

iii. Prove true for $n = k+1$

$$
\begin{aligned}
1+3+9+27+\ldots+3^{k-1}+3^{(k+1)-1} &= \frac{3^k-1}{2}+3^{(k+1)-1} \\
&= \frac{3^k-1}{2}+3^k \\
&= \frac{3^k-1+2\cdot 3^k}{2} \\
&= \frac{3\cdot 3^k-1}{2} \\
&= \frac{3^{k+1}-1}{2}
\end{aligned}
$$

15. A
16. E
17. D
18. B
19. E
20. B

Name: ______________________________ Date: _______________

1. Test the graph of the following equation for symmetry with respect to $\theta = \frac{\pi}{2}$, the polar axis, and the pole.

$$r = -6 + \cos(5\theta)$$

A) The graph is symmetric with respect to $\theta = \frac{\pi}{2}$ only.
B) The graph is symmetric with respect to the polar axis only.
C) The graph is symmetric with respect to the pole only.
D) The graph is symmetric with respect to all three.
E) The graph has no symmetries.

2. Which set of parametric equations represents the following line or conic? Use $x = h + a\sec\theta$ and $y = k + b\tan\theta$.

Hyperbola: vertices $(5,\ -3),\ (11,\ -3)$
foci $(0,\ -3),\ (16,\ -3)$

A) $x = -8 + \sqrt{55}\sec\theta$
$y = 3 + 3\tan\theta$

B) $x = 8 + \sqrt{55}\sec\theta$
$y = -3 + 3\tan\theta$

C) $x = -3 + \sqrt{55}\sec\theta$
$y = 8 + 3\tan\theta$

D) $x = 8 + 3\sec\theta$
$y = -3 + \sqrt{55}\tan\theta$

E) $x = -3 + 3\sec\theta$
$y = 8 + \sqrt{55}\tan\theta$

3. Find the eccentricity of the following ellipse. Round your answer to two decimals.

$$4x^2+9y^2-24x+18y-36=0$$

A) 1.25
B) 1.50
C) 329.06
D) 0.75
E) 0.67

4. Find three additional polar representations of the point $\left(5,-\frac{\pi}{3}\right)$, given in polar coordinates, using $-2\pi<\theta<2\pi$.

A) $\left(5,\frac{5\pi}{3}\right),\left(5,\frac{2\pi}{3}\right),\left(-5,-\frac{4\pi}{3}\right)$
B) $\left(5,\frac{5\pi}{3}\right),\left(5,\frac{2\pi}{3}\right),\left(5,-\frac{4\pi}{3}\right)$
C) $\left(5,\frac{5\pi}{3}\right),\left(-5,\frac{2\pi}{3}\right),\left(-5,-\frac{4\pi}{3}\right)$
D) $\left(-5,\frac{5\pi}{3}\right),\left(-5,\frac{2\pi}{3}\right),\left(5,-\frac{4\pi}{3}\right)$
E) $\left(-5,\frac{5\pi}{3}\right),\left(5,\frac{2\pi}{3}\right),\left(-5,-\frac{4\pi}{3}\right)$

5. Find the rectangular graph of the following polar equation.

$$\theta = \frac{5\pi}{4}$$

A)

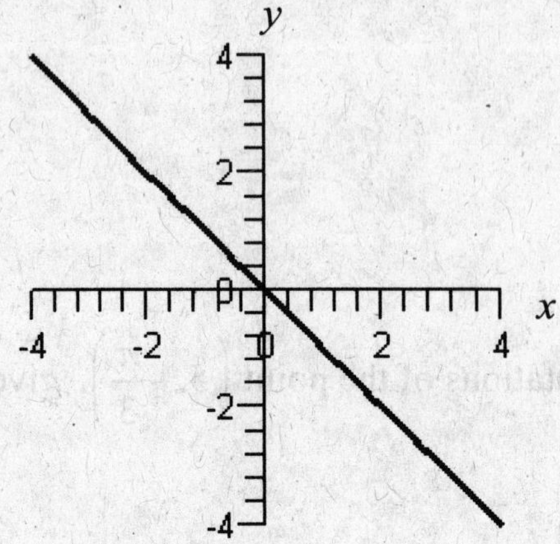

B)

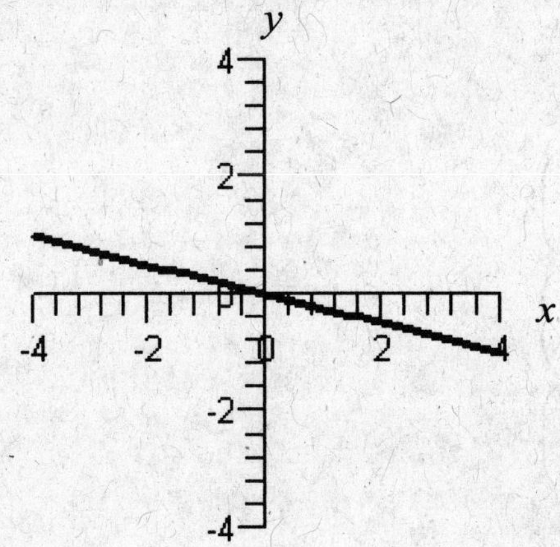

C)

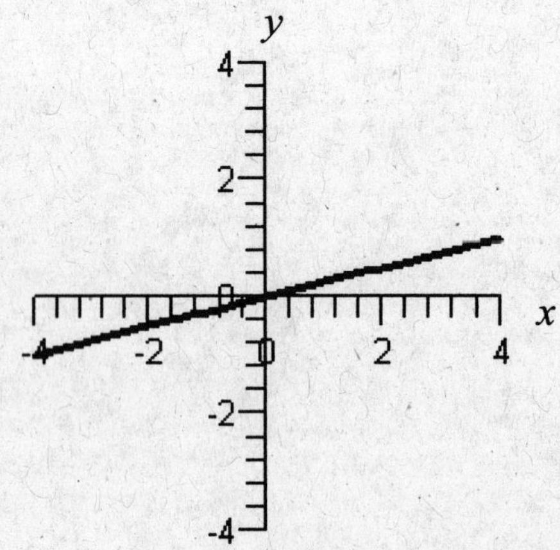

D)

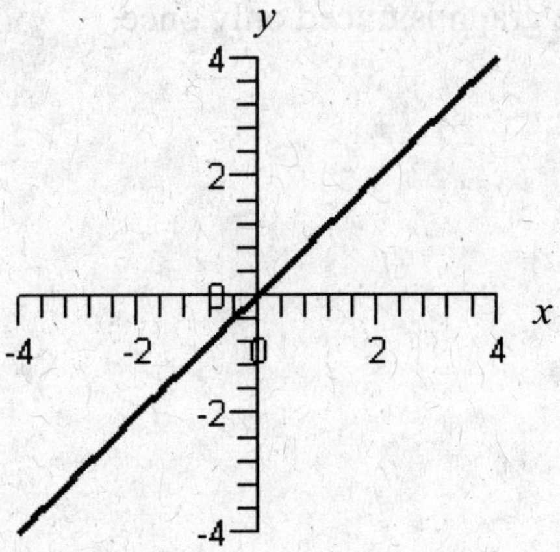

E)

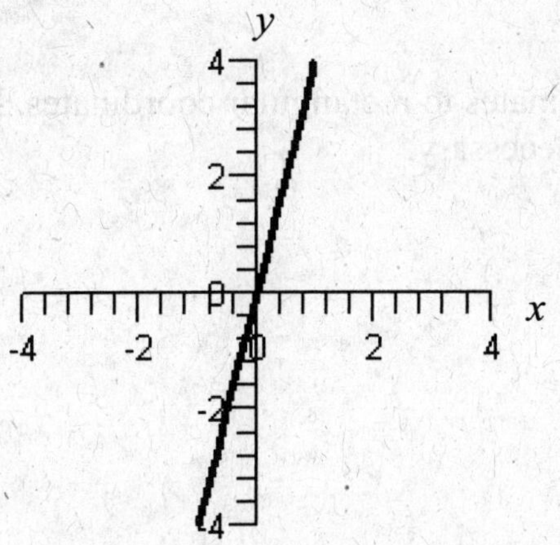

6. Find two sets of polar coordinates with $0 \le \theta < 2\pi$ for the point $(-2, 0)$, given in rectangular coordinates.
   A) $(2,0),(-2,\pi)$
   B) $(\sqrt{2},0),(-\sqrt{2},\pi)$
   C) $\left(-2,\frac{\pi}{2}\right),\left(2,\frac{3\pi}{2}\right)$
   D) $(-2,0),(2,\pi)$
   E) $\left(-\sqrt{2},\frac{\pi}{2}\right),\left(\sqrt{2},\frac{3\pi}{2}\right)$

7. Find an interval for θ for which the graph is traced only once.

$$r = 2\sin\left(\frac{3\theta}{2}\right)$$

A) $0 \le \theta < \frac{\pi}{3}$
B) $0 \le \theta < \pi$
C) $0 \le \theta < 4\pi$
D) $0 \le \theta < \frac{4\pi}{3}$
E) $-2\pi \le \theta < 2\pi$

8. Convert the point from polar coordinates to rectangular coordinates. Round answer to three decimal places, if necessary.

$$\left(3, -\frac{3\pi}{2}\right)$$

A) $(3, 0)$
B) $(0, 3)$
C) $(0, -3)$
D) $(-3, 0)$
E) $(3, -4.712)$

9. Find the graph of the following polar equation.

$$r = 6\sin(2\theta)$$

A)

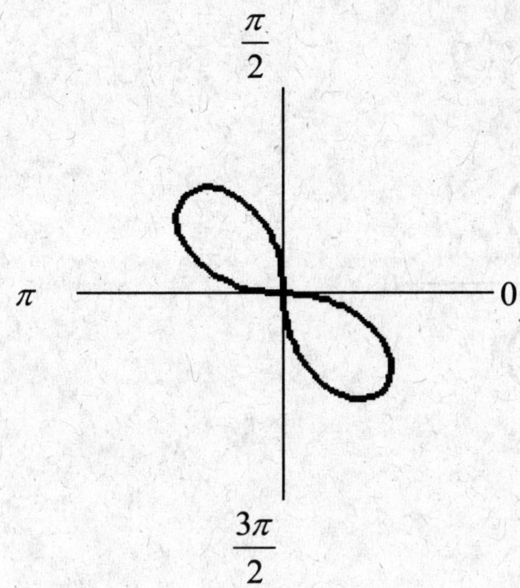

B)

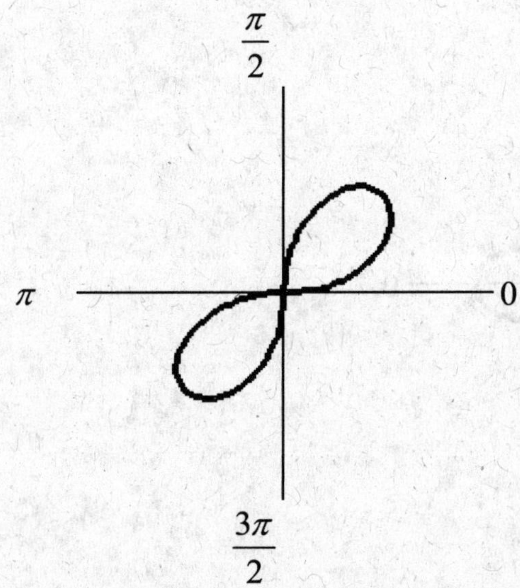

C)

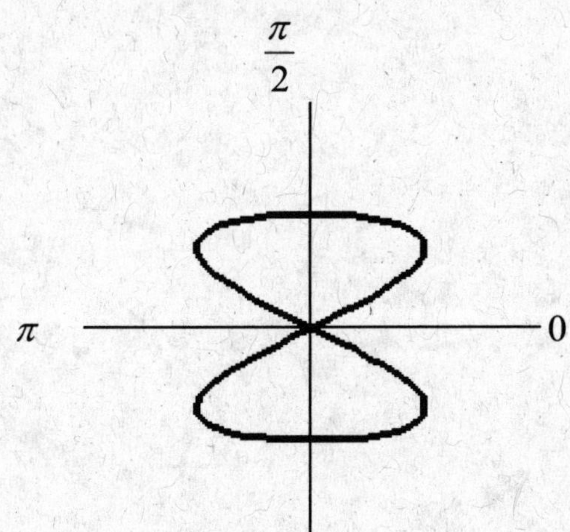

$\frac{3\pi}{2}$

D)

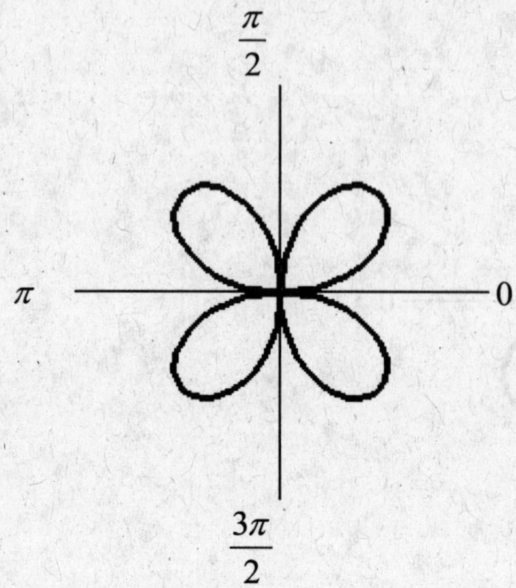

E)

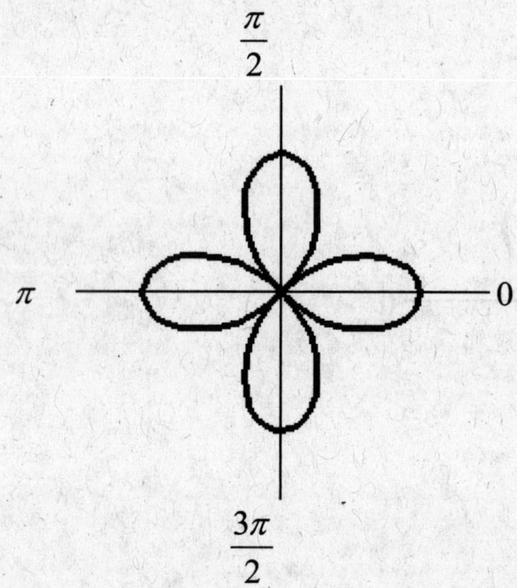

10. Find the center and foci of the ellipse.

$$\frac{(x-9)^2}{32}+\frac{(y+5)^2}{36}=1$$

A) center: (–9, 5) foci: (–9, 3), (–9, 7)
B) center: (9, –5) foci: (9, –7), (9, –3)
C) center: (9, –5) foci: (7, –5), (11, –5)
D) center: (–9, 5) foci: (–11, –5), (–7, –5)
E) center: (9, –5) foci: (–11, 5), (–7, 5)

11. Find the standard form of the parabola with the given characteristic and vertex at the origin.

directrix: $x = 5$

A) $x^2 = -20y$
B) $x^2 = 20y$
C) $x^2 = 5y$
D) $y^2 = 5x$
E) $y^2 = -20x$

12. Find the vertex and focus of the parabola.

$x^2 - 8y = 0$

A) vertex: $(2,\ 0)$ focus: (0, 0)
B) vertex: $(-2,\ 0)$ focus: (0, 0)
C) vertex: (0, 0) focus: $(2,\ 0)$
D) vertex: (0, 0) focus: $(0,\ 2)$
E) vertex: (0, 0) focus: $(0,\ -2)$

13. Find the graph of the following polar equation.
$r = -3\sin(3\theta)$

A)

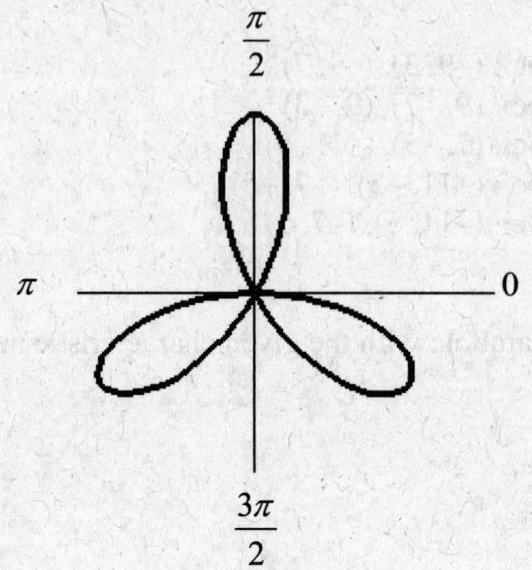

B)

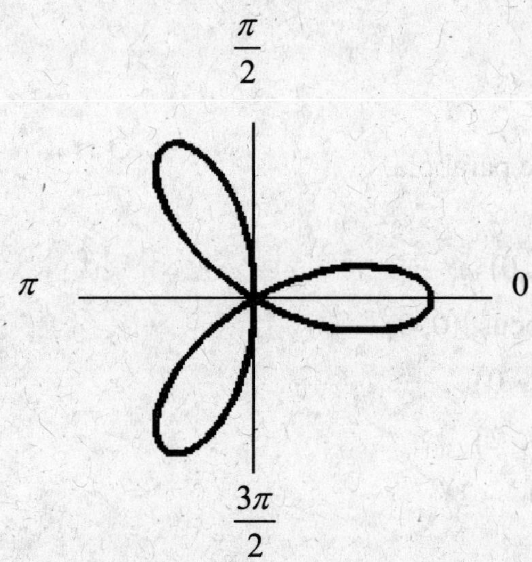

C)

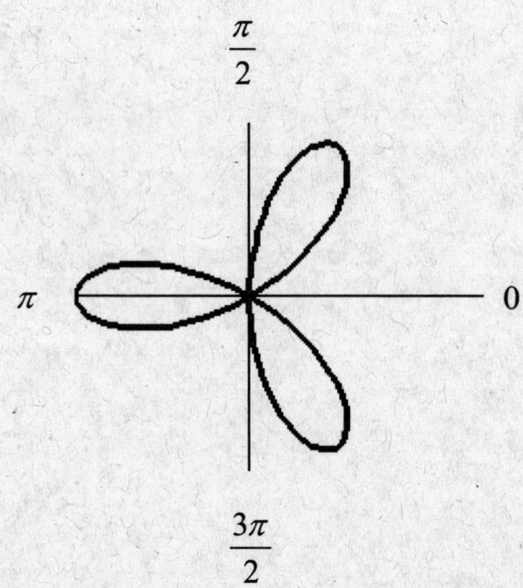

D)

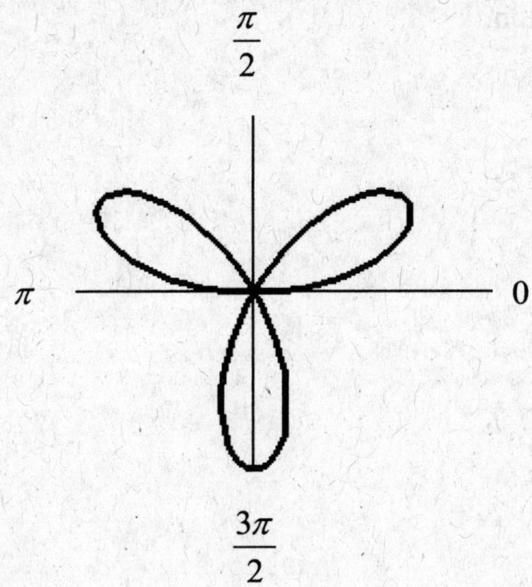

E)

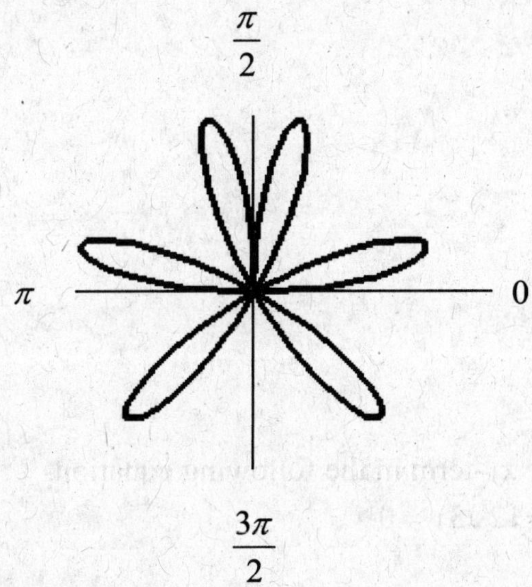

14. Find any zeros of r on the interval $0 \le \theta < 2\pi$.

$r = \sqrt{3} + 2\cos\theta$

A) zeros: $\frac{\pi}{6}, \frac{11\pi}{6}$

B) zeros: $\frac{5\pi}{6}, \frac{7\pi}{6}$

C) zeros: $\frac{5\pi}{3}, \frac{7\pi}{3}$

D) zeros: $\frac{\pi}{3}, \frac{11\pi}{3}$

E) zeros: $\frac{5\pi}{2}, \frac{7\pi}{2}$

15. Match the graph with its equation.

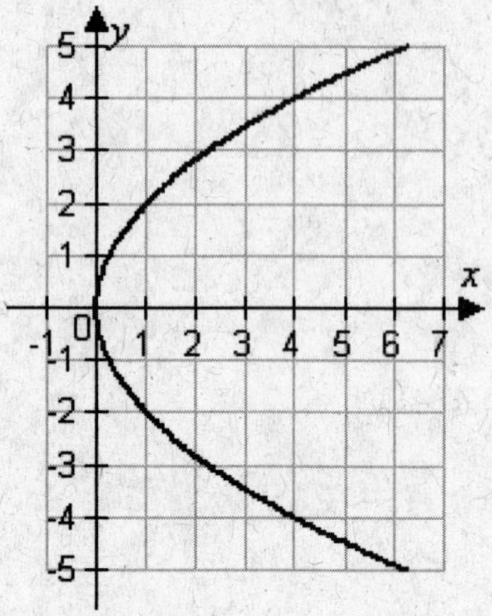

A) $x^2 = 4y$

B) $y^2 = 4x$

C) $y^2 = \dfrac{x}{4}$

D) $x^2 = \dfrac{y}{4}$

E) $(y+1)^2 = 4(x-2)$

16. Rotate the axes to eliminate the *xy*-term in the following equation.
$$18x^2 - 12\sqrt{3}xy + 6y^2 + 12x + 12\sqrt{3}y = 0$$

A) $(y')^2 + x' = 0$

B) $6\sqrt{3}(y')^2 + 6\sqrt{3}x' = 0$

C) $6(x')^2 - 6y' = 0$

D) $\dfrac{(y')^2}{6} + \dfrac{x'}{6} = 0$

E) $\dfrac{(x')^2}{12} + \dfrac{y'}{12} = 0$

17. Find the center and foci of the following ellipse.

$4x^2+8y^2-16x+32y+32=0$

A) Center: $(-2,2)$
   Foci: $(-2+\sqrt{2},2)$, $(-2-\sqrt{2},2)$
B) Center: $(2,-2)$
   Foci: $(2+\sqrt{2},-2)$, $(2-\sqrt{2},-2)$
C) Center: $(2,-2)$
   Foci: $(2,-2+\sqrt{2})$, $(2,-2-\sqrt{2})$
D) Center: $(-2,2)$
   Foci: $(-2,2+\sqrt{2})$, $(-2,2-\sqrt{2})$
E) Center: $(-2,2)$
   Foci: $(-2+\sqrt{6},2)$, $(-2-\sqrt{6},-2)$

18. A projectile is launched from ground level at an angle of θ with the horizontal. The initial velocity is v_0 feet per second and the path of the projectile is modeled by the parametric equations

$x=(v_0\cos\theta)t$ and $y=(v_0\sin\theta)t-16t^2$.

Use a graphing utility to graph the paths of a projectile launched from ground level with the values given for θ and v_0. Use the graph to approximate the maximum height and range of the projectile to the nearest foot.

$\theta=55°$, $v_0=56$ feet per second

A) maximum height ≈ 49 feet   range ≈ 4 feet
B) maximum height ≈ 4 feet   range ≈ 49 feet
C) maximum height ≈ 33 feet   range ≈ 92 feet
D) maximum height ≈ 92 feet   range ≈ 33 feet
E) maximum height ≈ 43 feet   range ≈ 8 feet

19. Find the vertices and asymptotes of the hyperbola.

$$9y^2 - 4x^2 = 36$$

A) vertices: $(0, \pm 2)$ asymptote: $y = \pm\frac{2}{3}x$

B) vertices: $(0, \pm 2)$ asymptote: $y = \pm\frac{3}{2}x$

C) vertices: $(\pm 2, 0)$ asymptote: $y = \pm\frac{2}{3}x$

D) vertices: $(\pm 2, 0)$ asymptote: $y = \pm\frac{3}{2}x$

E) vertices: $(\pm 2, 3)$ asymptote: $y = \pm\frac{2}{3}x$

20. Find the vertex and focus of the parabola below.

$$y^2 + 6y - 8x + 20 = 0$$

A) Vertex: $\left(-\frac{11}{8}, 3\right)$

Focus: $\left(\frac{5}{8}, 3\right)$

B) Vertex: $\left(\frac{11}{8}, -3\right)$

Focus: $\left(\frac{27}{8}, -3\right)$

C) Vertex: $\left(\frac{11}{8}, -3\right)$

Focus: $\left(\frac{75}{8}, -3\right)$

D) Vertex: $\left(-\frac{11}{8}, 3\right)$

Focus: $\left(\frac{53}{8}, 3\right)$

E) Vertex: $\left(-\frac{11}{8}, -3\right)$

Focus: $\left(\frac{53}{8}, -3\right)$

Answer Key

1. B
2. D
3. D
4. C
5. D
6. A
7. C
8. B
9. D
10. B
11. E
12. D
13. A
14. B
15. B
16. A
17. B
18. C
19. A
20. B

Name: ___________________________ Date: ____________

1. Test the graph of the following equation for symmetry with respect to $\theta = \frac{\pi}{2}$, the polar axis, and the pole.

   $r = 5 + 4\cos(\theta)$

   A) The graph is symmetric with respect to $\theta = \frac{\pi}{2}$ only.
   B) The graph is symmetric with respect to the polar axis only.
   C) The graph is symmetric with respect to the pole only.
   D) The graph is symmetric with respect to all three.
   E) The graph has no symmetries.

2. A projectile is launched from ground level at an angle of θ with the horizontal. The initial velocity is v_0 feet per second and the path of the projectile is modeled by the parametric equations
   $x = (v_0 \cos\theta)t$ and $y = (v_0 \sin\theta)t - 16t^2$.
   Use a graphing utility to graph the paths of a projectile launched from ground level with the values given for θ and v_0. Use the graph to approximate the maximum height and range of the projectile to the nearest foot.

   $\theta = 50°$, $v_0 = 104$ feet per second

   A) maximum height ≈ 12 feet range ≈ 171 feet
   B) maximum height ≈ 171 feet range ≈ 12 feet
   C) maximum height ≈ 99 feet range ≈ 333 feet
   D) maximum height ≈ 333 feet range ≈ 99 feet
   E) maximum height ≈ 38 feet range ≈ 8 feet

3. Sketch the graph of the polar equation using symmetry, zeros, maximum r-values, and any other additional points.
   $r = 2 - 3\cos\theta$

Use either grid below for your graph, whichever is more convenient.

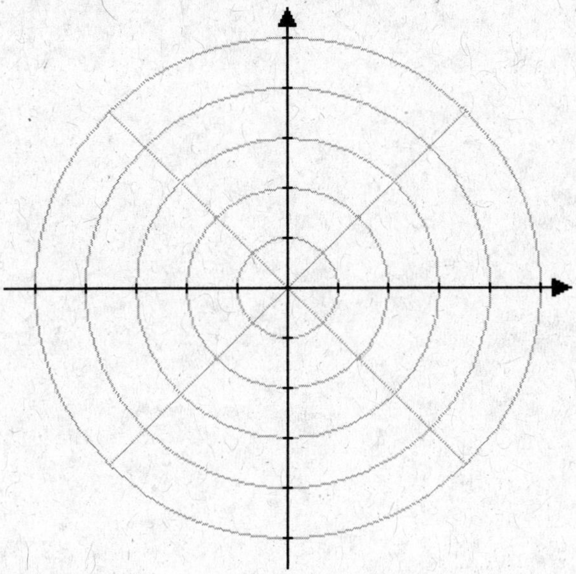

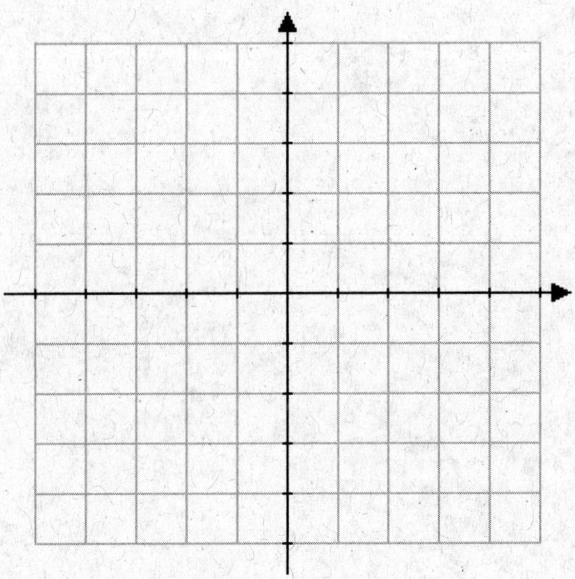

4. Which set of parametric equations represents the graph of the following rectangular equation using $t = 6 - x$?

$$y = x^2 + 9$$

A) $x = t + 6$
$y = (t + 6)^2 - 9$

B) $x = t - 6$
$y = (t - 6)^2 - 9$

C) $x = t - 6$
$y = (t - 6)^2 + 9$

D) $x = 6 + t$
$y = (6 + t)^2 + 9$

E) $x = 6 - t$
$y = (6 - t)^2 + 9$

5. Match the graph with its equation.

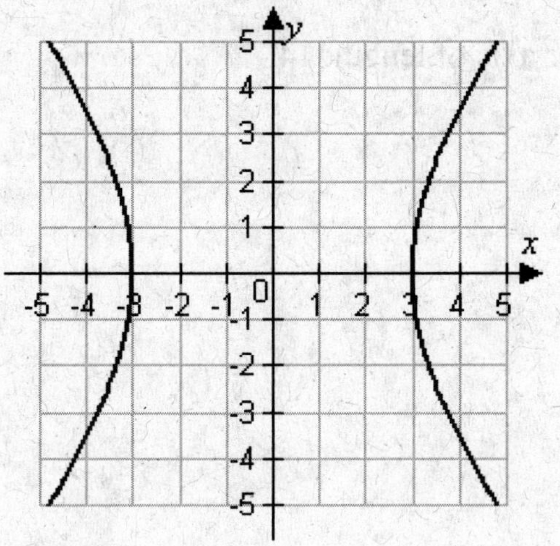

A) $\dfrac{x^2}{3} - \dfrac{y^2}{4} = 1$

B) $\dfrac{x^2}{4} + \dfrac{y^2}{3} = 1$

C) $\dfrac{x^2}{16} - \dfrac{y^2}{9} = 1$

D) $\dfrac{x^2}{9} - \dfrac{y^2}{16} = 1$

E) $\dfrac{x^2}{9} + \dfrac{y^2}{16} = 1$

6. Which answer is a rectangular form of the given polar equation?
$r = 18\cos\theta$

A) $(x-9)^2 + y^2 = 81$

B) $(x+9)^2 + y^2 = 81$

C) $(x+9)^2 + y^2 = 9$

D) $x^2 + (y+9)^2 = 81$

E) $x^2 + (y-9)^2 = 81$

7. Find the standard form of the equation of the ellipse with the following characteristics.

   foci: $(\pm 4, 0)$ major axis of length: 14

   A) $\frac{x^2}{49}+\frac{y^2}{33}=1$
   B) $\frac{x^2}{49}+\frac{y^2}{16}=1$
   C) $\frac{x^2}{16}+\frac{y^2}{49}=1$
   D) $\frac{x^2}{196}+\frac{y^2}{16}=1$
   E) $\frac{x^2}{196}+\frac{y^2}{180}=1$

8. Convert the following polar equation to rectangular form.

   $\theta=-\frac{5\pi}{3}$

   A) $y=\sqrt{3}$
   B) $y=\frac{\sqrt{3}}{3}$
   C) $y=-\sqrt{3}$
   D) $y=3$
   E) $y=-\frac{\sqrt{3}}{3}$

9. Classify the graph of the equation below as a circle, a parabola, an ellipse, or a hyperbola.

   $y^2+13x+y-79=0$

   A) ellipse
   B) parabola
   C) hyperbola
   D) circle

10. Rotate the axes to eliminate the *xy*-term in the following equation and then write the equation in standard form.

$3x^2-12xy+12y^2+15\sqrt{5}y+3=0$

A) $(x'-1)^2=3\left(y'-\frac{4}{5}\right)$

B) $(x'+3)^2=-5\left(y'+\frac{1}{3}\right)$

C) $(y'+1)^2=-\left(x'-\frac{4}{5}\right)$

D) $\left(y'-\frac{4}{5}\right)^2=3(x'-1)$

E) $\left(y'+\frac{4}{5}\right)^2=x'+\frac{1}{3}$

11. Find the standard form of the equation of the ellipse with the given characteristics.

vertices: (4, –6), (4, 10) minor axis of length: 4

A) $\frac{(x+4)^2}{4}+\frac{(y+2)^2}{64}=1$

B) $\frac{(x-4)^2}{4}+\frac{(y-2)^2}{64}=1$

C) $\frac{(x-4)^2}{64}+\frac{(y-2)^2}{4}=1$

D) $\frac{(x-2)^2}{64}+\frac{(y-4)^2}{4}=1$

E) $\frac{(x+2)^2}{64}+\frac{(y+4)^2}{4}=1$

12. Use the discriminant to classify the graph; then use the quadratic formula to solve for y.

$3x^2 - 2\sqrt{3}xy + y^2 + 10\sqrt{3}x - 6y - 24 = 0$

A) ellipse; $y = 3 + \sqrt{3}x \pm \sqrt{4\sqrt{3}x + 33}$

B) ellipse; $y = 3 + \sqrt{3}x \pm \sqrt{-4\sqrt{3}x + 33}$

C) parabola; $y = 3 + \sqrt{3}x \pm \sqrt{-4\sqrt{3}x + 33}$

D) parabola; $y = \sqrt{3}x \pm \sqrt{-4\sqrt{3}x + 33}$

E) hyperbola; $y = \sqrt{3}x \pm \sqrt{-4\sqrt{3}x + 33}$

13. Find a polar equation of the conic with the given characteristics and with one focus at the pole.

Conic	*Eccentricity*	*Directrix*
Hyperbola	$e = 5$	$x = 2$

A) $r = \dfrac{10}{1 - 5\cos\theta}$

B) $r = \dfrac{10}{1 + 5\cos\theta}$

C) $r = \dfrac{2}{1 + 5\cos\theta}$

D) $r = \dfrac{10}{1 + 5\sin\theta}$

E) $r = \dfrac{10}{1 - 5\sin\theta}$

14. Use the Quadratic Formula to solve for y in the following equation.

$36x^2 - 84xy + 49y^2 - 70x - 30y = 0$

A) $\dfrac{-42x - 15 \pm \sqrt{-84x + 3655}}{49}$

B) $\dfrac{2(42x + 15) \pm \sqrt{-84x + 14620}}{49}$

C) $\dfrac{42x + 15 \pm \sqrt{4690x + 225}}{49}$

D) $\dfrac{84x + 30 \pm \sqrt{-42x + 7085}}{98}$

E) $\dfrac{-84x - 30 \pm \sqrt{-42x + 4330}}{98}$

15. Find three additional polar representations of the point $\left(-5,-\frac{5\pi}{6}\right)$, given in polar coordinates, using $-2\pi < \theta < 2\pi$.

A) $\left(-5,\frac{7\pi}{6}\right),\left(-5,\frac{\pi}{6}\right),\left(5,-\frac{11\pi}{6}\right)$
B) $\left(-5,\frac{7\pi}{6}\right),\left(-5,\frac{\pi}{6}\right),\left(-5,-\frac{11\pi}{6}\right)$
C) $\left(-5,\frac{7\pi}{6}\right),\left(5,\frac{\pi}{6}\right),\left(5,-\frac{11\pi}{6}\right)$
D) $\left(5,\frac{7\pi}{6}\right),\left(5,\frac{\pi}{6}\right),\left(-5,-\frac{11\pi}{6}\right)$
E) $\left(5,\frac{7\pi}{6}\right),\left(-5,\frac{\pi}{6}\right),\left(5,-\frac{11\pi}{6}\right)$

16. Classify the graph of the equation below as a circle, a parabola, an ellipse, or a hyperbola.

$$2x^2+5y^2+8x-2y=0$$

A) ellipse
B) hyperbola
C) parabola
D) circle

17. Which set of parametric equations represents the following line or conic? Use $x = h + a\sec\theta$ and $y = k + b\tan\theta$.

    Hyperbola: vertices $(-5,\ 7),\ (-3,\ 7)$

    foci $(-8,\ 7),\ (0,\ 7)$

A) $x = 4 + \sqrt{15}\sec\theta$
$y = -7 + \tan\theta$

B) $x = -4 + \sqrt{15}\sec\theta$
$y = 7 + \tan\theta$

C) $x = 7 + \sqrt{15}\sec\theta$
$y = -4 + \tan\theta$

D) $x = -4 + \sec\theta$
$y = 7 + \sqrt{15}\tan\theta$

E) $x = 7 + \sec\theta$
$y = -4 + \sqrt{15}\tan\theta$

18. Match the graph with its equation.

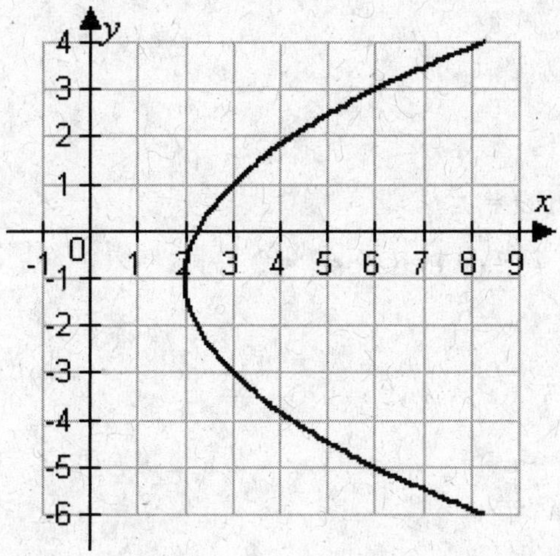

A) $y^2 = 4x$
B) $(y+1)^2 = 4(x-2)$
C) $x^2 = \frac{y}{4}$
D) $y^2 = \frac{x}{4}$
E) $x^2 = 4y$

19. Find the standard form of the parabola with the given characteristics.
focus: (8, 13) directrix: $y = -1$
A) $(x-8)^2 = 28(y-6)$
B) $(x+8)^2 = 28(y+6)$
C) $(x-6)^2 = 28(y-8)$
D) $(x+6)^2 = 28(y+8)$
E) $(x+8)^2 = 7(y+6)$

20. Test for symmetry with respect to $\theta = \pi/2$, the polar axis, and the pole.
$$r = \frac{2}{4+\sin\theta}$$
A) symmetric with respect to the pole
B) symmetric with respect to the polar axis
C) symmetric with respect to $\theta = \pi/2$
D) symmetric with respect to the pole and the polar axis
E) symmetric with respect to the pole and $\theta = \pi/2$

Answer Key

1. B
2. C
3.

symmetry: with respect to polar axis

Maximum value of $|r|$: $|r| = 5$ when $\theta = \pi$

zeros of r: $r = 0$ when $\theta \approx 0.841, 5.442$ (when $\cos\theta = 2/3$)

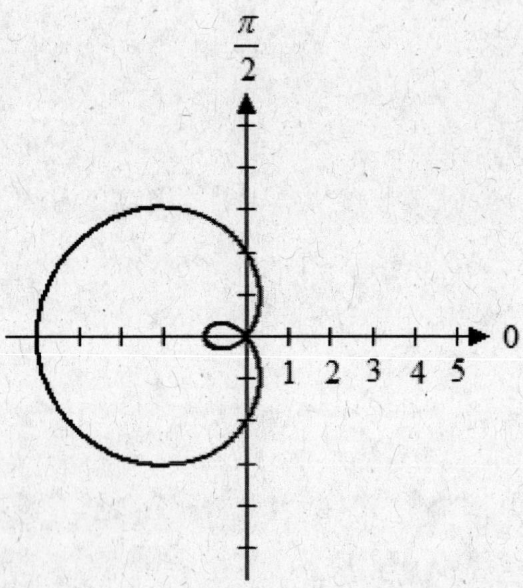

4. E
5. D
6. A
7. A
8. A
9. B
10. C
11. B
12. C
13. B
14. C
15. C
16. A
17. D
18. B
19. A
20. C

Name: ______________________________ Date: ____________

1. Rotate the axes to eliminate the *xy*-term in the following equation and then write the equation in standard form.

   $7x^2 - 28xy + 28y^2 + 35\sqrt{5}y + 7 = 0$

   A) $(x'-1)^2 = 7\left(y'-\frac{4}{5}\right)$

   B) $(x'+7)^2 = -5\left(y'+\frac{1}{7}\right)$

   C) $(y'+1)^2 = -\left(x'-\frac{4}{5}\right)$

   D) $\left(y'-\frac{4}{5}\right)^2 = 7(x'-1)$

   E) $\left(y'+\frac{4}{5}\right)^2 = x'+\frac{1}{7}$

2. Match the graph to a set of parametric equations.

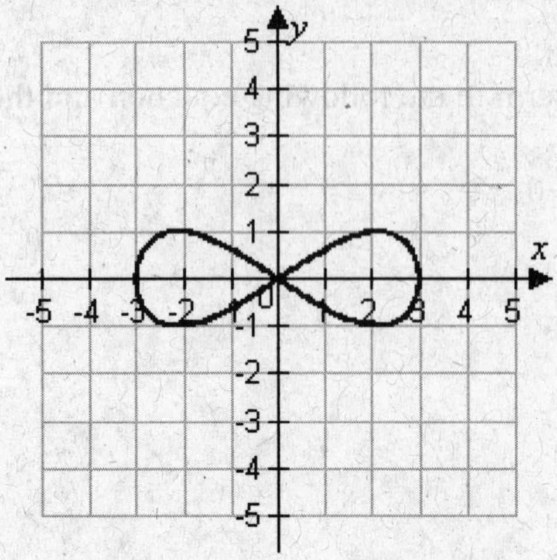

A) Witch of Agnesi: $x = 2\cot\theta$, $y = 2\sin^2\theta$

B) Lissajous curve: $x = 3\cos\theta$, $y = \sin 2\theta$

C) Involute of circle: $x = \cos\theta + \theta\sin\theta$, $y = \sin\theta - \theta\cos\theta$

D) Evolute of ellipse: $x = 3\cos^3\theta$, $y = 5\sin^3\theta$

E) Serpentine curve: $x = \cot\theta$, $y = 6\sin\theta\cos\theta$

3. Which answer is a rectangular form of the given polar equation?
$r = 10\cos\theta$

A) $(x-5)^2 + y^2 = 25$

B) $(x+5)^2 + y^2 = 25$

C) $(x+5)^2 + y^2 = 5$

D) $x^2 + (y+5)^2 = 25$

E) $x^2 + (y-5)^2 = 25$

4. Planets travel in elliptical orbits with the sun at one focus. Assume that the focus is at the pole, the major axis lies on the polar axis, and the length of the major axis is $2a$ (see figure). The polar equation of the orbit of a planet is given below, where e is the eccentricity. If $a = 88.908 \times 10^6$ miles and $e = 0.0260$, find the perihelion distance (the minimum distance from the sun to the planet). Round your answer to the nearest mile.

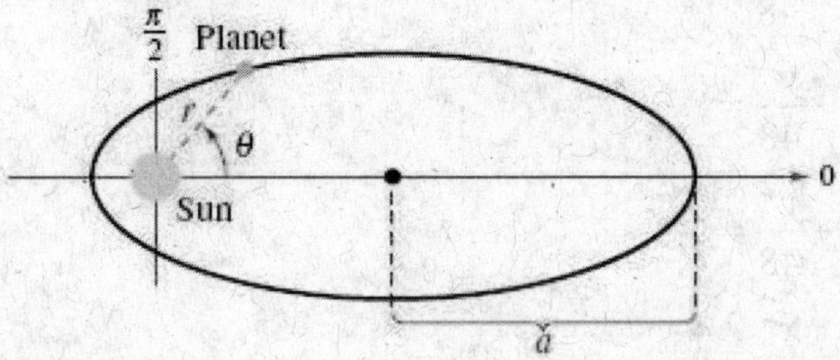

A) 86,596,392 miles
B) 91,219,608 miles
C) 8,659,639 miles
D) 9,121,961 miles
E) 865,963,920 miles

5. Find the standard form of the equation of the ellipse with the given characteristics.
center: (–8, –5) $a = 5c$ foci: (–11, –5), (–5, –5)

A) $\frac{(x-8)^2}{216} + \frac{(y-5)^2}{225} = 1$

B) $\frac{(x+8)^2}{216} + \frac{(y+5)^2}{225} = 1$

C) $\frac{(x+8)^2}{225} + \frac{(y+5)^2}{216} = 1$

D) $\frac{(x+5)^2}{225} + \frac{(y+8)^2}{216} = 1$

E) $\frac{(x-5)^2}{225} + \frac{(y-8)^2}{216} = 1$

6. Find the standard form of the equation of the hyperbola with the given characteristics.
   vertices: (0, ±4) foci: (0, ±8)

   A) $\dfrac{y^2}{16} - \dfrac{x^2}{64} = 1$

   B) $\dfrac{y^2}{16} - \dfrac{x^2}{48} = 1$

   C) $\dfrac{x^2}{16} - \dfrac{y^2}{48} = 1$

   D) $\dfrac{x^2}{16} - \dfrac{y^2}{48} = 64$

   E) $\dfrac{y^2}{16} - \dfrac{x^2}{48} = 64$

7. Use a graphing utility to graph the polar equation and show that the indicated line is an asymptote.

Name of Graph	*Polar Equation*	*Asymptote*
Hyperbolic Spiral	$r = -\dfrac{5}{\theta}$	$y = -5$

Use either grid below for your graph, whichever is more convenient.

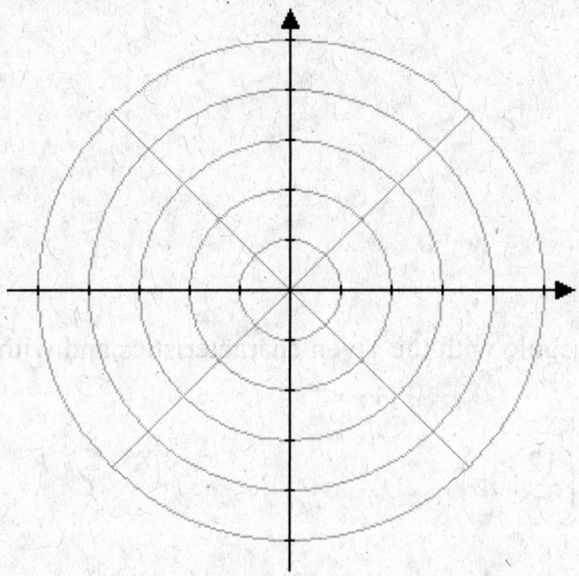

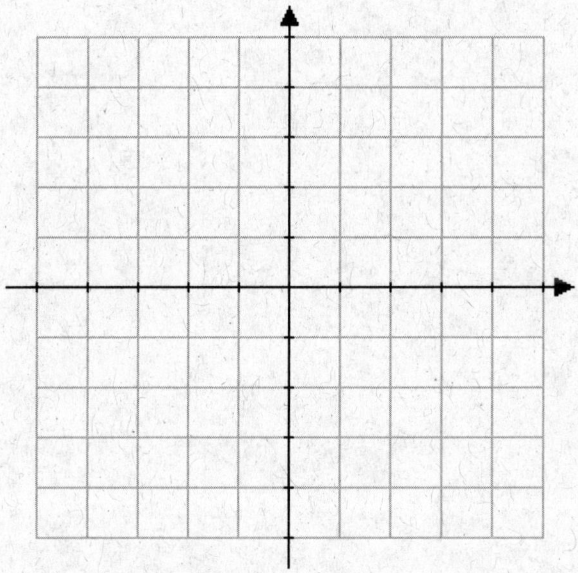

8. Find a polar equation of the conic with the given characteristics and with one focus at the pole.

Conic	*Eccentricity*	*Directrix*
Ellipse	$e = \frac{2}{5}$	$y = 2$

A) $r = \dfrac{2}{5 + 2\cos\theta}$

B) $r = \dfrac{4}{5 + 2\cos\theta}$

C) $r = \dfrac{4}{5 - 2\cos\theta}$

D) $r = \dfrac{4}{5 - 2\sin\theta}$

E) $r = \dfrac{4}{5 + 2\sin\theta}$

9. Find a polar equation of the conic with the given characteristics and with one focus at the pole.

Conic	*Vertices*
Ellipse	$(12, 0), \left(\frac{12}{7}, \pi\right)$

A) $r = \dfrac{12}{4 - 3\sin\theta}$

B) $r = \dfrac{12}{4 + 3\cos\theta}$

C) $r = \dfrac{12}{4 - 3\cos\theta}$

D) $r = \dfrac{12}{4 + 3\sin\theta}$

E) $r = \dfrac{12}{3 - 5\cos\theta}$

10. Find the standard form of the equation of the ellipse below.

$$8x^2+4y^2+64x-32y+32=0$$

A) $\dfrac{(x-4)^2}{\left(2\sqrt{5}\right)^2}+\dfrac{(y+4)^2}{\left(2\sqrt{10}\right)^2}=1$

B) $\dfrac{(x+4)^2}{\left(2\sqrt{5}\right)^2}+\dfrac{(y-4)^2}{\left(2\sqrt{10}\right)^2}=1$

C) $\dfrac{(x+4)^2}{\left(2\sqrt{10}\right)^2}-\dfrac{(y-4)^2}{\left(2\sqrt{5}\right)^2}=1$

D) $\dfrac{(x-4)^2}{\left(2\sqrt{10}\right)^2}+\dfrac{(y+4)^2}{\left(2\sqrt{5}\right)^2}=1$

E) $\dfrac{(x-4)^2}{\left(2\sqrt{5}\right)^2}-\dfrac{(y+4)^2}{\left(2\sqrt{10}\right)^2}=1$

11. Classify the graph of the equation below as a circle, a parabola, an ellipse, or a hyperbola.

$$3y^2+17x+y-103=0$$

A) ellipse
B) parabola
C) hyperbola
D) circle

12. Identify the center and radius of the circle below.
    $x^2 + y^2 + 6x + 6y - 5 = 0$
    A) Center: $(3,3)$
       Radius: $\sqrt{11}$
    B) Center: $(-3,-3)$
       Radius: $\sqrt{11}$
    C) Center: $(3,3)$
       Radius: $\sqrt{23}$
    D) Center: $(-3,-3)$
       Radius: $\sqrt{23}$
    E) Center: $(-6,-6)$
       Radius: $\sqrt{77}$

13. Find the standard form of the equation of the ellipse with the following characteristics.
    foci: $(\pm 8, 0)$ major axis of length: 22
    A) $\frac{x^2}{121} + \frac{y^2}{57} = 1$
    B) $\frac{x^2}{121} + \frac{y^2}{64} = 1$
    C) $\frac{x^2}{64} + \frac{y^2}{121} = 1$
    D) $\frac{x^2}{484} + \frac{y^2}{64} = 1$
    E) $\frac{x^2}{484} + \frac{y^2}{420} = 1$

14. Use the discriminant to classify the graph; then use the quadratic formula to solve for y.

$$3x^2 - 2\sqrt{3}xy + y^2 - 18\sqrt{3}x + 30y + 36 = 0$$

A) ellipse; $y = -15 + \sqrt{3}x \pm \sqrt{12\sqrt{3}x + 189}$

B) ellipse; $y = -15 + \sqrt{3}x \pm \sqrt{-12\sqrt{3}x + 189}$

C) parabola; $y = -15 + \sqrt{3}x \pm \sqrt{-12\sqrt{3}x + 189}$

D) parabola; $y = \sqrt{3}x \pm \sqrt{-12\sqrt{3}x + 189}$

E) hyperbola; $y = \sqrt{3}x \pm \sqrt{-12\sqrt{3}x + 189}$

15. Find the graph of the following polar equation.

$$r = 3 + 2\cos(\theta)$$

A)

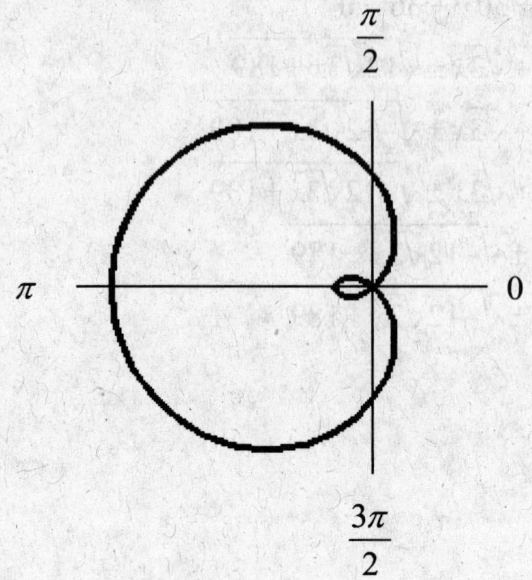

B)

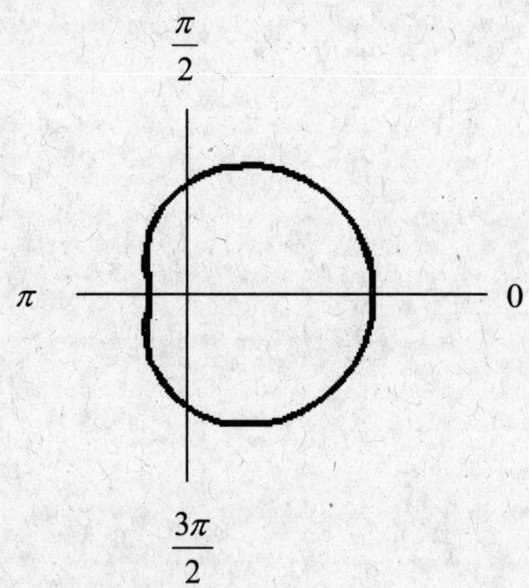

C)

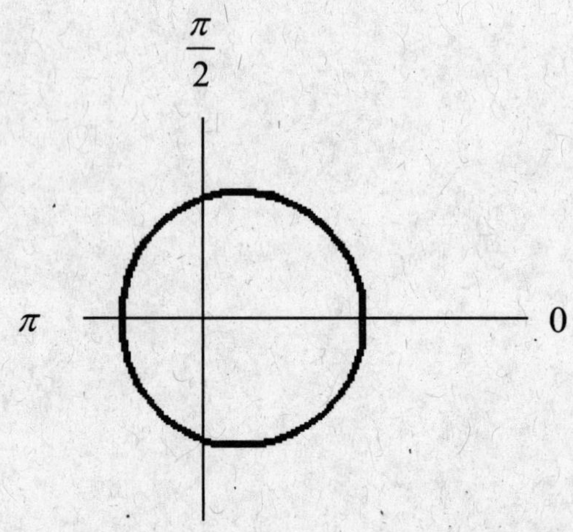

$\frac{3\pi}{2}$

D) 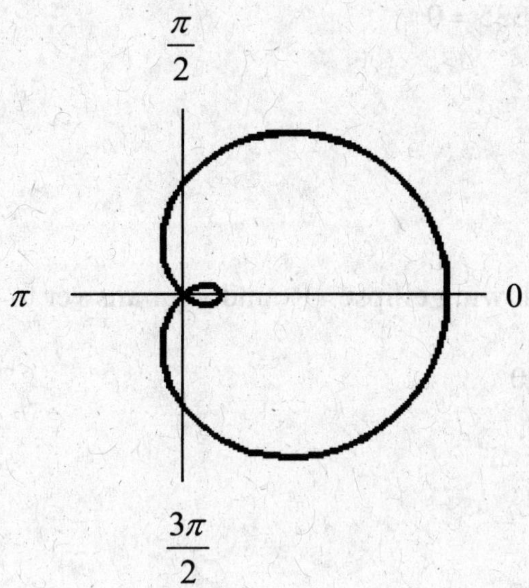

E) 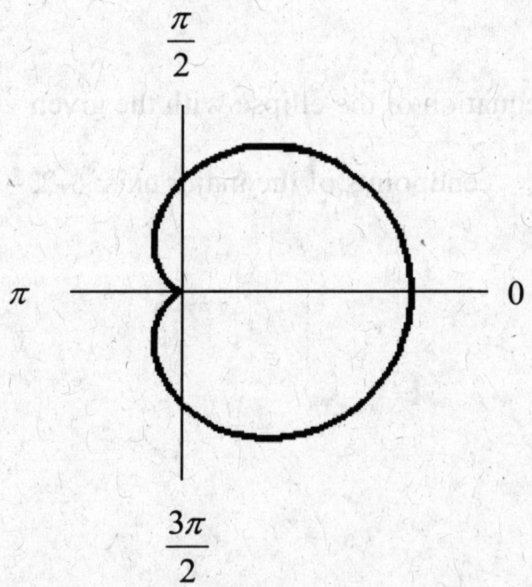

16. Classify the graph of the equation below as a circle, a parabola, an ellipse, or a hyperbola.

    $36x^2+16y^2+144x+20y-335=0$

    A) hyperbola
    B) parabola
    C) ellipse
    D) circle

17. Find the eccentricity of the following ellipse. Round your answer to two decimals.

    $2x^2+5y^2-4x+10y-10=0$

    A) 1.50
    B) 1.07
    C) 60.69
    D) 0.77
    E) 0.63

18. Find the standard form of the equation of the ellipse with the given characteristics.

    foci: (–2, 6), (–2, 10) endpoints of the major axis: (–2, –1), (–2, 17)

    A) $\frac{(x-2)^2}{77}+\frac{(y+8)^2}{81}=1$
    B) $\frac{(x+2)^2}{77}+\frac{(y-8)^2}{81}=1$
    C) $\frac{(x+2)^2}{81}+\frac{(y-8)^2}{77}=1$
    D) $\frac{(x-8)^2}{81}+\frac{(y+2)^2}{77}=1$
    E) $\frac{(x+8)^2}{81}+\frac{(y-2)^2}{77}=1$

19. Find the standard form of the equation of the hyperbola with the given characteristics.

    foci: $(\pm 3, 0)$ asymptotes: $y = \pm 4x$

    A) $\frac{x^2}{9} - \frac{y^2}{16} = 1$
    B) $\frac{y^2}{9} - \frac{x^2}{16} = 1$
    C) $\frac{y^2}{\frac{9}{17}} - \frac{x^2}{\frac{144}{17}} = 1$
    D) $\frac{x^2}{\frac{9}{17}} - \frac{y^2}{\frac{144}{17}} = 1$
    E) $\frac{x^2}{\frac{144}{17}} - \frac{y^2}{\frac{9}{17}} = 1$

20. Classify the graph of the equation as a circle, a parabola, an ellipse, or a hyperbola.

    $4x^2 - 3y^2 - 5x + 7y + 1 = 0$

    A) ellipse
    B) parabola
    C) hyperbola
    D) circle

Answer Key

1. C
2. B
3. A
4. A
5. C
6. B
7.

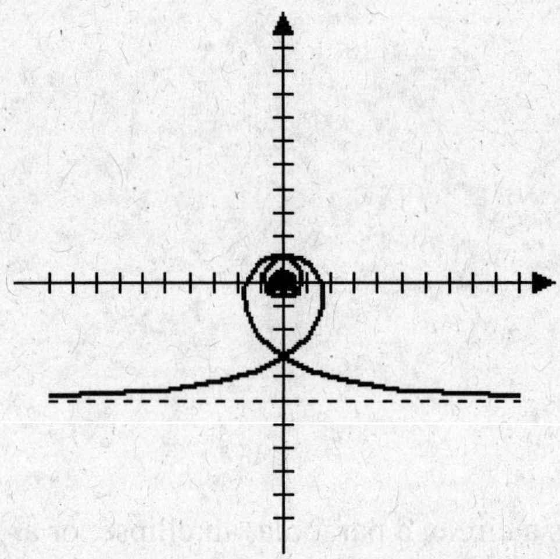

8. E
9. C
10. B
11. B
12. D
13. A
14. C
15. B
16. C
17. D
18. B
19. D
20. C

Name: ______________________________ Date: ____________

1. Find the standard form of the parabola with the given characteristic and vertex at the origin.
   vertical axis and passes through point $(10,\ -10)$

   A) $y^2 = -10x$
   B) $x^2 = -10y$
   C) $x^2 = \frac{5}{2}y$
   D) $y^2 = \frac{5}{2}x$
   E) $y^2 = \frac{5}{2}x^2$

2. Use the Quadratic Formula to solve for y in the following equation.
   $81x^2 - 108xy + 36y^2 - 40x - 50y = 0$

   A) $\dfrac{-54x - 25 \pm \sqrt{-108x + 2065}}{36}$
   B) $\dfrac{2(54x + 25) \pm \sqrt{-108x + 8260}}{36}$
   C) $\dfrac{54x + 25 \pm \sqrt{4140x + 625}}{36}$
   D) $\dfrac{108x + 50 \pm \sqrt{-54x + 3505}}{72}$
   E) $\dfrac{-108x - 50 \pm \sqrt{-54x + 3940}}{72}$

3. Test the graph of the following equation for symmetry with respect to $\theta = \frac{\pi}{2}$, the polar axis, and the pole.
   $r = -2 + \sin(2\theta)$

   A) The graph is symmetric with respect to $\theta = \frac{\pi}{2}$ only.
   B) The graph is symmetric with respect to the pole only.
   C) The graph is symmetric with respect to all three.
   D) The graph has no symmetries.
   E) The graph is symmetric with respect to the polar axis only.

4. Use a graphing utility to graph the rotated conic.

$$r = \frac{1}{1 + 2\cos(\theta + \pi/4)}$$

Use either grid below for your graph, whichever is more convenient.

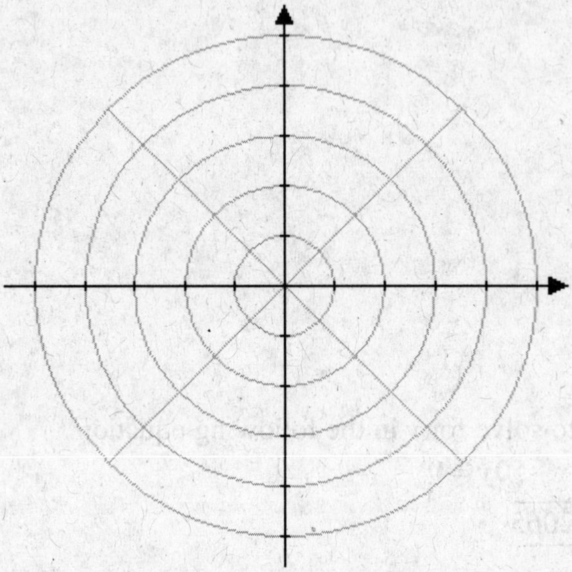

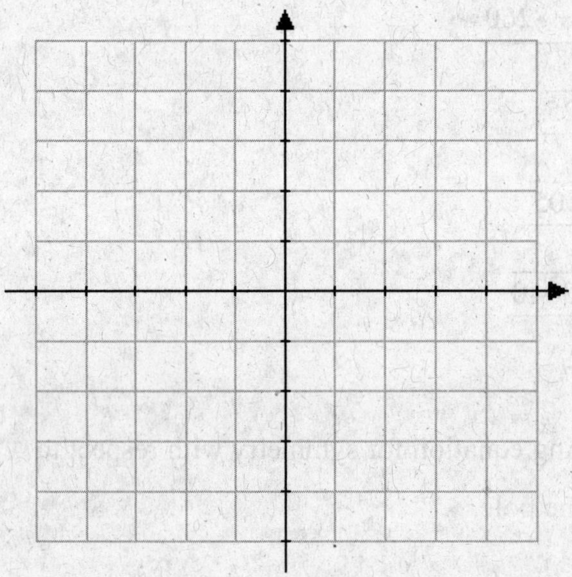

5. Find the standard form of the equation of the following hyperbola.

$$10x^2 - 6y^2 = 60$$

A) $\dfrac{x^2}{\left(\sqrt{6}\right)^2} - \dfrac{y^2}{\left(\sqrt{10}\right)^2} = 1$

B) $\dfrac{x^2}{10^2} - \dfrac{y^2}{6^2} = 1$

C) $\dfrac{y^2}{\left(\sqrt{10}\right)^2} - \dfrac{x^2}{\left(\sqrt{6}\right)^2} = 1$

D) $\dfrac{y^2}{\left(\sqrt{6}\right)^2} - \dfrac{x^2}{\left(\sqrt{10}\right)^2} = 1$

E) $\dfrac{x^2}{\left(\sqrt{10}\right)^2} - \dfrac{y^2}{\left(\sqrt{6}\right)^2} = 1$

6. Find the standard form of the parabola with the given characteristics.

focus: $(-9, 7)$ vertex: $(-9, 6)$

A) $(x + 9)^2 = 4(y - 6)$
B) $(x - 9)^2 = 4(y + 6)$
C) $(x - 6)^2 = 4(y + 9)$
D) $(x + 6)^2 = 4(y - 9)$
E) $(x - 9)^2 = (y + 6)$

7. Find the graph of the following polar equation.

$r = -3\cos(3\theta)$

A)

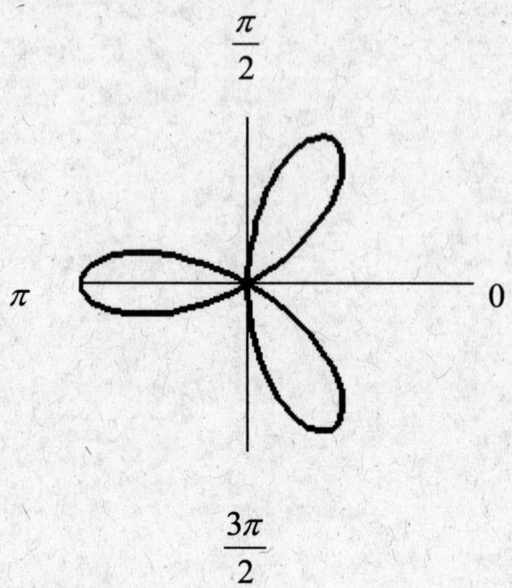

B)

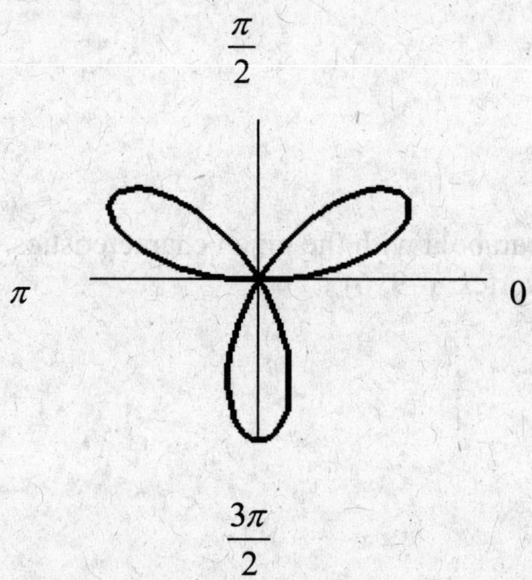

C)

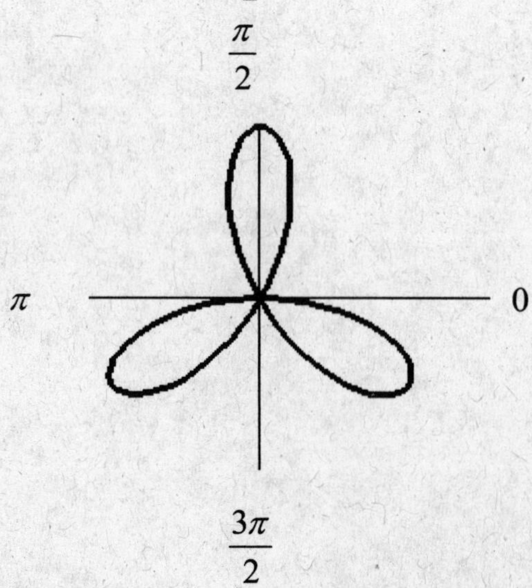

D)

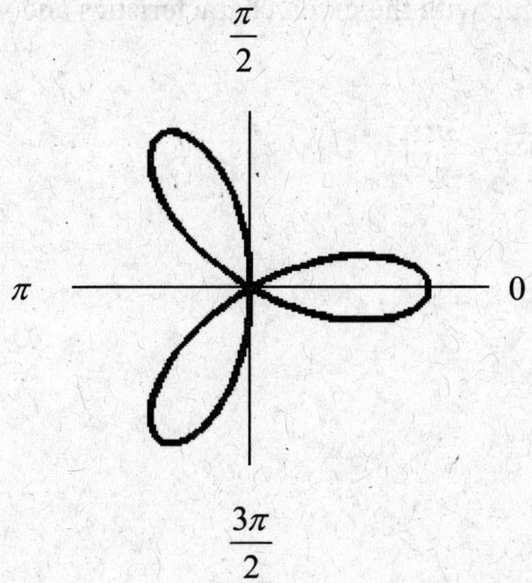

E)

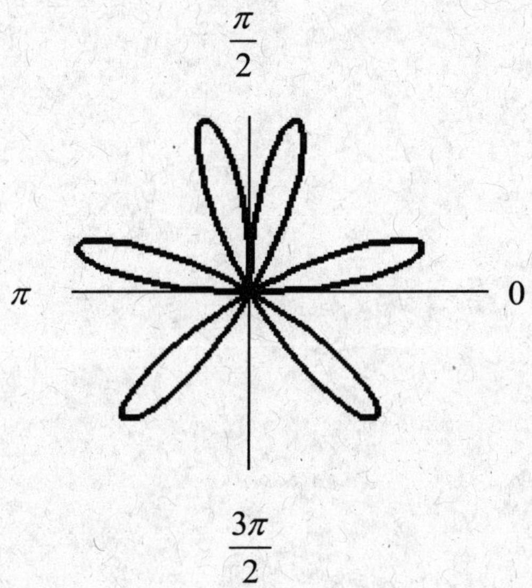

8. Find a polar equation of the conic with the given characteristics and with one focus at the pole.

Conic	*Vertices*
Hyperbola	$\left(\frac{2}{3}, \frac{\pi}{2}\right), \left(-2, \frac{3\pi}{2}\right)$

A) $r = \dfrac{2}{1-2\sin\theta}$

B) $r = \dfrac{2}{1+2\sin\theta}$

C) $r = \dfrac{2}{1+2\cos\theta}$

D) $r = \dfrac{2}{1-2\cos\theta}$

E) $r = \dfrac{2}{1+1\cos\theta}$

9. Sketch the graph of the polar equation using symmetry, zeros, maximum r-values, and any other additional points.

   $r^2 = 4\sin 2\theta$

   Use either grid below for your graph, whichever is more convenient.

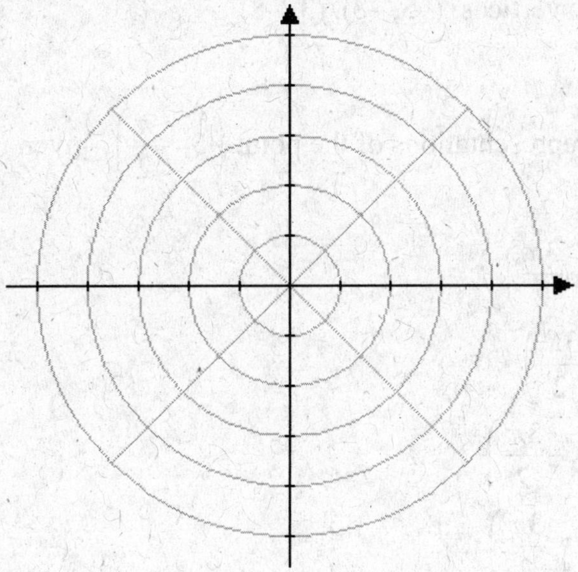

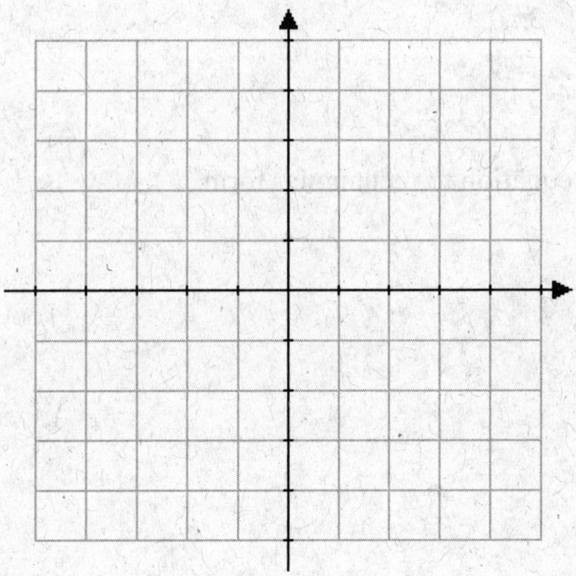

10. Find the center and vertices of the ellipse.

$$x^2 + 16y^2 - 2x - 256y + 1009 = 0$$

A) center: (8, 1) vertices: (4, 1), (12, 1)
B) center: (–1, –8) vertices: (–2, –8), (0, –8)
C) center: (1, 8) vertices: (0, 8), (2, 8)
D) center: (1, 8) vertices: (–3, 8), (5, 8)
E) center: (–1, –8) vertices: (–5, –8), (3, –8)

11. Find three additional polar representations of the point $\left(5, -\frac{\pi}{2}\right)$, given in polar coordinates, using $-2\pi < \theta < 2\pi$.

A) $\left(5, \frac{3\pi}{2}\right), \left(5, \frac{\pi}{2}\right), \left(-5, -\frac{3\pi}{2}\right)$
B) $\left(5, \frac{3\pi}{2}\right), \left(5, \frac{\pi}{2}\right), \left(5, -\frac{3\pi}{2}\right)$
C) $\left(5, \frac{3\pi}{2}\right), \left(-5, \frac{\pi}{2}\right), \left(-5, -\frac{3\pi}{2}\right)$
D) $\left(-5, \frac{3\pi}{2}\right), \left(-5, \frac{\pi}{2}\right), \left(5, -\frac{3\pi}{2}\right)$
E) $\left(-5, \frac{3\pi}{2}\right), \left(5, \frac{\pi}{2}\right), \left(-5, -\frac{3\pi}{2}\right)$

12. Convert the following polar equation to rectangular form.

$$\theta = \frac{2\pi}{3}$$

A) $y = -\sqrt{3}$
B) $y = -\frac{\sqrt{3}}{3}$
C) $y = \sqrt{3}$
D) $y = -3$
E) $y = \frac{\sqrt{3}}{3}$

13. Rotate the axes to eliminate the *xy*-term in the equation below and then write the equation in standard form.

$xy + 1 = 0$

A) $\dfrac{(x')^2}{(\sqrt{2})^2} - \dfrac{(y')^2}{(\sqrt{2})^2} = 1$

B) $\dfrac{(x')^2}{(\sqrt{2})^2} + \dfrac{(y')^2}{(\sqrt{2})^2} = 1$

C) $\dfrac{(x')^2}{2} + \dfrac{(y')^2}{2} = 1$

D) $\dfrac{(x')^2}{2} - \dfrac{(y')^2}{2} = 1$

E) $\dfrac{(y')^2}{(\sqrt{2})^2} - \dfrac{(x')^2}{(\sqrt{2})^2} = 1$

14. Find any zeros of r on the interval $0 \le \theta < 2\pi$.

$r = \sqrt{2} - 2\cos\theta$

A) zeros: $\dfrac{3\pi}{4}, \dfrac{5\pi}{4}$

B) zeros: $\dfrac{\pi}{4}, \dfrac{7\pi}{4}$

C) zeros: $\dfrac{\pi}{2}, \dfrac{7\pi}{2}$

D) zeros: $\dfrac{3\pi}{2}, \dfrac{5\pi}{2}$

E) zeros: $\dfrac{\pi}{8}, \dfrac{7\pi}{8}$

15. Classify the graph of the equation below as a circle, a parabola, an ellipse, or a hyperbola.

$3y^2 + 19x^2 + y - 115 = 0$

A) hyperbola
B) ellipse
C) circle
D) parabola

16. Solve the following system of quadratic equations algebraically by the method of substitution.

$$\begin{cases} -4x^2 - 3y^2 + 48 = 0 \\ -4x - 2y = 0 \end{cases}$$

A) $(\sqrt{3},-2\sqrt{3})$, $(\sqrt{5},-2\sqrt{5})$
B) $(-\sqrt{5},2\sqrt{5})$, $(\sqrt{5},-2\sqrt{5})$
C) $(-\sqrt{5},-2\sqrt{5})$, $(\sqrt{5},-2\sqrt{5})$
D) $(-\sqrt{3},-2\sqrt{3})$, $(\sqrt{3},2\sqrt{3})$
E) $(-\sqrt{3},2\sqrt{3})$, $(\sqrt{3},-2\sqrt{3})$

17. Which answer is a polar form of the given rectangular equation?

$$16xy = 144$$

A) $r^2 = 9\sec\theta\csc\theta$
B) $r^2 = 3\sec\theta\csc\theta$
C) $r^2 = 3\sin\theta\cos\theta$
D) $r^2 = 9\sin\theta\cos\theta$
E) $\theta = 9\sin\theta\cos\theta$

18. The graph of the following polar equation is a circle. Sketch its graph, and find its radius.

$r = 5\sin(\theta)$

A)

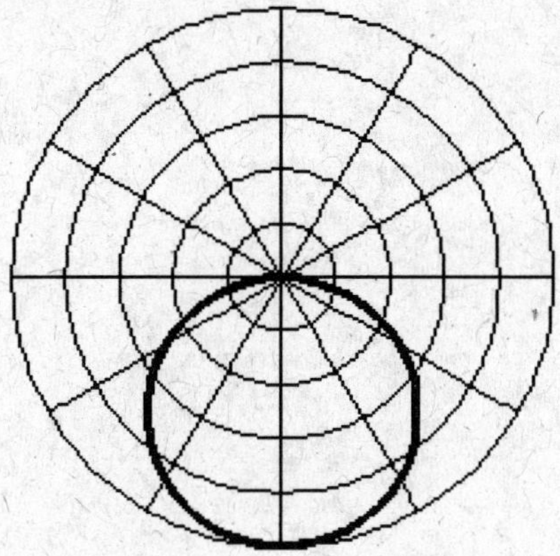

radius: 5

B)

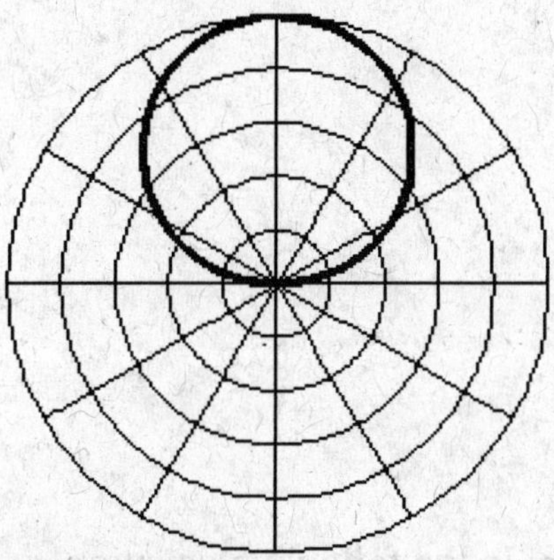

radius: 2.5

C)

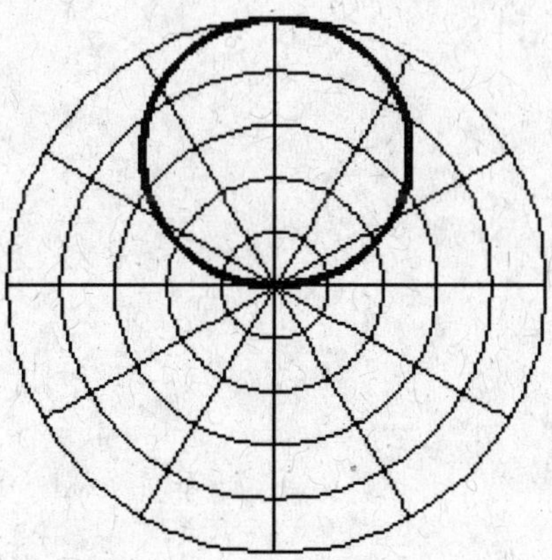

radius: 5

D)

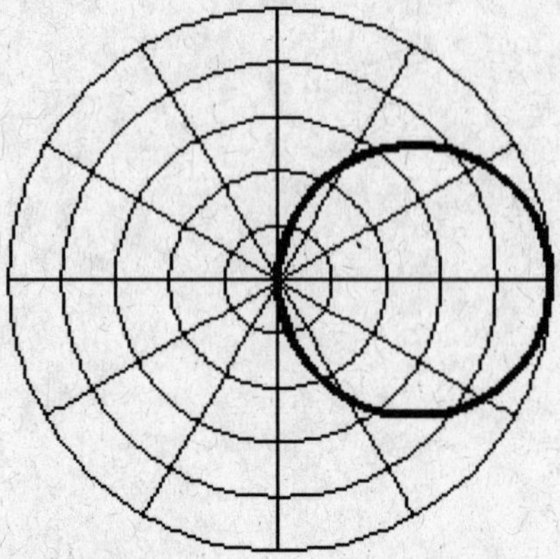

radius: 5

E)

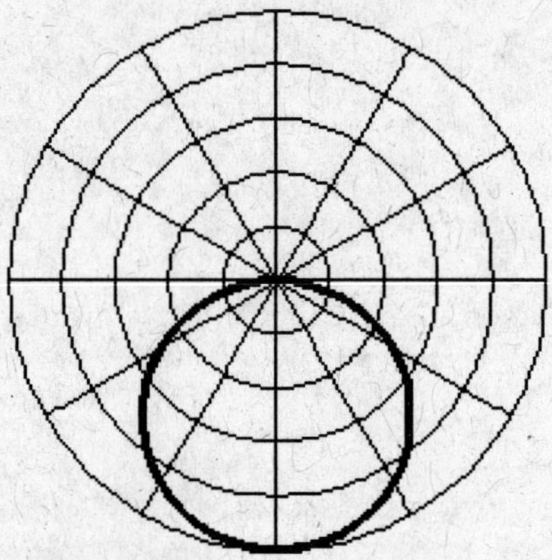

radius: 25

19. Rotate the axes to eliminate the *xy*-term in the following equation.

$$-30x^2+20\sqrt{3}xy-10y^2-20x-20\sqrt{3}y=0$$

A) $(y')^2+x'=0$

B) $-10\sqrt{3}(y')^2-10\sqrt{3}x'=0$

C) $-10(x')^2+10y'=0$

D) $-\dfrac{(y')^2}{10}-\dfrac{x'}{10}=0$

E) $-\dfrac{(x')^2}{20}-\dfrac{y'}{20}=0$

20. Which answer is a rectangular form of the given polar equation?

$$r = \frac{5}{8 + \sin\theta}$$

A) $63x^2 - 64y^2 - 10x + 25 = 0$

B) $65x^2 + 64y^2 - 10x + 25 = 0$

C) $63x^2 + 64y^2 + 10x - 25 = 0$

D) $64x^2 + 63y^2 + 10y - 25 = 0$

E) $64x^2 + 65y^2 - 10y + 25 = 0$

Answer Key

1. B
2. C
3. B
4.

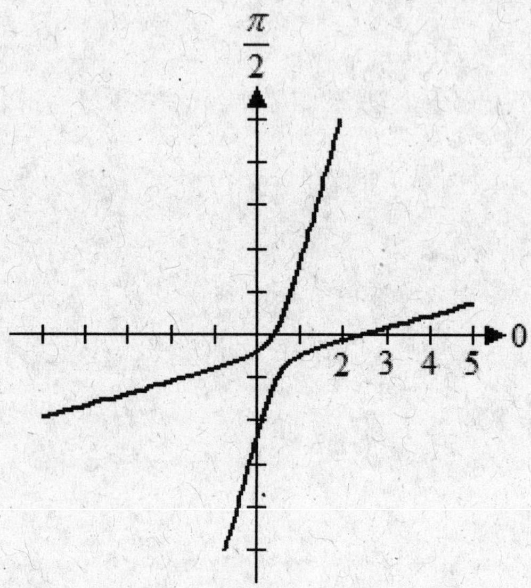

5. A
6. A
7. A
8. B
9.

symmetry: with respect to pole

maximum value of $|r|$: $|r| = 2$ when $\theta = \frac{5\pi}{4}$

zeros of r: $r = 0$ when $\theta = 0, \frac{\pi}{2}, \pi, \frac{3\pi}{2}$

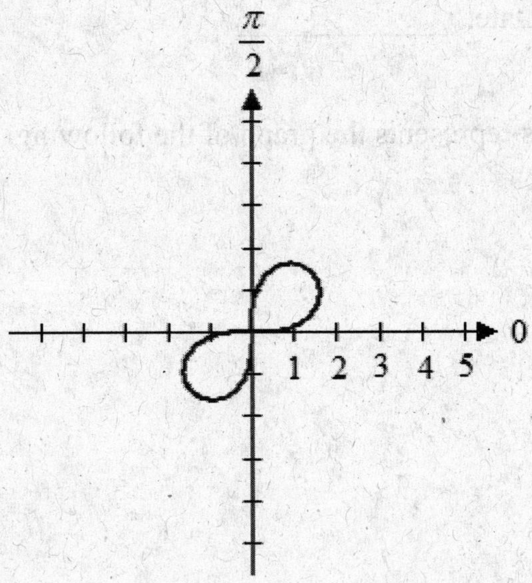

10. D
11. C
12. A
13. E
14. B
15. B
16. E
17. A
18. B
19. A
20. D

Name: ____________________________ Date: ______________

1. Which set of parametric equations represents the graph of the following rectangular equation using $t = 6 - x$?

$$y = x^2 + 2$$

A) $x = t + 6$
$y = (t+6)^2 - 2$

B) $x = t - 6$
$y = (t-6)^2 - 2$

C) $x = t - 6$
$y = (t-6)^2 + 2$

D) $x = 6 + t$
$y = (6+t)^2 + 2$

E) $x = 6 - t$
$y = (6-t)^2 + 2$

2. Which set of parametric equations represents the following line or conic? Use $x = x_1 + t(x_2 - x_1)$ and $y = y_1 + t(y_2 - y_1)$.

Line: passes through $(7, 6)$ and $(-4, 3)$

A) $x = -11t + 7$
$y = -3t + 6$

B) $x = -11t + 7$
$y = 3t + 6$

C) $x = -11t - 7$
$y = -3t - 6$

D) $x = -3t + 6$
$y = -11t + 7$

E) $x = -3t - 6$
$y = -11t - 7$

3. Use a graphing utility to graph the rotated conic.

$$r = \frac{1}{3 - 3\sin(\theta - 2\pi/3)}$$

Use either grid below for your graph, whichever is more convenient.

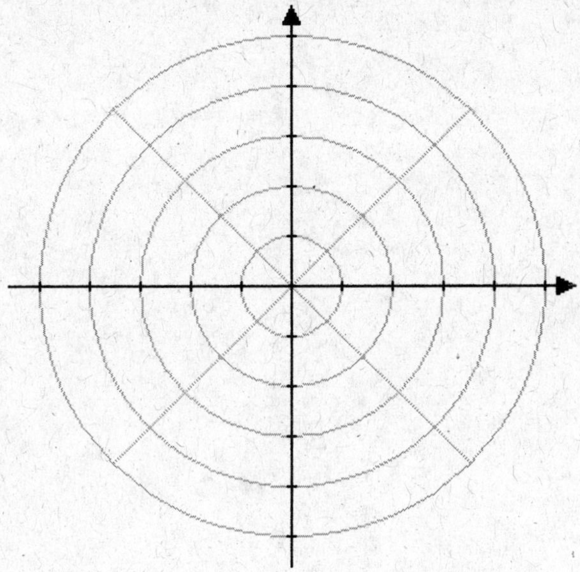

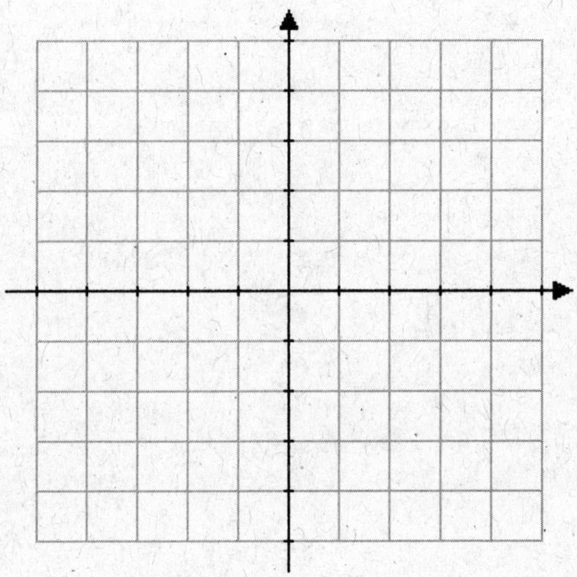

4. Find a polar equation of the conic with the given characteristics and with one focus at the pole.

Conic	*Eccentricity*	*Directrix*
Parabola	$e = 1$	$x = -4$

A) $r = \dfrac{4}{1 + \sin\theta}$

B) $r = \dfrac{4}{1 - \sin\theta}$

C) $r = \dfrac{4}{1 - \cos\theta}$

D) $r = \dfrac{4}{1 + \cos\theta}$

E) $r = \dfrac{1}{1 - 4\cos\theta}$

5. Sketch the curve represented by the following parametric equations. Indicate the orientation of the curve.

   $x = e^t, y = e^{-2t}$

   A)

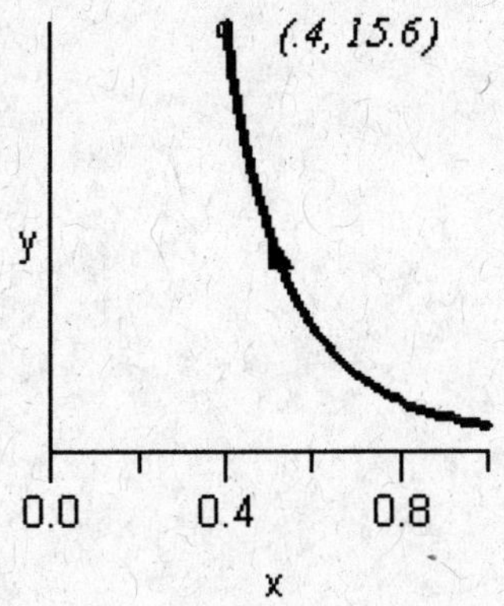

   B)

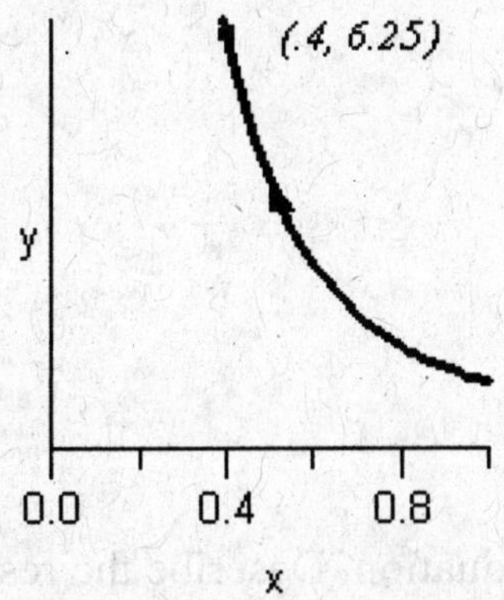

   C)

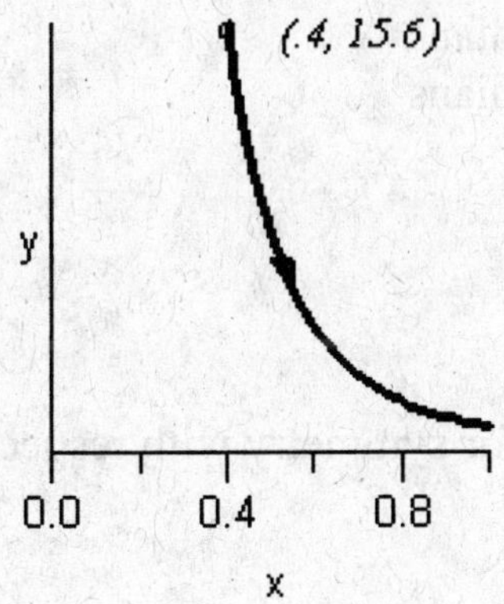

D)

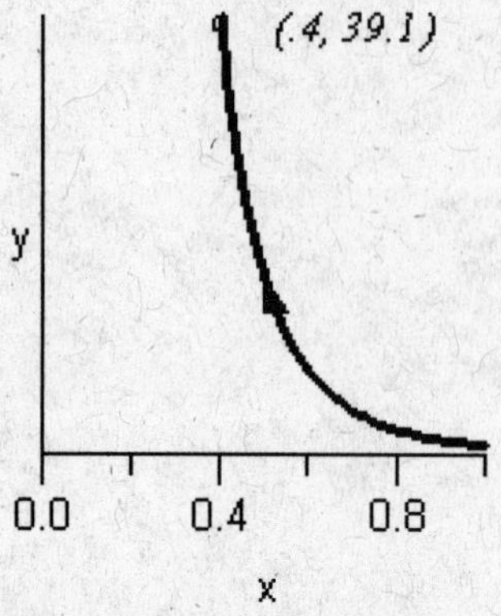

E)

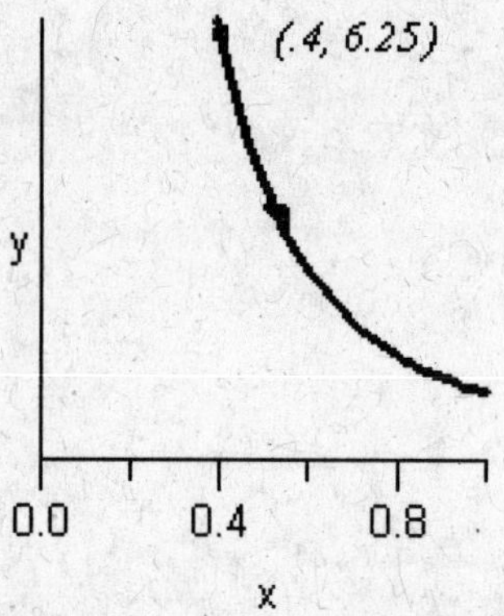

6. Use a graphing utility to graph the polar equation. Describe the resulting graph.

$$r = 4\cos(7\theta - 2)$$

A) rose curve with 7 petals, rotated 2 radians
B) rose curve with 7 petals, rotated 4 radians
C) lemniscate rotated 4 radians
D) limaçon rotated 4 radians
E) cardioid rotated 2 radians

7. Test the graph of the following equation for symmetry with respect to $\theta = \frac{\pi}{2}$, the polar axis, and the pole.

$$r = 2 - 2\sin(4\theta)$$

A) The graph is symmetric with respect to $\theta = \frac{\pi}{2}$ only.
B) The graph is symmetric with respect to the pole only.
C) The graph is symmetric with respect to all three.
D) The graph has no symmetries.
E) The graph is symmetric with respect to the polar axis only.

8. Find the y-intercepts of the graph of the circle below.

$$(x-5)^2+(y+2)^2=36$$

A) $5+4\sqrt{2},\ 5-4\sqrt{2}$
B) $5+2\sqrt{10},\ 5-2\sqrt{10}$
C) $2+\sqrt{61},\ 2-\sqrt{61}$
D) $-2+\sqrt{11},\ -2-\sqrt{11}$
E) $-2+\sqrt{61},\ -2-\sqrt{61}$

9. Find the standard form of the equation of the hyperbola with the given characteristics.

foci: $(\pm 5, 0)$ asymptotes: $y=\pm 3x$

A) $\dfrac{x^2}{25}-\dfrac{y^2}{9}=1$
B) $\dfrac{y^2}{25}-\dfrac{x^2}{9}=1$
C) $\dfrac{y^2}{\frac{5}{2}}-\dfrac{x^2}{\frac{45}{2}}=1$
D) $\dfrac{x^2}{\frac{5}{2}}-\dfrac{y^2}{\frac{45}{2}}=1$
E) $\dfrac{x^2}{\frac{45}{2}}-\dfrac{y^2}{\frac{5}{2}}=1$

10. Find the center, vertices, and foci of the ellipse below.
$x^2 + 8y^2 = 80$

A) Center: $(0,0)$
Vertices: $(4\sqrt{5},0), (-4\sqrt{5},0)$
Foci: $(\sqrt{70},0),\ (-\sqrt{70},0)$

B) Center: $(0,0)$
Vertices: $(0,10),\ (0,-10)$
Foci: $(0,24\sqrt{11}),\ (0,-24\sqrt{11})$

C) Center: $(80,10)$
Vertices: $(160,10),\ (-160,10)$
Foci: $(80+24\sqrt{11},10),\ (80-24\sqrt{11},10)$

D) Center: $(80,10)$
Vertices: $(80,20),\ (80,-20)$
Foci: $(80,10+24\sqrt{11}),\ (80,10-24\sqrt{11})$

E) Center: $(0,0)$
Vertices: $(80,0),\ (-80,0)$
Foci: $(6336,0),\ (-6336,0)$

11. Find the standard form of the equation of the hyperbola below.
$8x^2 - y^2 - 80x - 16y + 96 = 0$

A) $\dfrac{(x-5)^2}{(\sqrt{5})^2} - \dfrac{(y+8)^2}{(2\sqrt{10})^2} = 1$

B) $\dfrac{(x+5)^2}{(\sqrt{5})^2} + \dfrac{(y-8)^2}{(\sqrt{5})^2} = 1$

C) $\dfrac{(y+5)^2}{(2\sqrt{10})^2} + \dfrac{(x-8)^2}{(\sqrt{5})^2} = 1$

D) $\dfrac{(y+5)^2}{(\sqrt{5})^2} - \dfrac{(x-8)^2}{(2\sqrt{10})^2} = 1$

E) $\dfrac{(y-5)^2}{(\sqrt{5})^2} - \dfrac{(x+8)^2}{(2\sqrt{2})^2} = 1$

12. Find the standard form of the equation of the ellipse centered at the origin having vertices at $(-4,0)$ and $(4,0)$ and foci at $(-1,0)$ and $(1,0)$.

A) $\dfrac{x^2}{(\sqrt{15})^2} + \dfrac{y^2}{4^2} = 1$

B) $\dfrac{x^2}{4} + \dfrac{y^2}{\sqrt{15}} = 1$

C) $\dfrac{x^2}{4} + \dfrac{y^2}{\sqrt{17}} = 1$

D) $\dfrac{x^2}{\sqrt{15}} + \dfrac{y^2}{4} = 1$

E) $\dfrac{x^2}{4^2} + \dfrac{y^2}{(\sqrt{15})^2} = 1$

13. Find the standard form of the parabola with the given characteristic and vertex at the origin.
    focus: (0, –5)
    A) $x^2 = -20y$
    B) $x^2 = -5y$
    C) $x^2 = 5y$
    D) $y^2 = -20x$
    E) $y^2 = -5x$

14. Test for symmetry with respect to $\theta = \pi/2$, the polar axis, and the pole.

$$r = \frac{6}{5 + \sin\theta}$$

    A) symmetric with respect to the pole
    B) symmetric with respect to the polar axis
    C) symmetric with respect to $\theta = \pi/2$
    D) symmetric with respect to the pole and the polar axis
    E) symmetric with respect to the pole and $\theta = \pi/2$

15. Find the vertex and focus of the parabola below.

$y^2+10y-8x+30=0$

A) Vertex: $\left(-\frac{5}{8},5\right)$

Focus: $\left(\frac{11}{8},5\right)$

B) Vertex: $\left(\frac{5}{8},-5\right)$

Focus: $\left(\frac{21}{8},-5\right)$

C) Vertex: $\left(\frac{5}{8},-5\right)$

Focus: $\left(\frac{69}{8},-5\right)$

D) Vertex: $\left(-\frac{5}{8},5\right)$

Focus: $\left(\frac{59}{8},5\right)$

E) Vertex: $\left(-\frac{5}{8},-5\right)$

Focus: $\left(\frac{59}{8},-5\right)$

16. Rotate the axes to eliminate the *xy*-term in the equation. Then write the equation in standard form.

$$113x^2 - 30xy + 113y^2 - 6272 = 0$$

A) $\frac{(x')^2}{81} + \frac{(y')^2}{64} = 1$

B) $\frac{(x')^2}{49} + \frac{(y')^2}{36} = 1$

C) $\frac{(x')^2}{64} + \frac{(y')^2}{64} = 1$

D) $\frac{(x')^2}{49} + \frac{(y')^2}{49} = 1$

E) $\frac{(x')^2}{64} + \frac{(y')^2}{49} = 1$

17. Identify the center and radius of the circle below.

$$x^2 + y^2 + 10x + 10y - 5 = 0$$

A) Center: $(5, 5)$
Radius: $\sqrt{15}$

B) Center: $(-5, -5)$
Radius: $\sqrt{15}$

C) Center: $(5, 5)$
Radius: $\sqrt{55}$

D) Center: $(-5, -5)$
Radius: $\sqrt{55}$

E) Center: $(-10, -10)$
Radius: $\sqrt{205}$

18. Identify the center and radius of the circle below.
$$(x+1)^2+(y-8)^2=9$$
A) Center: $(-1,8)$
Radius: 3
B) Center: $(-1,8)$
Radius: 9
C) Center: $(1,-8)$
Radius: 3
D) Center: $(1,-8)$
Radius: 9
E) Center: $(8,-1)$
Radius: 3

19. Which answer is a polar form of the given rectangular equation?
$$25xy=225$$
A) $r^2=9\sec\theta\csc\theta$
B) $r^2=3\sec\theta\csc\theta$
C) $r^2=3\sin\theta\cos\theta$
D) $r^2=9\sin\theta\cos\theta$
E) $\theta=9\sin\theta\cos\theta$

20. Find the standard form of the parabola with the given characteristics.
directrix: $x=-1$ vertex: $(7,-1)$
A) $(x+7)^2=32(y-1)$
B) $(x-7)^2=32(y+1)$
C) $(x-1)^2=32(y+7)$
D) $(y-1)^2=32(x+7)$
E) $(y+1)^2=32(x-7)$

Answer Key

1. E
2. A
3. 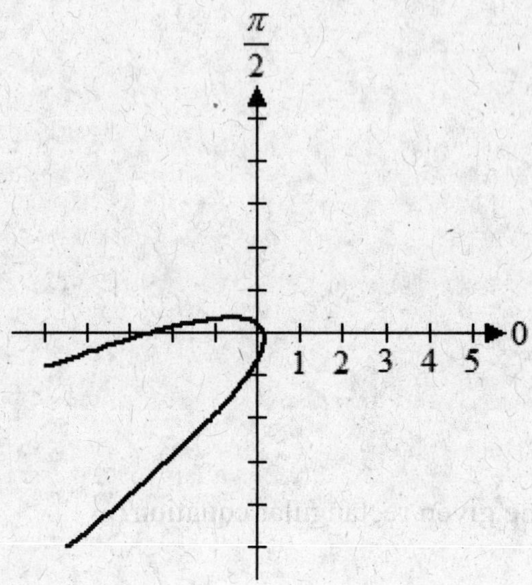

4. C
5. E
6. A
7. B
8. D
9. D
10. A
11. A
12. E
13. A
14. C
15. B
16. E
17. D
18. A
19. A
20. E

Name: ____________________________ Date: _____________

1. Find the y-intercepts of the graph of the circle below.

   $(x-3)^2+(y+2)^2=36$

   A) $3+4\sqrt{2},\ 3-4\sqrt{2}$
   B) $3+2\sqrt{10},\ 3-2\sqrt{10}$
   C) $2+3\sqrt{5},\ 2-3\sqrt{5}$
   D) $-2+3\sqrt{3},\ -2-3\sqrt{3}$
   E) $-2+3\sqrt{5},\ -2-3\sqrt{5}$

2. Find the center and foci of the hyperbola.

   $$\frac{(y+1)^2}{25}-\frac{(x+2)^2}{24}=1$$

   A) center: (2, 1) foci: (2, –6), (2, 8)
   B) center: (–2, –1) foci: (–2, –8), (–2, 6)
   C) center: (–1, –2) foci: (–8, –2), (6, –2)
   D) center: (–2, –1) foci: (–9, –1), (5, –1)
   E) center: (1, 2) foci: (1, –5), (1, 9)

3. Classify the graph of the equation below as a circle, a parabola, an ellipse, or a hyperbola.

   $3y^2+17x^2+y-103=0$

   A) hyperbola
   B) ellipse
   C) circle
   D) parabola

4. Rotate the axes to eliminate the xy-term in the equation. Then write the equation in standard form.

$$41x^2 + 18xy + 41y^2 - 800 = 0$$

A) $\frac{(x')^2}{25} + \frac{(y')^2}{36} = 1$

B) $\frac{(x')^2}{9} + \frac{(y')^2}{16} = 1$

C) $\frac{(x')^2}{16} + \frac{(y')^2}{36} = 1$

D) $\frac{(x')^2}{9} + \frac{(y')^2}{25} = 1$

E) $\frac{(x')^2}{16} + \frac{(y')^2}{25} = 1$

5. Sketch the graph of the ellipse, using the latera recta.

$$4x^2 + 9y^2 = 36$$

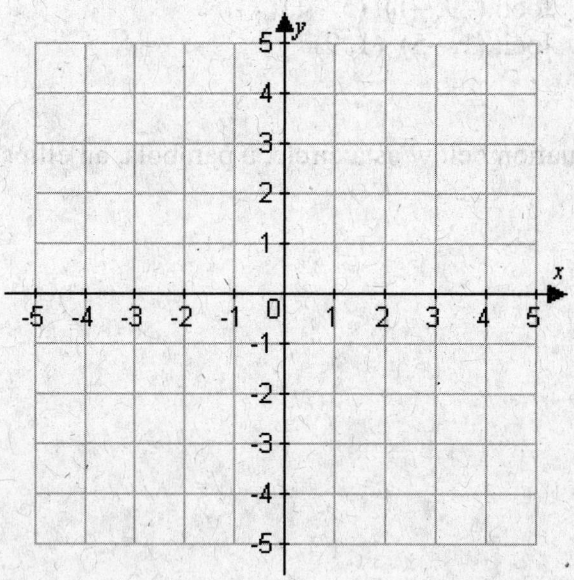

6. Consider the parametric equations $x = 5\sqrt{t}$ and $y = 2 + 7t$. Find the rectangular equation by eliminating the parameter.
   A) $y = \frac{7}{25}x^2 + \frac{2}{25}$
   B) $y = 175x^2 + 2$
   C) $y = \frac{1}{25}x^2 - \frac{2}{7}$
   D) $y = \frac{7}{25}x^2 + 2$
   E) $y = \frac{1}{175}x^2 + 14$

7. Classify the graph of the equation below as a circle, a parabola, an ellipse, or a hyperbola.
   $64x^2 + 64y^2 + 192x + 24y - 327 = 0$
   A) hyperbola
   B) ellipse
   C) circle
   D) parabola

8. Which set of parametric equations represents the following line or conic?
Use $x = h + a\sec\theta$ and $y = k + b\tan\theta$.

Hyperbola: vertices $(-10, -1), (2, -1)$

foci $(-11, -1), (3, -1)$

A) $x = 4 + \sqrt{13}\sec\theta$
$y = 1 + 6\tan\theta$

B) $x = -4 + \sqrt{13}\sec\theta$
$y = -1 + 6\tan\theta$

C) $x = -1 + \sqrt{13}\sec\theta$
$y = -4 + 6\tan\theta$

D) $x = -4 + 6\sec\theta$
$y = -1 + \sqrt{13}\tan\theta$

E) $x = -1 + 6\sec\theta$
$y = -4 + \sqrt{13}\tan\theta$

9. Which answer is a polar form of the given rectangular equation?
$9xy = 144$

A) $r^2 = 16\sec\theta\csc\theta$
B) $r^2 = 4\sec\theta\csc\theta$
C) $r^2 = 4\sin\theta\cos\theta$
D) $r^2 = 16\sin\theta\cos\theta$
E) $\theta = 16\sin\theta\cos\theta$

10. Match the graph with its equation.

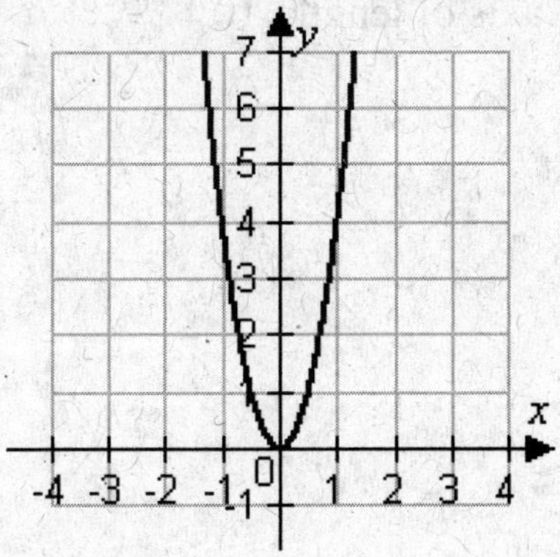

A) $x^2 = 4y$

B) $x^2 = \frac{y}{4}$

C) $y^2 = \frac{x}{4}$

D) $y^2 = 4x$

E) $(y+1)^2 = 4(x-2)$

11. Which answer is a polar form of the given rectangular equation?

$x^2 + y^2 = 25$

A) $r = 5$
B) $r = 25$
C) $r^2 + \theta^2 = 25$
D) $\theta = 25$
E) $\theta = 5$

12. Find the standard form of the equation of the ellipse with the given characteristics.
    vertices: (–6, –8), (–6, 6) minor axis of length: 10

    A) $\frac{(x-6)^2}{25}+\frac{(y-1)^2}{49}=1$

    B) $\frac{(x+6)^2}{25}+\frac{(y+1)^2}{49}=1$

    C) $\frac{(x+6)^2}{49}+\frac{(y+1)^2}{25}=1$

    D) $\frac{(x+1)^2}{49}+\frac{(y+6)^2}{25}=1$

    E) $\frac{(x-1)^2}{49}+\frac{(y-6)^2}{25}=1$

13. Find the standard form of the equation of the following hyperbola.
    $5x^2-3y^2=15$

    A) $\frac{x^2}{\left(\sqrt{3}\right)^2}-\frac{y^2}{\left(\sqrt{5}\right)^2}=1$

    B) $\frac{x^2}{5^2}-\frac{y^2}{3^2}=1$

    C) $\frac{y^2}{\left(\sqrt{5}\right)^2}-\frac{x^2}{\left(\sqrt{3}\right)^2}=1$

    D) $\frac{y^2}{\left(\sqrt{3}\right)^2}-\frac{x^2}{\left(\sqrt{5}\right)^2}=1$

    E) $\frac{x^2}{\left(\sqrt{5}\right)^2}-\frac{y^2}{\left(\sqrt{3}\right)^2}=1$

14. Find a polar equation of the conic with the given characteristics and with one focus at the pole.

Conic	*Eccentricity*	*Directrix*
Ellipse	$e=\dfrac{1}{2}$	$y=5$

A) $r=\dfrac{1}{2+\cos\theta}$

B) $r=\dfrac{5}{2+\cos\theta}$

C) $r=\dfrac{5}{2-\cos\theta}$

D) $r=\dfrac{5}{2-\sin\theta}$

E) $r=\dfrac{5}{2+\sin\theta}$

15. Solve the following system of quadratic equations algebraically by the method of substitution.

$$\begin{cases} 8x^2-6y^2+80=0 \\ 8x+4y=0 \end{cases}$$

A) $\left(\sqrt{5},-2\sqrt{5}\right)$, $\left(\sqrt{7},-2\sqrt{7}\right)$

B) $\left(-\sqrt{7},2\sqrt{7}\right)$, $\left(\sqrt{7},-2\sqrt{7}\right)$

C) $\left(-\sqrt{7},-2\sqrt{7}\right)$, $\left(\sqrt{7},-2\sqrt{7}\right)$

D) $\left(-\sqrt{5},-2\sqrt{5}\right)$, $\left(\sqrt{5},2\sqrt{5}\right)$

E) $\left(-\sqrt{5},2\sqrt{5}\right)$, $\left(\sqrt{5},-2\sqrt{5}\right)$

16. Find the graph of the following polar equation.

$r = 2\sin(2\theta)$

A)

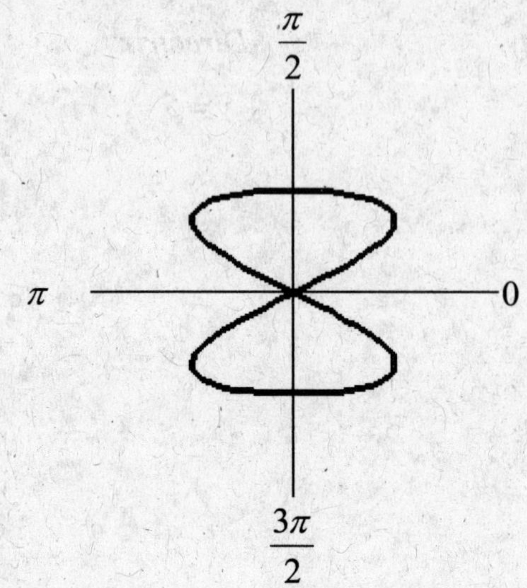

B)

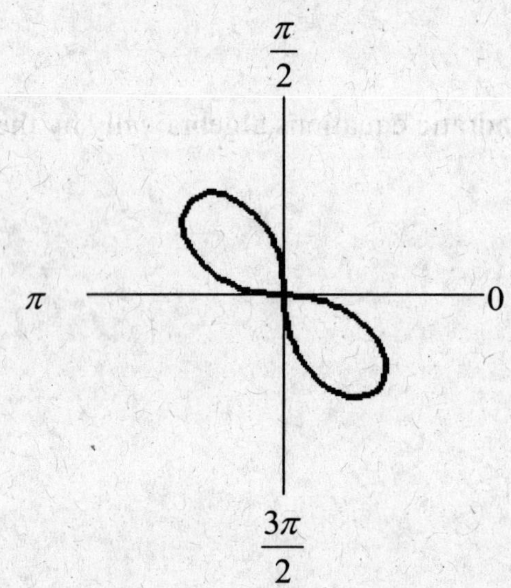

C)

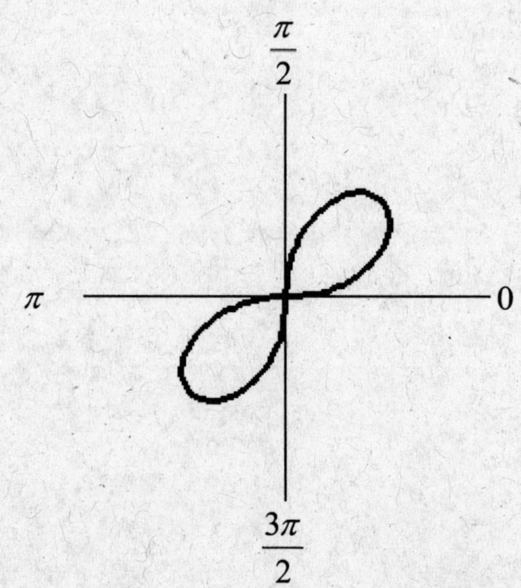

D)

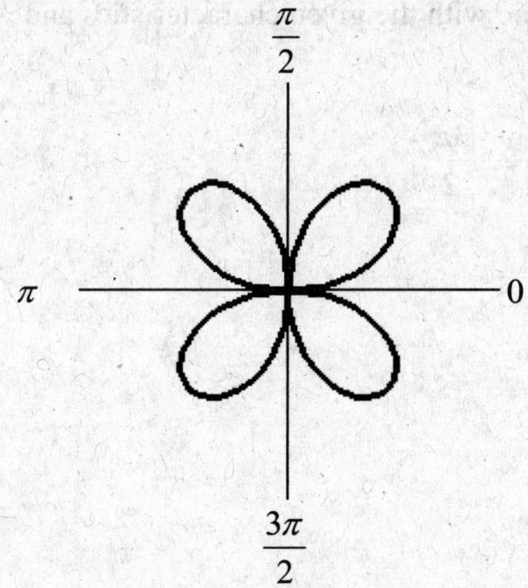

E)

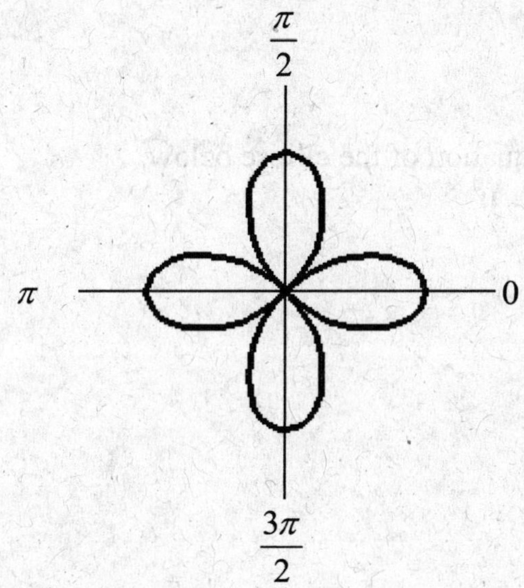

17. Find a polar equation of the conic with the given characteristics and with one focus at the pole.

Conic	*Vertices*
Hyperbola	$\left(\frac{2}{3}, \frac{\pi}{2}\right), \left(-2, \frac{3\pi}{2}\right)$

A) $r = \dfrac{2}{1-2\sin\theta}$

B) $r = \dfrac{2}{1+2\sin\theta}$

C) $r = \dfrac{2}{1+2\cos\theta}$

D) $r = \dfrac{2}{1-2\cos\theta}$

E) $r = \dfrac{2}{1+1\cos\theta}$

18. Find the standard form of the equation of the ellipse below.

$$2x^2 + 8y^2 + 24x - 64y + 16 = 0$$

A) $\dfrac{(x-6)^2}{\left(2\sqrt{23}\right)^2} + \dfrac{(y+4)^2}{\left(\sqrt{23}\right)^2} = 1$

B) $\dfrac{(x+6)^2}{\left(2\sqrt{23}\right)^2} + \dfrac{(y-4)^2}{\left(\sqrt{23}\right)^2} = 1$

C) $\dfrac{(x+4)^2}{\left(\sqrt{23}\right)^2} - \dfrac{(y-6)^2}{\left(2\sqrt{23}\right)^2} = 1$

D) $\dfrac{(x-4)^2}{\left(\sqrt{23}\right)^2} + \dfrac{(y+6)^2}{\left(2\sqrt{23}\right)^2} = 1$

E) $\dfrac{(x-4)^2}{\left(2\sqrt{23}\right)^2} - \dfrac{(y+6)^2}{\left(\sqrt{23}\right)^2} = 1$

19. Classify the graph of the equation as a circle, a parabola, an ellipse, or a hyperbola.

    $7x^2 + 7y^2 - 7x + 9y - 2 = 0$

    A) ellipse
    B) parabola
    C) circle
    D) hyperbola

20. Use the Quadratic Formula to solve for y in the following equation.

    $49x^2 - 42xy + 9y^2 - 90x - 10y = 0$

    A) $\dfrac{-21x - 5 \pm \sqrt{-42x + 835}}{9}$
    B) $\dfrac{2(21x + 5) \pm \sqrt{-42x + 3340}}{9}$
    C) $\dfrac{21x + 5 \pm \sqrt{1020x + 25}}{9}$
    D) $\dfrac{42x + 10 \pm \sqrt{-21x + 1645}}{18}$
    E) $\dfrac{-42x - 10 \pm \sqrt{-21x + 910}}{18}$

Answer Key

1. D
2. B
3. B
4. E
5.

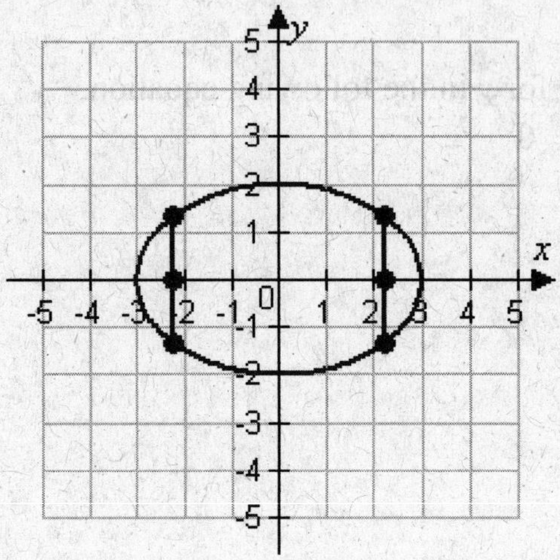

6. D
7. C
8. D
9. A
10. B
11. A
12. B
13. A
14. E
15. E
16. D
17. B
18. B
19. C
20. C

Name: ________________________ Date: ____________

1. Write the component form of the vector described below.
   Initial point: $(-1,-5,-6)$
   Terminal point: $(4,-2,-4)$
   A) $\langle 5,3,2\rangle$
   B) $\langle -5,-3,-2\rangle$
   C) $\langle 4,-2,-4\rangle$
   D) $\langle 5,3,-10\rangle$
   E) $\langle -4,10,24\rangle$

2. Find the coordinates of the point located five units in front of the *yz*-plane, seven units to the right of the *xz*-plane, and nine units below the *xy*-plane.
   A) $(5,-9,7)$
   B) $(5,7,-9)$
   C) $(5,-7,-9)$
   D) $(5,-9,-7)$
   E) $(5,12,3)$

3. Find the area of the parallelogram that has the vectors as adjacent sides.
   $\mathbf{u}=\langle -4,-4,-1\rangle, \mathbf{v}=\langle -1,4,-4\rangle$
   A) 12
   B) $\sqrt{41}$
   C) $2\sqrt{2}$
   D) 15
   E) $5\sqrt{41}$

4. Find the angle between the two planes in degrees. Round to a tenth of a degree.

$$x+3y+6z=4$$

$$x-6y-z=4$$

A) 9.4°
B) 33.4°
C) 123.4°
D) 28.8°
E) 2.2°

5. Find the magnitude of the vector **v**.

$$\mathbf{v}=\langle -7,1,-8\rangle$$

A) –14
B) 16
C) $\sqrt{14}$
D) $2\sqrt{114}$
E) $\sqrt{114}$

6. Find the general form of the equation of the plane passing through the three points. [Be sure to reduce the coefficients in your answer to lowest terms by dividing out any common factor.]

$$(-1,-4,-2),(-6,5,4),(1,-2,-6)$$

A) $12x+2y+7z+34=0$
B) $12x+2y+7z=0$
C) $x+4y+2z=0$
D) $12x+2y+7z-34=0$
E) $x+4y+2z-136=0$

7. Find a set of symmetric equations of the line that passes through the points $(5,0,5)$ and $(7,8,-6)$.
   A) $\frac{x}{7}=\frac{y}{8}=\frac{z}{-6}$
   B) $\frac{x-5}{2}=\frac{y-8}{8}=\frac{z-6}{-11}$
   C) $\frac{x-5}{-2}=\frac{y}{8}=\frac{z-5}{11}$
   D) $\frac{x+2}{5}=y-8=\frac{z-11}{5}$
   E) $\frac{x-5}{2}=\frac{y}{8}=\frac{z-5}{-11}$

8. Find a set of parametric equations for the line through the point and parallel to the specified vector. Show all your work.
   $(-1,-5,9)$, parallel to $\langle -7,6,-3\rangle$

9. Find the angle between the vectors **u** and **v**. Express your answer in degrees and round to the nearest tenth of a degree.
   $\mathbf{u}=-6\mathbf{i}+9\mathbf{j}+2\mathbf{k}, \mathbf{v}=-3\mathbf{i}-7\mathbf{j}-4\mathbf{k}$
   A) 29.3°
   B) 34.1°
   C) 60.7°
   D) 90°
   E) 124.1°

10. Find the general form of the equation of the plane with the given characteristics.
    The plane passes through the points $(-4,4,3)$ and $(2,7,-7)$ and is perpendicular to the plane $2x+3y+3z=1$.
    A) $39x-38y+12z+272=0$
    B) $2x+3y+3z+13=0$
    C) $39x+38y+12z+32=0$
    D) $2x+3y+3z=0$
    E) $38x-39y+14z+266=0$

11. Determine whether the planes are parallel, orthogonal, or neither.

$2x-3y-6z=-4$

$8x-12y-24z=-14$

A) parallel
B) orthogonal
C) neither

12. Find the distance between the points.

$(1,9,3),(5,-6,8)$

A) 20
B) 24
C) $2\sqrt{6}$
D) $2\sqrt{266}$
E) $\sqrt{266}$

13. Determine the values of c such that $\|c\mathbf{u}\|=2$, where $\mathbf{u}=3\mathbf{i}-4\mathbf{j}+2\mathbf{k}$.

A) $c=\pm 2$
B) $c=\pm\dfrac{2\sqrt{29}}{29}$
C) $c=\pm\dfrac{\sqrt{29}}{2}$
D) $c=\pm\dfrac{1}{2}$
E) $c=\pm\dfrac{\sqrt{29}}{29}$

14. Find the area of the parallelogram formed by the points $A(2,-3,5)$, $B(7,-2,7)$, $C(3,-2,11)$, and $D(8,-1,13)$.

A) 816
B) $2\sqrt{5}$
C) 20
D) $4\sqrt{51}$
E) 18

15. Find a set of parametric equations for the line that passes through the given points. Show all your work.

    $(4,-7,2),(-7,-1,-5)$

16. Find the triple scalar product $\mathbf{u}\cdot(\mathbf{v}\times\mathbf{w})$ for the vectors

    $\mathbf{u}=-\mathbf{i}+2\mathbf{j}+\mathbf{k}, \mathbf{v}=7\mathbf{i}+8\mathbf{j}+3\mathbf{k}, \mathbf{w}=4\mathbf{i}+7\mathbf{j}-3\mathbf{k}$

    A) 75
    B) 128
    C) –128
    D) –4
    E) 0

17. Find the dot product of **u** and **v**.

    $\mathbf{u}=9\mathbf{i}-5\mathbf{j}-9\mathbf{k}, \mathbf{v}=-2\mathbf{i}+6\mathbf{j}+5\mathbf{k}$

    A) –93
    B) 4
    C) $-18\mathbf{i}-30\mathbf{j}-45\mathbf{k}$
    D) –3
    E) $7\mathbf{i}+\mathbf{j}-4\mathbf{k}$

18. Find the vector **z**, given $\mathbf{u}=\langle 0,-8,-4\rangle$, $\mathbf{v}=\langle -5,0,-6\rangle$, and $\mathbf{w}=\langle -17,25,-1\rangle$.

    $-2\mathbf{u}+4\mathbf{v}-3\mathbf{z}=\mathbf{w}$

    A) $\langle 3,9,15\rangle$
    B) $\langle -3,-1,-5\rangle$
    C) $\langle 1,3,-5\rangle$
    D) $\langle -1,-3,-5\rangle$
    E) $\langle 0,-3,-4\rangle$

19. Find the lengths of the sides of the right triangle whose vertices are located at the given points. Show that these lengths satisfy the Pythagorean Theorem. Show all of your work.

    $(-6,-3,4),(-7,-1,3),(6,4,0)$

20. Find the torque on the crankshaft **V** using the data shown in the figure. Round to the nearest tenth of a foot-pound.

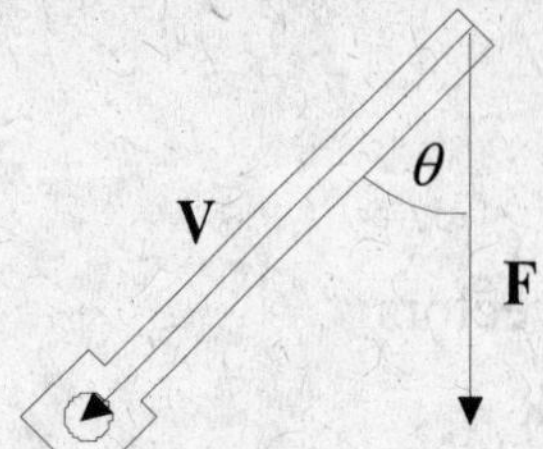

$\|\mathbf{V}\| = 1.6$ ft

$\|\mathbf{F}\| = 20$ lb

$\theta = 60°$

A) 27.7 ft-lb
B) 16.0 ft-lb
C) 55.4 ft-lb
D) 32.0 ft-lb
E) 0

Answer Key

1. A
2. B
3. E
4. C
5. E
6. A
7. E
8. Answers may vary. One possible answer is shown below.
   $x=-1-7t, y=-5+6t, z=9-3t$
9. E
10. A
11. A
12. E
13. B
14. D
15. Answers may vary. One possible answer is shown below.
    $x=4-11t, y=-7+6t, z=2-7t$
16. B
17. A
18. D
19. Let point $P=(-6,-3,4)$, point $Q=(-7,-1,3)$, and point $R=(6,4,0)$.
    $$|\overline{PQ}|^2=(-7+6)^2+(-1+3)^2+(3-4)^2=6; \Rightarrow |\overline{PQ}|=\sqrt{6}$$
    $$|\overline{QR}|^2=(6+7)^2+(4+1)^2+(0-3)^2=203; \Rightarrow |\overline{QR}|=\sqrt{203}$$
    $$|\overline{PR}|^2=(6+6)^2+(4+3)^2+(0-4)^2=209; \Rightarrow |\overline{PR}|=\sqrt{209}$$
    Note that 6 + 203 = 209 in accordance with the Pythagorean Theorem.
20. A

Name: ____________________________ Date: _____________

1. Use the scalar triple product to find the volume of the parallelepiped having adjacent edges $\langle 1,2,2\rangle$, $\langle 2,3,1\rangle$, and $\langle 1,3,4\rangle$.
   A) 29
   B) 15
   C) 1
   D) 51
   E) 28

2. Find the center and radius of the sphere.
   $x^2+y^2+z^2-16x+6y+6z+18=0$
   A) center: $(-8,3,3)$; radius: 8
   B) center: $(8,-3,-3)$; radius: 8
   C) center: $(8,3,3)$; radius: 8
   D) center: $(-8,3,3)$; radius: 64
   E) center: $(8,-3,-3)$; radius: 64

3. Determine whether **u** and **v** are parallel, orthogonal, or neither.
   $\mathbf{u}=\langle 5,6,-8\rangle, \mathbf{v}=\langle 10,12,-16\rangle$
   A) parallel
   B) orthogonal
   C) neither

4. Find the vector **z**, given $\mathbf{u}=\langle -3,3,9\rangle$ and $\mathbf{v}=\langle -7,5,7\rangle$.
   $\mathbf{z}=-3\mathbf{u}-5\mathbf{v}$
   A) $\langle 2,-4,-20\rangle$
   B) $\langle 44,-34,-62\rangle$
   C) $\langle -10,8,16\rangle$
   D) $\langle 36,-30,-66\rangle$
   E) $\langle -26,16,8\rangle$

5. Find the vector **z,** given $\mathbf{u} = \langle 5, -8, 2 \rangle$, $\mathbf{v} = \langle -6, -2, 9 \rangle$, and $\mathbf{w} = \langle -13, 34, -10 \rangle$.
   $-3\mathbf{u} - 2\mathbf{v} - 2\mathbf{z} = \mathbf{w}$
   A) $\langle -10, 6, 14 \rangle$
   B) $\langle -3, 5, -7 \rangle$
   C) $\langle -5, 3, -7 \rangle$
   D) $\langle 5, -3, -7 \rangle$
   E) $\langle 6, -3, -6 \rangle$

6. Find the coordinates of the point located three units in front of the *yz*-plane, eight units to the right of the *xz*-plane, and five units above the *xy*-plane.
   A) $(3,5,8)$
   B) $(3,8,5)$
   C) $(3,-8,5)$
   D) $(3,5,-8)$
   E) $(3,11,16)$

7. Determine whether the planes are parallel, orthogonal, or neither.
   $6x - y - z = 4$
   $24x - 4y - 4z = 18$
   A) parallel
   B) orthogonal
   C) neither

8. Find a unit vector orthogonal to **u** and **v.**
   u = leadcoeff(a)**i** coeff(b)**j** coeff(c)**k**, **v** = leadcoeff(d)**i** coeff(f)**j** coeff(g)**k**
   A) $\frac{1}{\text{leadcoeff(da 1 fac)}\sqrt{\text{da 1 rad}}}$(leadcoeff(a* d)**i** coeff(b* f)**j** coeff(c* g)**k**)
   B) $\frac{1}{\text{leadcoeff(da 2 fac)}\sqrt{\text{da 2 rad}}}$(leadcoeff(a^ 2)**i** coeff(b^ 2)**j** coeff(c^ 2)**k**)
   C) leadcoeff(b* g– c* f)**i** coeff(– a* g+ c* d)**j** coeff(a* f– d* b)**k**
   D) $\frac{1}{\text{leadcoeff(wfac)}\sqrt{\text{wrad}}}$(leadcoeff(b* g– c* f)**i** coeff(– a* g+ c* d)**j** coeff(a* f– d* b)**k**)
   E) leadcoeff(a* d)**i** coeff(b* f)**j** coeff(c* g)**k**

9. Find the distance between the point and the plane.

   $(-2,-1,-4)$

   $-3x-2y-z=1$

   A) 11
   B) $\frac{11}{14}$
   C) $\frac{13}{\sqrt{14}}$
   D) 0
   E) $\frac{11}{\sqrt{14}}$

10. Find a set of parametric equations for the line that passes through the given points. Show all your work.

    $\left(4,\frac{9}{2},\frac{-3}{2}\right),\left(\frac{1}{2},\frac{-3}{2},-2\right)$

11. Find the area of the parallelogram that has the vectors as adjacent sides.

    $\mathbf{u}=\langle 4,2,-5\rangle, \mathbf{v}=\langle -4,1,2\rangle$

    A) 14
    B) $\sqrt{41}$
    C) $2\sqrt{6}$
    D) 11
    E) $3\sqrt{41}$

12. The weight of a crate is 300 newtons. Find the tension in each of the supporting cables shown in the figure. The coordinates of the points A, B, C, and D are given below the figure. Round to the nearest newton.

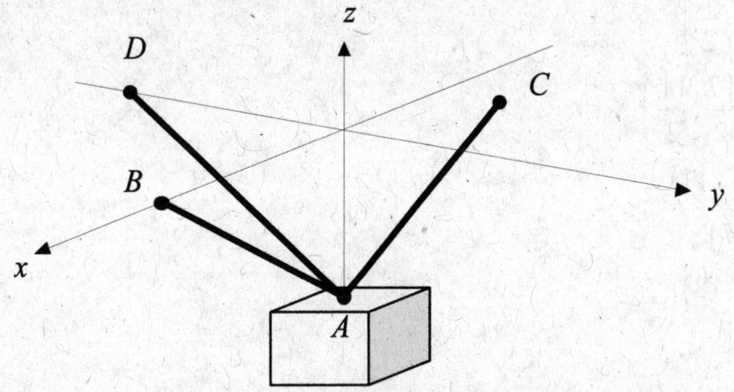

[Figure not necessarily to scale.]

point $A=(0,0,-130)$, point $B=(90,0,0)$, point $C=(-40,40,0)$, point $D=(0,-1$

A) cable AB = 196; cable AC = 97; cable AD = 68
B) cable AB = 97; cable AC = 68; cable AD = 196
C) cable AB = 97; cable AC = 196; cable AD = 68
D) cable AB = 68; cable AC = 196; cable AD = 97
E) cable AB = 196; cable AC = 68; cable AD = 97

13. Find a set of parametric equations for the line that passes through the given points. Show all your work.

$(9,8,-2),(-3,-6,4)$

14. Find the area of the parallelogram that has the vectors as adjacent sides.

$\mathbf{u}=2\mathbf{i}+\mathbf{j}+\mathbf{k}, \mathbf{v}=5\mathbf{i}-5\mathbf{j}+5\mathbf{k}$

A) 5
B) $\sqrt{14}$
C) $\sqrt{10}$
D) 4
E) $5\sqrt{14}$

15. Find the lengths of the sides of the right triangle whose vertices are located at the given points. Show that these lengths satisfy the Pythagorean Theorem. Show all of your work.

$(8,-4,4),(9,5,3),(-3,6,0)$

16. Find the midpoint of the line segment joining the points.
    $(-5,8,9),(-8,9,2)$
    A) $\left(\frac{3}{2},\frac{-1}{2},\frac{7}{2}\right)$
    B) $\left(\frac{-13}{2},\frac{17}{2},\frac{11}{2}\right)$
    C) $(-13,17,11)$
    D) $(3,-1,7)$
    E) $(40,72,18)$

17. Find the magnitude of the vector **v**.
    $\mathbf{v}=\langle 0,-5,-3\rangle$
    A) –8
    B) 8
    C) $2\sqrt{2}$
    D) $2\sqrt{34}$
    E) $\sqrt{34}$

18. Find the general form of the equation of the plane passing through the three points. [Be sure to reduce the coefficients in your answer to lowest terms by dividing out any common factor.]
    $(5,-5,-6),(6,-1,4),(-2,-5,5)$
    A) $44x-81y+28z-457=0$
    B) $44x-81y+28z=0$
    C) $5x-5y-6z=0$
    D) $44x-81y+28z+457=0$
    E) $5x-5y-6z+457=0$

19. Find the angle between the vectors **u** and **v**. Express your answer in degrees and round to the nearest tenth of a degree.

$\mathbf{u} = \langle -3,6,3 \rangle, \mathbf{v} = \langle 9,9,9 \rangle$

A) 25.2°
B) 28.1°
C) 64.8°
D) 90°
E) 61.9°

20. Find symmetric equations for the line through the point and parallel to the specified line. Show all your work.

$$(-5,-3,3), \text{ parallel to } \begin{aligned} x &= -8+7t \\ y &= 7+3t \\ z &= -9-5t \end{aligned}$$

Answer Key

1. C
2. B
3. A
4. B
5. D
6. B
7. A
8. D
9. E
10. Answers may vary. One possible answer is shown below.
    $x = 4 - \frac{7}{2}t, y = \frac{9}{2} - 6t, z = -\frac{3}{2} - \frac{1}{2}t$
11. E
12. C
13. Answers may vary. One possible answer is shown below.
    $x = 9 - 12t, y = 8 - 14t, z = -2 + 6t$
14. E
15. Let point $P = (8, -4, 4)$, point $Q = (9, 5, 3)$, and point $R = (-3, 6, 0)$.

    $\left|\overline{PQ}\right|^2 = (9-8)^2 + (5+4)^2 + (3-4)^2 = 83; \Rightarrow \left|\overline{PQ}\right| = \sqrt{83}$

    $\left|\overline{QR}\right|^2 = (-3-9)^2 + (6-5)^2 + (0-3)^2 = 154; \Rightarrow \left|\overline{QR}\right| = \sqrt{154}$

    $\left|\overline{PR}\right|^2 = (-3-8)^2 + (6+4)^2 + (0-4)^2 = 237; \Rightarrow \left|\overline{PR}\right| = \sqrt{237}$

    Note that 83 + 154 = 237 in accordance with the Pythagorean Theorem.
16. B
17. E
18. A
19. E
20. Answers may vary. One possible answer is shown below.
    $\frac{x+5}{7} = \frac{y+3}{3} = \frac{z-3}{-5}$

Name: ______________________ Date: ___________

1. Find the angle of intersection of the planes in degrees. Round to a tenth of a degree.
   $3x-2y+5z=3$
   $-x+3y+3z=3$
   A) 4.8°
   B) 12.9°
   C) 77.1°
   D) 12.6°
   E) 1.3°

2. Find the magnitude of the vector **v**.
   $\mathbf{v}=\langle 0,-4,-4\rangle$
   A) –8
   B) 8
   C) $2\sqrt{2}$
   D) $8\sqrt{2}$
   E) $4\sqrt{2}$

3. Find a unit vector orthogonal to 3**i** +2**j** and **j** +5**k**.
   A) $10\mathbf{i}-15\mathbf{j}+3\mathbf{k}$
   B) $\frac{10}{\sqrt{334}}\mathbf{i}-\frac{15}{\sqrt{334}}\mathbf{j}+\frac{3}{\sqrt{334}}\mathbf{k}$
   C) $10\mathbf{i}+15\mathbf{j}+3\mathbf{k}$
   D) $-\frac{10}{\sqrt{334}}\mathbf{i}+\frac{15}{\sqrt{334}}\mathbf{j}-\frac{3}{\sqrt{334}}\mathbf{k}$
   E) $\frac{10}{\sqrt{334}}\mathbf{i}+\frac{15}{\sqrt{334}}\mathbf{j}+\frac{3}{\sqrt{334}}\mathbf{k}$

4. Find the area of the parallelogram formed by the points $A(2,-3,5)$, $B(8,-2,7)$, $C(3,-2,8)$, and $D(9,-1,10)$.
   A) 282
   B) $\sqrt{10}$
   C) 10
   D) $\sqrt{282}$
   E) 13

5. Find the standard form of the equation of the sphere with the given characteristics.
   Endpoints of a diameter: $(-8,4,-2),(-8,-2,6)$

   A) $(x+16)^2+(x-2)^2+(x-4)^2=25$
   B) $(x+8)^2+(x-1)^2+(x-2)^2=50$
   C) $(x+8)^2+(x-1)^2+(x-2)^2=100$
   D) $(x+8)^2+(x-1)^2+(x-2)^2=25$
   E) $(x-8)^2+(x+1)^2+(x+2)^2=25$

6. Find $\mathbf{u}\times\mathbf{v}$.
   $\mathbf{u}=\langle 7,-5,1\rangle, \mathbf{v}=\langle -3,4,8\rangle$

   A) $\langle -44,-59,13\rangle$
   B) -33
   C) 7
   D) $\langle -44,59,13\rangle$
   E) $\langle -21,-20,8\rangle$

7. Find the general form of the equation of the plane passing through the three points. [Be sure to reduce the coefficients in your answer to lowest terms by dividing out any common factor.]
   $(-2,-1,4),(3,-1,2),(-6,-5,-6)$

   A) $4x-29y+10z-61=0$
   B) $4x-29y+10z=0$
   C) $2x+y-4z=0$
   D) $4x-29y+10z+61=0$
   E) $2x+y-4z+122=0$

8. Find the distance between the points.
   $(2,-3,3),(-6,2,7)$

   A) 5
   B) 17
   C) $\sqrt{17}$
   D) $2\sqrt{105}$
   E) $\sqrt{105}$

9. Find the vector **z**, given $\mathbf{u}=\langle -4,9,8\rangle$, $\mathbf{v}=\langle 5,4,4\rangle$, and $\mathbf{w}=\langle -3,23,4\rangle$.

   $3\mathbf{u}-\mathbf{v}+2\mathbf{z}=\mathbf{w}$

   A) $\langle 14,0,-16\rangle$
   B) $\langle 0,7,-8\rangle$
   C) $\langle -7,0,-8\rangle$
   D) $\langle 7,0,-8\rangle$
   E) $\langle 8,0,-7\rangle$

10. Find the dot product of **u** and **v**.

    $\mathbf{u}=\langle 1,-6,-4\rangle, \mathbf{v}=\langle 6,4,9\rangle$

    A) –54
    B) 10
    C) $\langle 6,-24,-36\rangle$
    D) 18
    E) $\langle 7,-2,5\rangle$

11. Find the angle between the vectors **u** and **v**. Express your answer in degrees and round to the nearest tenth of a degree.

    $\mathbf{u}=\langle -2,6,6\rangle, \mathbf{v}=\langle -3,7,7\rangle$

    A) 44.9°
    B) 86.4°
    C) 45.1°
    D) 90°
    E) 3.6°

12. Find a set of parametric equations for the line that passes through the given points. Show all your work.

    $\left(9,\frac{9}{2},\frac{7}{2}\right),\left(\frac{3}{2},\frac{1}{2},9\right)$

13. Find the area of the parallelogram that has the vectors as adjacent sides.
$\mathbf{u} = \langle 2,3,5 \rangle, \mathbf{v} = \langle -3,3,-5 \rangle$
A) 3
B) $\sqrt{46}$
C) $\sqrt{22}$
D) 34
E) $5\sqrt{46}$

14. Find the magnitude of the vector **v** described below.
Initial point: $(1,-8,-6)$
Terminal point: $(-9,-1,-6)$
A) –29
B) 17
C) $\sqrt{17}$
D) $2\sqrt{149}$
E) $\sqrt{149}$

15. For the points $A(1,-2,-1), B(2,-7,4), C(1,-6,-5), D(2,-11,0)$:

a. Verify that the points are vertices of a parallelogram. Show all work.

b. Find the area of the parallelogram. Show all work.

c. Determine whether the parallelogram is a rectangle.

16. Find the area of the triangle with the given vertices.
$(-4,-2,2),(-9,-7,0),(-4,-3,5)$
A) 0
B) 1
C) $\frac{7\sqrt{11}}{2}$
D) $\frac{7\sqrt{11}}{4}$
E) $7\sqrt{11}$

17. Find a set of parametric equations for the line through the point and parallel to the specified line. Show all your work.

$$(3,-3,9), \text{ parallel to } \begin{aligned} x &= 2-5t \\ y &= -9-3t \\ z &= 8-2t \end{aligned}$$

18. Find the magnitude of the vector described below.

Initial point: $(5,6,-4)$

Terminal point: $(-4,2,7)$

A) 2
B) 36
C) $2\sqrt{6}$
D) $2\sqrt{218}$
E) $\sqrt{218}$

19. The lights in an auditorium are 30-pound disks of radius 24 inches. Each disk is supported by three equally spaced 45-inch wires attached to the ceiling. Find the tension in each wire. Round your answer to two decimals.

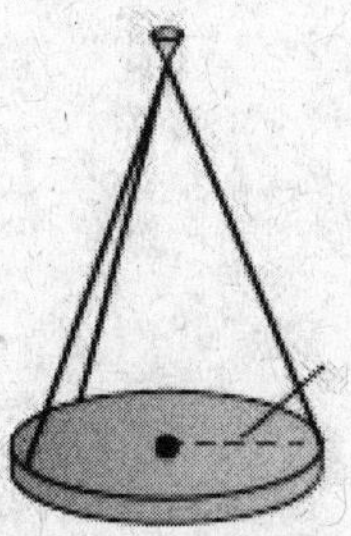

A) 6.30 pounds
B) 35.46 pounds
C) 11.82 pounds
D) 18.91 pounds
E) 10.00 pounds

20. Find the angle between the two planes in degrees. Round to a tenth of a degree.

$$x + 2y + 6z = 0$$

$$-4x - y - 3z = 0$$

A) 10.8°
B) 47.3°
C) 137.3°
D) 36.3°
E) 2.4°

Answer Key

1. C
2. E
3. B
4. D
5. D
6. A
7. A
8. E
9. D
10. A
11. E
12. Answers may vary. One possible answer is shown below.
    $x = 9 - \frac{15}{2}t, y = \frac{9}{2} - 4t, z = \frac{7}{2} + \frac{11}{2}t$
13. E
14. E
15. Answer details may vary.
    a.
    $\overrightarrow{AB} = \langle 1,-5,5\rangle = \overrightarrow{CD} \Rightarrow \overrightarrow{AB} \parallel \overrightarrow{CD}$
    $\overrightarrow{AC} = \langle 0,-4,-4\rangle = \overrightarrow{BD} \Rightarrow \overrightarrow{AC} \parallel \overrightarrow{BD}$

    b. Area $= \left\|\overrightarrow{AB} \times \overrightarrow{AC}\right\| = \left\|\begin{vmatrix} \mathbf{i} & \mathbf{j} & \mathbf{k} \\ 1 & -5 & 5 \\ 0 & -4 & -4 \end{vmatrix}\right\| = \left\|\langle 40,4,-4\rangle\right\| = 4\sqrt{102}$

    c. $\theta = \sin^{-1}\left(\frac{\left\|\overrightarrow{AB} \times \overrightarrow{AC}\right\|}{\left\|\overrightarrow{AB}\right\|\left\|\overrightarrow{AC}\right\|}\right) = \sin^{-1}\left(\frac{4\sqrt{102}}{\sqrt{51}\sqrt{32}}\right) \approx 90.0°$
    Since $\theta \neq 90°$, the parallelogram is not a rectangle.
16. C
17. Answers may vary. One possible answer is shown below.
    $x = 3 - 5t, y = -3 - 3t, z = 9 - 2t$
18. E
19. C
20. C

Name: ______________________________ Date: ______________

1. Find the general form of the equation of the plane passing through the three points. [Be sure to reduce the coefficients in your answer to lowest terms by dividing out any common factor.]

   $(5,2,-1),(3,3,2),(4,4,1)$

   A) $4x-y+3z-15=0$
   B) $4x-y+3z=0$
   C) $5x+2y-z=0$
   D) $4x-y+3z+15=0$
   E) $5x+2y-z-15=0$

2. Find symmetric equations for the line through the point and parallel to the specified vector. Show all your work.

   $(1,8,6)$, parallel to $\langle 9,-6,5\rangle$

3. Find symmetric equations for the line through the point and parallel to the specified line. Show all your work.

   $(-6,-2,5)$, parallel to $\begin{aligned} x&=-1-3t \\ y&=6-8t \\ z&=4-3t \end{aligned}$

4. Find the angle, in degrees, between two adjacent *sides* of the pyramid shown below. Round to the nearest tenth of a degree. [Note: The *base* of the pyramid is not considered a *side*.]

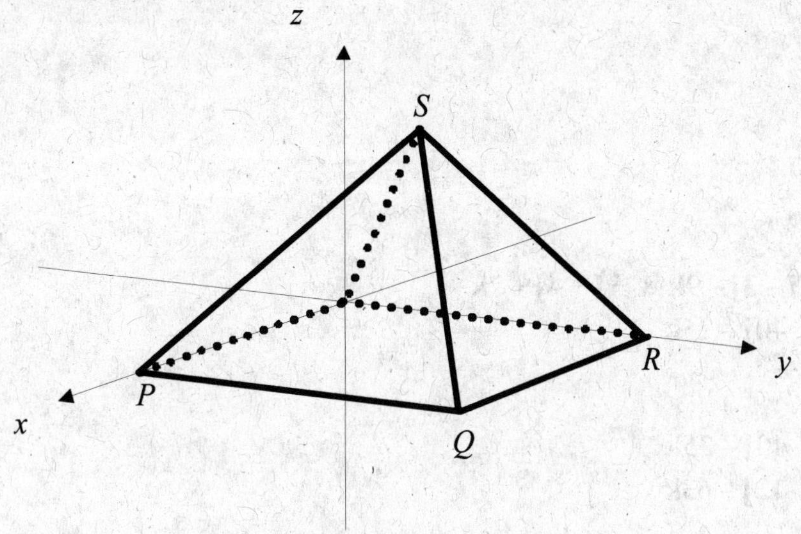

$P(8,0,0), Q(8,8,0), R(0,8,0), S(4,4,3)$

A) 2.3°
B) 129.8°
C) 90°
D) 45°
E) 39.8°

5. Find the angle of intersection of the planes in degrees. Round to a tenth of a degree.

$3x - 2y + 4z = 3$

$5x - 3y + 3z = -6$

A) –0.8°
B) 69.1°
C) 20.9°
D) 43.1°
E) 0.4°

6. Find the area of the parallelogram that has the vectors as adjacent sides.
$\mathbf{u} = \langle 4, -1, 4 \rangle, \mathbf{v} = \langle 2, -3, 0 \rangle$
   A) 11
   B) $\sqrt{77}$
   C) $\sqrt{11}$
   D) 4
   E) $2\sqrt{77}$

7. Find $\mathbf{u} \times \mathbf{v}$.
$\mathbf{u} = -7\mathbf{i} - 3\mathbf{j} - 9\mathbf{k}, \mathbf{v} = \mathbf{i} + 4\mathbf{j} + 7\mathbf{k}$
   A) $15\mathbf{i} + 40\mathbf{j} - 25\mathbf{k}$
   B) -82
   C) -58
   D) $15\mathbf{i} - 40\mathbf{j} - 25\mathbf{k}$
   E) $-7\mathbf{i} - 12\mathbf{j} - 63\mathbf{k}$

8. Find the midpoint of the line segment joining the points.
$(-2, -7, 3), (-3, -4, -6)$
   A) $\left(\frac{1}{2}, \frac{-3}{2}, \frac{9}{2}\right)$
   B) $\left(\frac{-5}{2}, \frac{-11}{2}, \frac{-3}{2}\right)$
   C) $(-5, -11, -3)$
   D) $(1, -3, 9)$
   E) $(6, 28, -18)$

9. Find the magnitude of the vector $\mathbf{v}$.
$\mathbf{v} = \langle -3, -3, -5 \rangle$
   A) -11
   B) 11
   C) $\sqrt{11}$
   D) $2\sqrt{43}$
   E) $\sqrt{43}$

10. Find the angle between the vectors **u** and **v**. Express your answer in degrees and round to the nearest tenth of a degree.

    $\mathbf{u} = 4\mathbf{i} + 2\mathbf{j} - \mathbf{k}, \mathbf{v} = 7\mathbf{i} - 7\mathbf{j} + 4\mathbf{k}$

    A) 11.6°
    B) 11.8°
    C) 78.4°
    D) 90°
    E) 78.2°

11. Find the angle between the two planes in degrees. Round to a tenth of a degree.

    $2x - y - 4z = 4$

    $-3x - 2y + 4z = 4$

    A) 11.5°
    B) 54.1°
    C) 144.1°
    D) 39.0°
    E) 2.5°

12. Find the distance between the points.

    $(1,5,5),(-2,6,-1)$

    A) 14
    B) 10
    C) $\sqrt{10}$
    D) $2\sqrt{46}$
    E) $\sqrt{46}$

13. Find the volume of the parallelpiped with the given vertices.

    $A(-3,-8,-6), B(-7,0,2), C(-12,-3,3), D(-16,5,11),$

    $E(-9,-7,2), F(-13,1,10), G(-18,-2,11), H(-22,6,19)$

    A) 400
    B) 188
    C) 100
    D) 278
    E) 228

14. Find the coordinates of the point located four units in front of the *yz*-plane, nine units to the right of the *xz*-plane, and three units below the *xy*-plane.
    A) $(4,-3,9)$
    B) $(4,9,-3)$
    C) $(4,-9,-3)$
    D) $(4,-3,-9)$
    E) $(4,13,10)$

15. Find a set of symmetric equations of the line that passes through the points $(3,0,3)$ and $(7,4,-2)$.
    A) $\frac{x}{7}=\frac{y}{4}=\frac{z}{-2}$
    B) $\frac{x-3}{4}=\frac{y-4}{4}=\frac{z-2}{-5}$
    C) $\frac{x-3}{-4}=\frac{y}{4}=\frac{z-3}{5}$
    D) $\frac{x+4}{3}=y-4=\frac{z-5}{3}$
    E) $\frac{x-3}{4}=\frac{y}{4}=\frac{z-3}{-5}$

16. Find the cross product of the unit vectors k × i.
    A) j
    B) -j
    C) i
    D) -i
    E) k

17. Write the component form of the vector described below.

    Initial point: $(-7,2,-1)$

    Terminal point: $(-5,-9,-6)$

    A) $\langle 2,-11,-5\rangle$
    B) $\langle -2,11,5\rangle$
    C) $\langle -5,-9,-6\rangle$
    D) $\langle 2,-11,-7\rangle$
    E) $\langle 35,-18,6\rangle$

18. Find the triple scalar product $\mathbf{u}\cdot(\mathbf{v}\times\mathbf{w})$ for the vectors

    $\mathbf{u}=\langle -3,-5,-8\rangle, \mathbf{v}=\langle -1,-6,3\rangle, \mathbf{w}=\langle -8,-7,6\rangle$

    A) –378
    B) 463
    C) –463
    D) 283
    E) 0

19. Find the triple scalar product $\mathbf{u}\cdot(\mathbf{v}\times\mathbf{w})$ for the vectors

    $\mathbf{u}=-\mathbf{i}-5\mathbf{j}+9\mathbf{k}, \mathbf{v}=-2\mathbf{i}-3\mathbf{j}-8\mathbf{k}, \mathbf{w}=-\mathbf{i}-6\mathbf{j}+\mathbf{k}$

    A) –164
    B) 82
    C) –82
    D) 182
    E) 0

20. Determine whether **u** and **v** are parallel, orthogonal, or neither.

    $\mathbf{u}=\langle 6,-5,2\rangle, \mathbf{v}=\langle 30,-25,10\rangle$

    A) parallel
    B) orthogonal
    C) neither

Answer Key

1. A
2. Answers may vary. One possible answer is shown below.
   $$\frac{x-1}{9}=\frac{y-8}{-6}=\frac{z-6}{5}$$
3. Answers may vary. One possible answer is shown below.
   $$\frac{x+6}{-3}=\frac{y+2}{-8}=\frac{z-5}{-3}$$
4. B
5. C
6. E
7. A
8. B
9. E
10. E
11. C
12. E
13. B
14. B
15. E
16. A
17. A
18. B
19. B
20. A

Name: ______________________ Date: ___________

1. Determine whether **u** and **v** are parallel, orthogonal, or neither.

   $\mathbf{u} = \langle -1, 6, -3 \rangle, \mathbf{v} = \langle -3, 18, -9 \rangle$

   A) parallel
   B) orthogonal
   C) neither

2. Find the general form of the equation of the plane passing through the point and perpendicular to the specified line.

   $$(5, -6, -4), \quad \begin{aligned} x &= 2 + 3t \\ y &= -3 + 4t \\ z &= -6 + 6t \end{aligned}$$

   A) $3x + 4y + 6z = 0$
   B) $3x + 4y + 6z - 33 = 0$
   C) $3x + 4y + 6z + 33 = 0$
   D) $5x - 6y - 4z - 33 = 0$
   E) $5x - 6y - 4z + 33 = 0$

3. Find symmetric equations for the line through the point and parallel to the specified line. Show all your work.

   $$(5, -9, -1), \text{ parallel to } \begin{aligned} x &= -8 - 9t \\ y &= 5 - 5t \\ z &= -8 + 3t \end{aligned}$$

4. Find the acute interior angle of the parallelogram formed by $A(2, -2, 2)$, $B(5, -1, 4)$, $C(6, 0, 10)$, and $D(3, -1, 8)$. Round your answer to two decimals.

   A) 43.92°
   B) 35.76°
   C) 46.08°
   D) 1.79°
   E) 31.25°

5. Determine the values of c such that $\|c\mathbf{u}\| = 6$, where $\mathbf{u} = -3\mathbf{i} + 2\mathbf{j} - 5\mathbf{k}$.
   A) $c = \pm 6$
   B) $c = \pm \dfrac{3\sqrt{38}}{19}$
   C) $c = \pm \dfrac{\sqrt{38}}{6}$
   D) $c = \pm \dfrac{1}{6}$
   E) $c = \pm \dfrac{\sqrt{38}}{38}$

6. Find the volume of the parallelpiped with the given vertices.
   $A(0,-5,3), B(3,-1,-4), C(7,-3,-1), D(10,1,-8),$
   $E(-3,-13,6), F(0,-9,-1), G(4,-11,2), H(7,-7,-5)$
   A) 43
   B) 236
   C) 308
   D) 296
   E) 256

7. Find the distance between the point and the plane.
   $(-3,-5,-2)$
   $4x + 6y + 6z = -18$
   A) 36
   B) $\dfrac{9}{22}$
   C) $\dfrac{36}{\sqrt{22}}$
   D) 0
   E) $\dfrac{18}{\sqrt{22}}$

8. Find $\mathbf{u}\times\mathbf{v}$.

$\mathbf{u}=\langle -1,-2,1\rangle, \mathbf{v}=\langle -8,-1,-8\rangle$

A) $\langle 17,-16,-15\rangle$
B) 2
C) –2
D) $\langle 17,16,-15\rangle$
E) $\langle 8,2,-8\rangle$

9. The lights in an auditorium are 25-pound disks of radius 16 inches. Each disk is supported by three equally spaced 60-inch wires attached to the ceiling. Find the tension in each wire. Round your answer to two decimals.

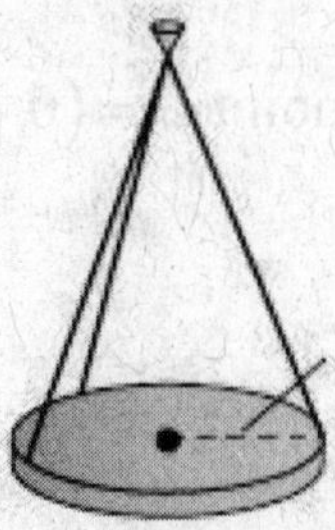

A) 2.31 pounds
B) 25.94 pounds
C) 8.65 pounds
D) 6.92 pounds
E) 8.33 pounds

10. For the points $A(4,-1,-1), B(-1,-2,4), C(5,-4,4), D(0,-5,9)$:

a. Verify that the points are vertices of a parallelogram. Show all work.

b. Find the area of the parallelogram. Show all work.

c. Determine whether the parallelogram is a rectangle.

11. The weight of a crate is 200 newtons. Find the tension in each of the supporting cables shown in the figure. The coordinates of the points A, B, C, and D are given below the figure. Round to the nearest newton.

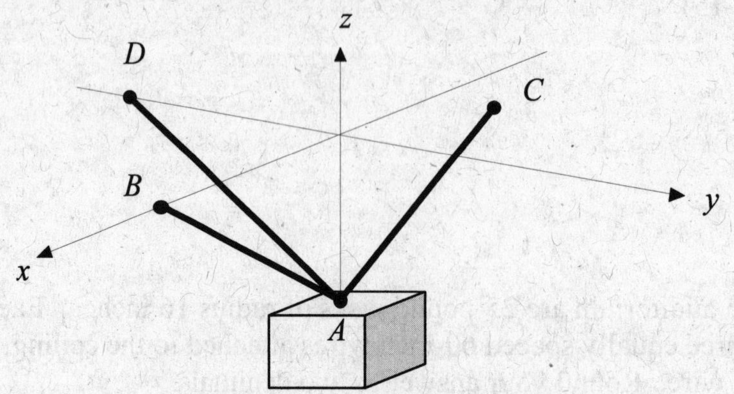

[Figure not necessarily to scale.]

point $A=(0,0,-180)$, point $B=(110,0,0)$, point $C=(-40,50,0)$, point $D=(0,-130,0)$

A) cable AB = 121; cable AC = 49; cable AD = 54
B) cable AB = 49; cable AC = 54; cable AD = 121
C) cable AB = 49; cable AC = 121; cable AD = 54
D) cable AB = 54; cable AC = 121; cable AD = 49
E) cable AB = 121; cable AC = 54; cable AD = 49

12. Find the general form of the equation of the plane with the given characteristics.
The plane passes through the point $(-3,6,-2)$ and is parallel to the yz-plane.

A) $y=6$
B) $z=-2$
C) $x+y+z=1$
D) $y+z=4$
E) $x=-3$

13. Find the magnitude of the vector **v**.
$\mathbf{v}=\langle 7,-4,0\rangle$

A) 3
B) 11
C) $\sqrt{3}$
D) $2\sqrt{65}$
E) $\sqrt{65}$

14. Find the angle between the vectors **u** and **v**. Express your answer in degrees and round to the nearest tenth of a degree.

$$\mathbf{u} = 6\mathbf{i} - 3\mathbf{j} - 9\mathbf{k}, \mathbf{v} = -\mathbf{i} - \mathbf{j} - 6\mathbf{k}$$

A) 36.4°
B) 47.5°
C) 53.6°
D) 90°
E) 42.5°

15. Find the dot product of **u** and **v**.

$$\mathbf{u} = \langle -8, 7, -4 \rangle, \mathbf{v} = \langle 8, 8, 8 \rangle$$

A) –40
B) 19
C) $\langle -64, 56, -32 \rangle$
D) 24
E) $\langle 0, 15, 4 \rangle$

16. Find the magnitude of the vector described below.

Initial point: $(0, -6, -4)$

Terminal point: $(-1, 5, 5)$

A) 19
B) 50
C) $\sqrt{21}$
D) $2\sqrt{203}$
E) $\sqrt{203}$

17. Find the lengths of the sides of the right triangle whose vertices are located at the given points. Show that these lengths satisfy the Pythagorean Theorem. Show all of your work.

$$(-2, -3, 6), (-3, -8, 5), (2, -8, 0)$$

18. Find the midpoint of the line segment joining the points.
$(-3,7,-8),(8,2,5)$
A) $\left(\frac{-11}{2},\frac{5}{2},\frac{-13}{2}\right)$
B) $\left(\frac{5}{2},\frac{9}{2},\frac{-3}{2}\right)$
C) $(5,9,-3)$
D) $(-11,5,-13)$
E) $(-24,14,-40)$

19. Find the vector **z**, given $\mathbf{u}=\langle -8,1,5\rangle$ and $\mathbf{v}=\langle -6,-3,6\rangle$.
$\mathbf{z}=-3\mathbf{u}-4\mathbf{v}$
A) $\langle 18,-6,-9\rangle$
B) $\langle 48,9,-39\rangle$
C) $\langle -14,-2,11\rangle$
D) $\langle 50,5,-38\rangle$
E) $\langle 0,-15,9\rangle$

20. Find a set of parametric equations for the line through the point and parallel to the specified line. Show all your work.

$(-8,-4,3)$, parallel to $\begin{aligned} x &= -7-7t \\ y &= 2-8t \\ z &= -4+9t \end{aligned}$

Answer Key

1. A
2. C
3. Answers may vary. One possible answer is shown below.
   $\frac{x-5}{-9}=\frac{y+9}{-5}=\frac{z+1}{3}$
4. C
5. B
6. B
7. E
8. A
9. C
10. Answer details may vary.

    a.

    $\overrightarrow{AB}=\langle -5,-1,5\rangle=\overrightarrow{CD}\Rightarrow\overrightarrow{AB}\parallel\overrightarrow{CD}$

    $\overrightarrow{AC}=\langle 1,-3,5\rangle=\overrightarrow{BD}\Rightarrow\overrightarrow{AC}\parallel\overrightarrow{BD}$

    b. Area $=\left\|\overrightarrow{AB}\times\overrightarrow{AC}\right\|=\left\|\begin{matrix}\mathbf{i} & \mathbf{j} & \mathbf{k}\\ -5 & -1 & 5\\ 1 & -3 & 5\end{matrix}\right\|=\left\|\langle 10,30,16\rangle\right\|=2\sqrt{314}$

    c. $\theta=\sin^{-1}\left(\frac{\left\|\overrightarrow{AB}\times\overrightarrow{AC}\right\|}{\left\|\overrightarrow{AB}\right\|\left\|\overrightarrow{AC}\right\|}\right)=\sin^{-1}\left(\frac{2\sqrt{314}}{\sqrt{51}\sqrt{35}}\right)\approx 57.0°$

    Since $\theta\neq 90°$, the parallelogram is not a rectangle.
11. C
12. E
13. E
14. E
15. A
16. E
17. Let point $P=(-2,-3,6)$, point $Q=(-3,-8,5)$, and point $R=(2,-8,0)$.

    $\left|\overline{PQ}\right|^2=(-3+2)^2+(-8+3)^2+(5-6)^2=27;\Rightarrow\left|\overline{PQ}\right|=3\sqrt{3}$

    $\left|\overline{QR}\right|^2=(2+3)^2+(-8+8)^2+(0-5)^2=50;\Rightarrow\left|\overline{QR}\right|=5\sqrt{2}$

    $\left|\overline{PR}\right|^2=(2+2)^2+(-8+3)^2+(0-6)^2=77;\Rightarrow\left|\overline{PR}\right|=\sqrt{77}$

    Note that 27 + 50 = 77 in accordance with the Pythagorean Theorem.
18. B

19. B
20. Answers may vary. One possible answer is shown below.
    $x = -8 - 7t, y = -4 - 8t, z = 3 + 9t$

Name: ______________________ Date: __________

1. Find a set of parametric equations for the line through the point and parallel to the specified line. Show all your work.

$$(-3,-7,-6), \text{ parallel to } \begin{aligned} x &= 9-8t \\ y &= 3-4t \\ z &= -2+6t \end{aligned}$$

2. Find the general form of the equation of the plane with the given characteristics.
   The plane passes through the points $(4,5,-2)$ and $(-1,-4,-1)$ and is perpendicular to the plane $2x+5y+3z=-3$.
   A) $32x-17y+7z-29=0$
   B) $2x+5y+3z+27=0$
   C) $32x+17y+7z+199=0$
   D) $2x+5y+3z=0$
   E) $33x-16y+5z-42=0$

3. Find the acute interior angle of the parallelogram formed by $A(1,-1,2)$, $B(6,0,4)$, $C(7,1,9)$, and $D(2,0,7)$. Round your answer to two decimals.
   A) 34.21°
   B) 39.59°
   C) 55.79°
   D) 1.67°
   E) 43.15°

4. The lights in an auditorium are 25-pound disks of radius 16 inches. Each disk is supported by three equally spaced 50-inch wires attached to the ceiling. Find the tension in each wire. Round your answer to two decimals.

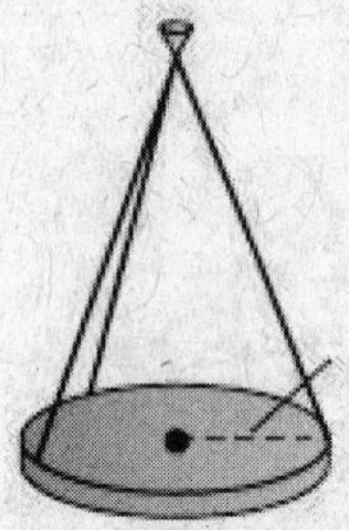

A) 2.81 pounds
B) 26.39 pounds
C) 8.80 pounds
D) 8.44 pounds
E) 8.33 pounds

5. Determine the values of c such that $\|c\mathbf{u}\| = 6$, where $\mathbf{u} = 6\mathbf{i} + 5\mathbf{j} + 5\mathbf{k}$.

A) $c = \pm 6$
B) $c = \pm\dfrac{3\sqrt{86}}{43}$
C) $c = \pm\dfrac{\sqrt{86}}{6}$
D) $c = \pm\dfrac{1}{6}$
E) $c = \pm\dfrac{\sqrt{86}}{86}$

6. Find the lengths of the sides of the right triangle whose vertices are located at the given points. Show that these lengths satisfy the Pythagorean Theorem. Show all of your work.

$(-9,6,-1),(-8,3,-2),(3,6,0)$

7. Find a set of parametric equations for the line that passes through the given points. Show all your work.

$(-4,2,8),(5,-2,-7)$

8. Find the angle between the vectors **u** and **v.** Express your answer in degrees and round to the nearest tenth of a degree.

   $\mathbf{u} = -5\mathbf{i} - 6\mathbf{j} - \mathbf{k}, \mathbf{v} = 6\mathbf{i} + \mathbf{j} + \mathbf{k}$

   A) 37.3°
   B) 49.7°
   C) 52.7°
   D) 90°
   E) 139.7°

9. Find the dot product of **u** and **v.**

   $\mathbf{u} = \langle -3, -2, 4 \rangle, \mathbf{v} = \langle -4, -5, -7 \rangle$

   A) –6
   B) –17
   C) $\langle 12, 10, -28 \rangle$
   D) 50
   E) $\langle -7, -7, -3 \rangle$

10. Find $\mathbf{u} \times \mathbf{v}$.

    $\mathbf{u} = \langle 3, -7, -1 \rangle, \mathbf{v} = \langle -9, 4, -4 \rangle$

    A) $\langle 32, 21, -51 \rangle$
    B) –51
    C) 5
    D) $\langle 32, -21, -51 \rangle$
    E) $\langle -27, -28, 4 \rangle$

11. Find the volume of the parallelpiped with the given vertices.

    $A(8,-5,-4), B(2,2,5), C(9,4,-12), D(3,11,-3),$
    $E(8,-6,-10), F(2,1,-1), G(9,3,-18), H(3,10,-9)$

    A) 369
    B) 405
    C) 321
    D) 455
    E) 430

12. Find the angle, in degrees, between two adjacent *sides* of the pyramid shown below. Round to the nearest tenth of a degree. [Note: The *base* of the pyramid is not considered a *side*.]

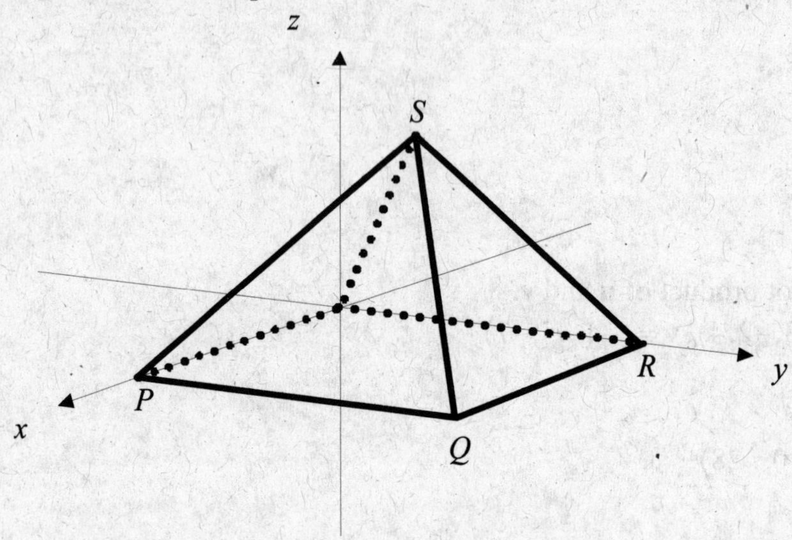

$$P(2,0,0), Q(2,2,0), R(0,2,0), S(1,1,7)$$

A) 1.6°
B) 91.1°
C) 90°
D) 45°
E) 1.1°

13. Find the general form of the equation of the plane passing through the point and perpendicular to the specified line.

$$(6,-8,9), \quad \begin{aligned} x &= -1+5t \\ y &= -3+t \\ z &= -2-6t \end{aligned}$$

A) $5x+y-6z=0$
B) $5x+y-6z-32=0$
C) $5x+y-6z+32=0$
D) $6x-8y+9z-32=0$
E) $6x-8y+9z+32=0$

14. Find the vector **z**, given $\mathbf{u} = \langle 3, -3, 8 \rangle$, $\mathbf{v} = \langle -7, -5, 2 \rangle$, and $\mathbf{w} = \langle 8, -24, 18 \rangle$.

$-\mathbf{u} + 3\mathbf{v} - 4\mathbf{z} = \mathbf{w}$

A) $\langle 32, -12, 20 \rangle$
B) $\langle 3, -8, -5 \rangle$
C) $\langle 8, -3, -5 \rangle$
D) $\langle -8, 3, -5 \rangle$
E) $\langle -7, 3, -4 \rangle$

15. Find symmetric equations for the line through the point and parallel to the specified vector. Show all your work.

$(-5, 3, 8)$, parallel to $\langle -4, 5, 7 \rangle$

16. Find a set of symmetric equations of the line that passes through the points $(6, 0, 6)$ and $(2, 3, -7)$.

A) $\dfrac{x}{2} = \dfrac{y}{3} = \dfrac{z}{-7}$
B) $\dfrac{x-6}{-4} = \dfrac{y-3}{3} = \dfrac{z-7}{-13}$
C) $\dfrac{x-6}{4} = \dfrac{y}{3} = \dfrac{z-6}{13}$
D) $\dfrac{x-4}{6} = y-3 = \dfrac{z-13}{6}$
E) $\dfrac{x-6}{-4} = \dfrac{y}{3} = \dfrac{z-6}{-13}$

17. Find the area of the parallelogram that has the vectors as adjacent sides.

$\mathbf{u} = -3\mathbf{i} - 2\mathbf{j} + 5\mathbf{k}, \mathbf{v} = \mathbf{i} - 2\mathbf{j} + \mathbf{k}$

A) 1
B) $\sqrt{3}$
C) $\sqrt{6}$
D) 16
E) $8\sqrt{3}$

18. Find $\mathbf{u} \times \mathbf{v}$.

$\mathbf{u} = -8\mathbf{i} - 5\mathbf{j} + 3\mathbf{k}, \mathbf{v} = -6\mathbf{i} - 7\mathbf{j} - 3\mathbf{k}$

A) $36\mathbf{i} - 42\mathbf{j} + 26\mathbf{k}$
B) 74
C) 4
D) $36\mathbf{i} + 42\mathbf{j} + 26\mathbf{k}$
E) $48\mathbf{i} + 35\mathbf{j} - 9\mathbf{k}$

19. Find the magnitude of the vector **v**.

$\mathbf{v} = \langle -4, 0, -8 \rangle$

A) -12
B) 12
C) $2\sqrt{3}$
D) $8\sqrt{5}$
E) $4\sqrt{5}$

20. Use the scalar triple product to find the volume of the parallelepiped having adjacent edges $\langle 1,3,5 \rangle$, $\langle 2,3,3 \rangle$, and $\langle 3,3,4 \rangle$.

A) 15
B) 45
C) 9
D) 147
E) 93

Answer Key

1. Answers may vary. One possible answer is shown below.
   $x=-3-8t, y=-7-4t, z=-6+6t$
2. A
3. C
4. C
5. B
6. Let point $P=(-9,6,-1)$, point $Q=(-8,3,-2)$, and point $R=(3,6,0)$.
   $$\left|\overline{PQ}\right|^2=(-8+9)^2+(3-6)^2+(-2+1)^2=11; \Rightarrow \left|\overline{PQ}\right|=\sqrt{11}$$
   $$\left|\overline{QR}\right|^2=(3+8)^2+(6-3)^2+(0+2)^2=134; \Rightarrow \left|\overline{QR}\right|=\sqrt{134}$$
   $$\left|\overline{PR}\right|^2=(3+9)^2+(6-6)^2+(0+1)^2=145; \Rightarrow \left|\overline{PR}\right|=\sqrt{145}$$
   Note that 11 + 134 = 145 in accordance with the Pythagorean Theorem.
7. Answers may vary. One possible answer is shown below.
   $x=-4+9t, y=2-4t, z=8-15t$
8. E
9. A
10. A
11. B
12. B
13. C
14. D
15. Answers may vary. One possible answer is shown below.
   $$\frac{x+5}{-4}=\frac{y-3}{5}=\frac{z-8}{7}$$
16. E
17. E
18. A
19. E
20. C

Name: ____________________________ Date: ____________

1. A union contract guarantees a 15% salary increase yearly for 3 years. For a current salary of \$31,000, the salary $f(t)$ (in thousands of dollars) for the next 3 years is given by

$$f(t) = \begin{cases} 31.00, & 0 < t \le 1 \\ 35.65, & 1 < t \le 2 \\ 41.00, & 2 < t \le 3 \end{cases}$$

where t represents the time in years. Find the limit of f as $t \to 1.00$, if it exists.
A) 31.00
B) 35.65
C) 1.00
D) 41.00
E) The limit does not exist.

2. Use the graph to find $\lim\limits_{x \to 3} \dfrac{4x^2 - 36}{x - 3}$.

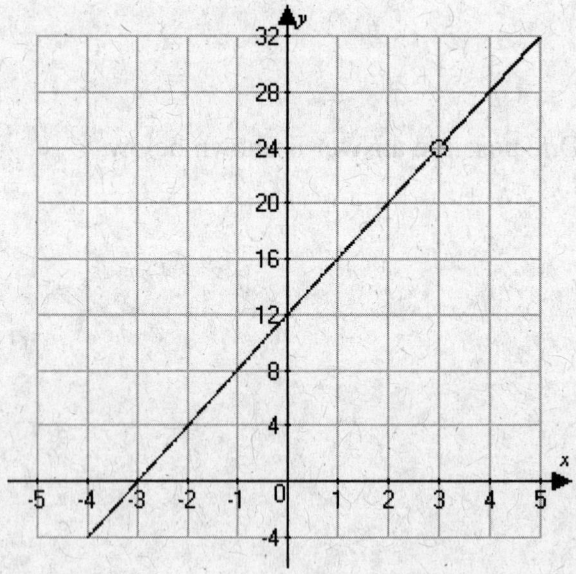

A) 24
B) ∞
C) 12
D) 0
E) limit does not exist

3. Find
$$\lim_{x\to 3}\left[g(x)-f(x)\right]$$
for $f(x)=4x^3$ and $g(x)=\dfrac{\sqrt{x^2+9}}{5x^2}$.
A) $\dfrac{\sqrt{2}}{15}-108$
B) $\dfrac{\sqrt{10}}{5}-108$
C) $\dfrac{36\sqrt{2}}{5}$
D) $\dfrac{\sqrt{10}}{5}+108$
E) limit does not exist

4. Use the first six terms to predict the limit of the sequence $a_n=\dfrac{5n^3+8}{n+3}$ (assume n begins with 1).
A) $\dfrac{5}{3}$
B) 111
C) 79
D) 0
E) the sequence diverges

5. The cost function for a certain model of a digital camera given by $C=12.00x+48,450$, where C is the cost (in dollars) and x is the number of cameras produced. Find the average cost per unit when $x=100$. Round your answer to the nearest cent.
A) \$49,650.00
B) \$484.50
C) \$12.00
D) \$496.50
E) \$484.62

6. Using the summation formulas and properties, evaluate the following expression.

$$\sum_{i=1}^{30} 8$$

A) 3720
B) 465
C) 232
D) 240
E) 38

7. Find an equation of the tangent line to the graph of the following function at the point $(3,-30)$.

$$-3x^2 - 3$$

A) $y = -18x - 84$
B) $y = -18x + 84$
C) $y = -18x + 24$
D) $y = 180x + 570$
E) $y = 180x - 570$

8. Use the graph to find $\lim_{x \to -1} \frac{1}{x+1}$, if it exists.

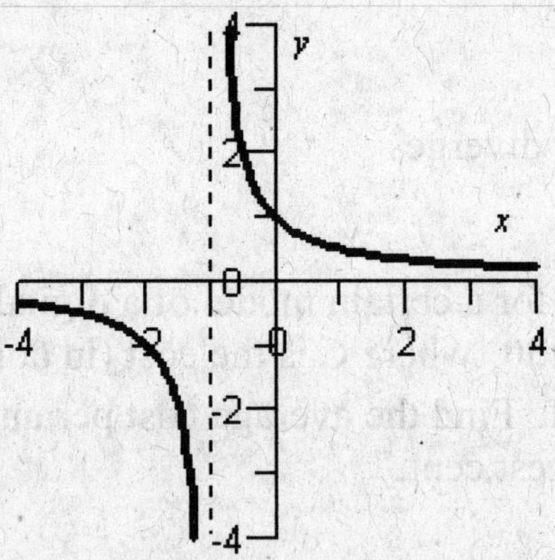

A) 0
B) 2
C) 4
D) –1
E) The limit does not exist.

9. Find
$$\lim_{x\to 6} \frac{4x}{\sqrt{x+4}}$$
by direct substitution.

A) $\dfrac{6}{\sqrt{10}}$
B) $4\sqrt{6}$
C) $24\sqrt{10}$
D) 1
E) $\dfrac{24}{\sqrt{10}}$

10. Use the graph to determine $\lim_{x\to 0} \dfrac{x^2 - x}{x}$ (if it exists).

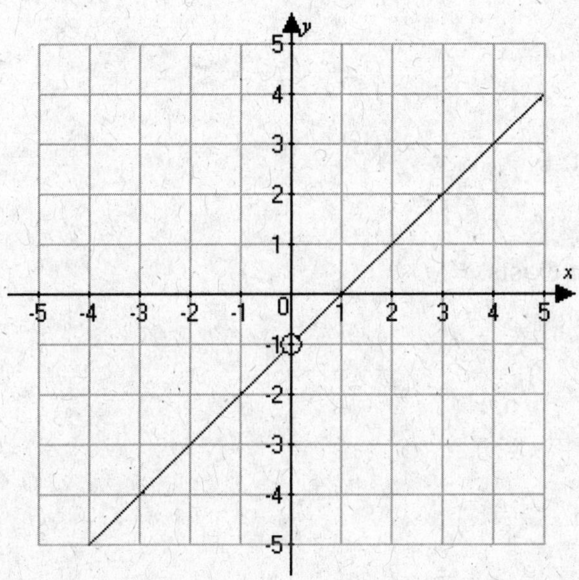

A) 0
B) 1
C) −1
D) 2
E) limit does not exist

11. Find $\lim_{x\to-\infty}\left[\frac{x}{(5+x)^2}+7\right]$ (if it exists).

A) $\frac{8}{25}$
B) –7
C) 7
D) ∞
E) limit does not exist

12. Estimate the following limit numerically, if it exists.

$$\lim_{x\to1}\frac{x-1}{x^2+5x-6}$$

A) $\frac{1}{7}$
B) 0
C) $-\frac{1}{5}$
D) $\frac{1}{5}$
E) The limit does not exist.

13. Find the following limit, if it exists.

$$\lim_{x\to3}x^7$$

A) 3
B) 21
C) –3
D) 2187
E) The limit does not exist.

14. Use the graph below to find $\lim_{x \to -10} \frac{|x+10|}{x+10}$, if it exists.

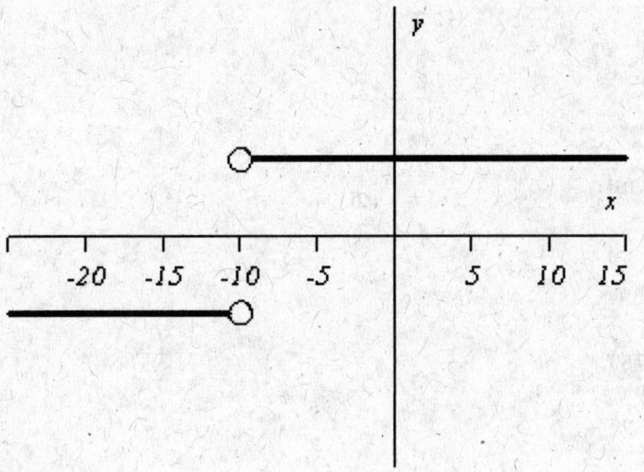

A) 1
B) −10
C) 10
D) −1
E) The limit does not exist.

15. Use the function below and its derivative to determine any points on the graph of *f* at which the tangent line is horizontal.

$f(x) = -3x^4 + 6x^2,\ f'(x) = -12x^3 + 12x$

A) $(-1,3),\ (0,0),\ (1,3)$
B) $(-1,-3),\ (1,-12)$
C) $(-1,3),\ (1,3)$
D) $(0,0),\ (1,0)$
E) $(0,0),\ (1,3)$

16. Find the slope of the graph of the following function at the point $(1,-3)$.

$-x^2 - 2$

A) 4
B) −2
C) 6
D) −3
E) −4

17. Find the following limit, if it exists.

$$\lim_{x\to\infty} \frac{3}{x^{-6}}$$

A) $\frac{1}{2}$
B) -1
C) 3
D) The limit does not exist.
E) 0

18. Find $\lim_{x\to\infty} \frac{1-6x}{1+8x}$ (if it exists).

A) 0
B) $-\frac{3}{4}$
C) $\frac{3}{4}$
D) $-\infty$
E) limit does not exist

19. Consider the graph of the function and approximate $\lim_{x\to 0} \frac{\sin(-8x)}{x}$, if it exists.

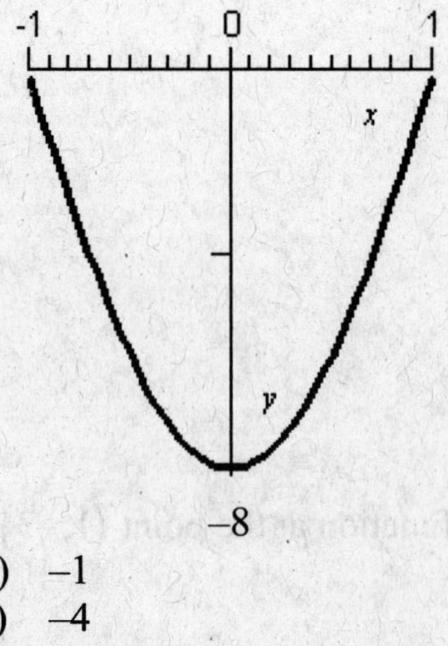

A) -1
B) -4
C) -8
D) 1
E) The limit does not exist.

20. Use the derivative of $f(x)=3x^3+9x$ to determine any points on the graph of $f(x)$ at which the tangent line is horizontal.
   A) $(1,12)$
   B) $(1,12)$ and $(-1,-12)$
   C) $(3,108)$ and $(-3,108)$
   D) $(0,0)$
   E) $f(x)$ has no points with a horizontal tangent line.

Answer Key

1. E
2. A
3. A
4. E
5. D
6. D
7. C
8. E
9. E
10. C
11. C
12. A
13. D
14. E
15. A
16. B
17. D
18. B
19. C
20. E

Name: ______________________________ Date: _______________

1. Use the function below and its derivative to determine any points on the graph of f at which the tangent line is horizontal.
   $f(x) = -2x^4 + 4x^2,\ f'(x) = -8x^3 + 8x$
   A) $(-1,2),\ (0,0),\ (1,2)$
   B) $(-1,-2),\ (1,-8)$
   C) $(-1,2),\ (1,2)$
   D) $(0,0),\ (1,0)$
   E) $(0,0),\ (1,2)$

2. Find $\lim\limits_{x \to 9^-} \dfrac{x-9}{x^2-81}$.
   A) $\dfrac{1}{18}$
   B) $\dfrac{1}{9}$
   C) $\dfrac{9}{2}$
   D) $\dfrac{2}{9}$
   E) limit does not exist

3. Find $\lim\limits_{x \to 0^-} \dfrac{x}{\sqrt{x+7}-\sqrt{7}}$
   A) 14
   B) $\dfrac{7}{2}$
   C) $\dfrac{\sqrt{7}}{2}$
   D) $2\sqrt{7}$
   E) limit does not exist

4. Find the following limit.

$$\lim_{x \to -4} 3x^2 + 3x + 1$$

A) 52
B) 36
C) 37
D) –23
E) –35

5. Use the figure below to approximate the slope of the curve at the point (x, y).

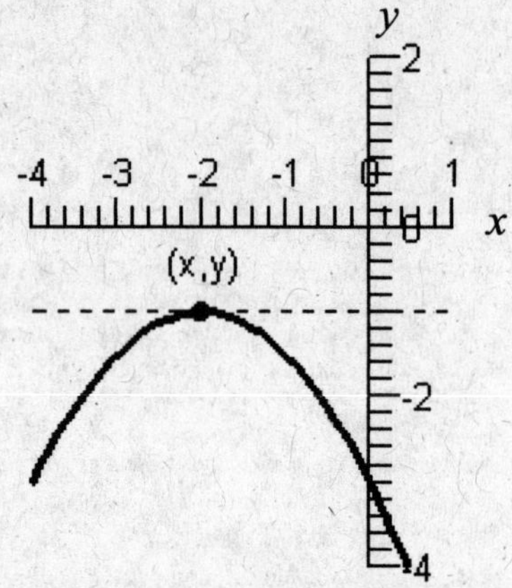

A) $-\frac{1}{2}$
B) The slope is undefined.
C) 1
D) $\frac{1}{2}$
E) 0

6. Use the graph to find $\lim_{x \to 3} \frac{4x^2 - 36}{x-3}$.

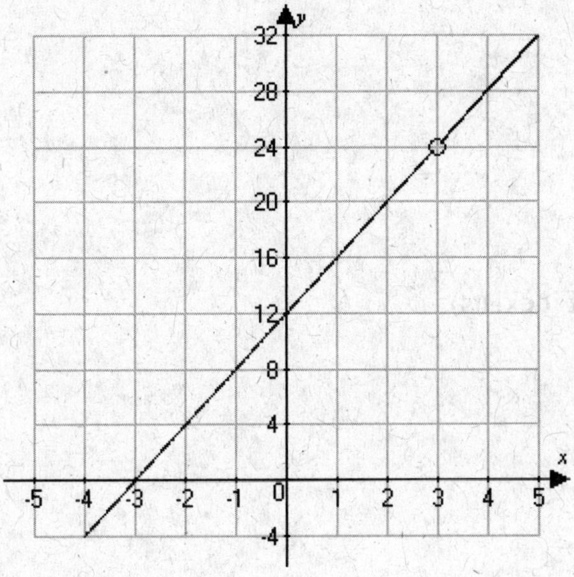

A) 24
B) ∞
C) 12
D) 0
E) limit does not exist

7. The cost function for a certain model of a digital camera given by $C = 13.50x + 48{,}950$, where C is the cost (in dollars) and x is the number of cameras produced. Find the average cost per unit when $x = 200$. Round your answer to the nearest cent.
A) \$51,650.00
B) \$244.75
C) \$13.50
D) \$258.25
E) \$244.82

8. Find the following limit, if it exists.
$\lim_{x \to -3} x^5$
A) –3
B) –15
C) 3
D) –243
E) The limit does not exist.

9. Find the slope of the graph of the following function at the point $(-1,1)$.

   $3x^2 - 2$

   A) 4
   B) –6
   C) 6
   D) 1
   E) –8

10. Find $\lim_{x \to -\infty} \left[\frac{x}{(5+x)^2} + 8 \right]$ (if it exists).

   A) $\frac{9}{25}$
   B) –8
   C) 8
   D) ∞
   E) limit does not exist

11. Find the following limit, if it exists.

   $\lim_{x \to -3} \frac{9 - 3x - 2x^2}{3 + x}$

   A) –3
   B) 9
   C) 7
   D) –11
   E) The limit does not exist.

12. A union contract guarantees a 16% salary increase yearly for 3 years. For a current salary of $31,500, the salary $f(t)$ (in thousands of dollars) for the next 3 years is given by

$$f(t) = \begin{cases} 31.50, & 0 < t \le 1 \\ 36.54, & 1 < t \le 2 \\ 42.39, & 2 < t \le 3 \end{cases}$$

where t represents the time in years. Find the limit of f as $t \to 1.00$, if it exists.
A) 31.50
B) 36.54
C) 1.00
D) 42.39
E) The limit does not exist.

13. Use the limit process to find the area of the region between $f(x) = 6x + 7$ and the x-axis on the interval $[0,5]$.
A) 185
B) 110
C) 65
D) 75
E) 82

14. Use the limit process to find the area of the region between $f(x) = \frac{1}{4}\left(x^2 + 4x\right)$ and the x-axis on the interval $[1,4]$.
A) $\frac{143}{2}$
B) $\frac{81}{4}$
C) $\frac{47}{2}$
D) 78
E) $\frac{51}{4}$

15. Use the limit process to find the slope of the graph of $\sqrt{x+4}$ at $(5,3)$.
   A) $\frac{1}{3}$
   B) 3
   C) ∞
   D) $\frac{1}{6}$
   E) the slope is undefined at this point

16. Complete the table and numerically estimate the limit as x approaches infinity for $f(x)=x-\sqrt{x^2+4}$.

x	10^0	10^1	10^2	10^3	10^4	10^5	10^6
$f(x)$							

   A) 0
   B) 4
   C) –4
   D) ∞
   E) limit does not exist

17. Find the derivative of $f(x)=4x^2+2x+3$.
   A) $8x+2$
   B) $4x+2$
   C) $8x+5$
   D) $4x+5$
   E) $8x$

18. Find the following limit. Round your answer to two decimals.

$$\lim_{x\to 3}\frac{\sqrt{x^2+7}}{2x^2}$$

   A) 0.22
   B) 1.32
   C) 0.44
   D) 0.18
   E) 1.41

19. Find

$$\lim_{x\to 8}\frac{5x}{\sqrt{x+2}}$$

by direct substitution.

A) $\frac{8}{\sqrt{10}}$
B) $10\sqrt{2}$
C) $40\sqrt{10}$
D) 1
E) $\frac{40}{\sqrt{10}}$

20. Use the limit process to find the slope of the graph of the following function at the point $(5,-27)$.

$$g(x)=-7-4x$$

A) –11
B) –27
C) –3
D) –4
E) 5

Answer Key

1. A
2. A
3. D
4. C
5. E
6. A
7. D
8. D
9. B
10. C
11. B
12. E
13. B
14. E
15. D
16. A
17. A
18. A
19. E
20. D

Name: ______________________________ Date: ______________

1. Find $\lim\limits_{h\to 0}\dfrac{f(x+h)-f(x)}{h}$ for $f(x)=2+6x$.
   A) –6
   B) –4
   C) 6
   D) 2
   E) limit does not exist

2. Find the following limit, if it exists.
   $$\lim_{x\to\infty}\frac{4x^2+3}{-2-x^2}$$
   A) $-\dfrac{7}{3}$
   B) –4
   C) 4
   D) 7
   E) The limit does not exist.

3. Use the figure below to approximate the slope of the curve at the point (x, y).

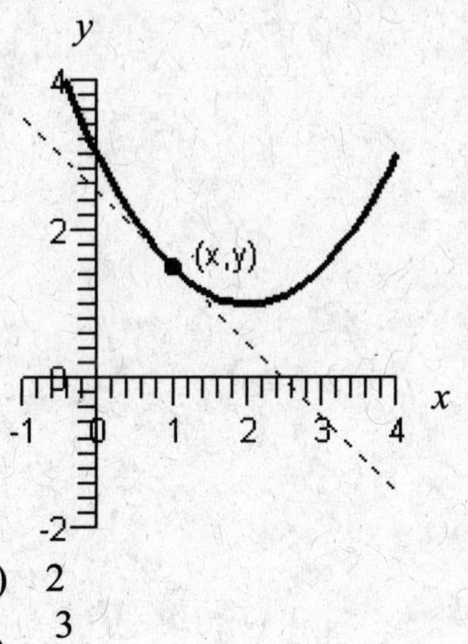

   A) 2
   B) $\dfrac{3}{2}$
   C) –2
   D) –1
   E) 1

4. Find

$$\lim_{x \to 7} \frac{x-8}{x^2+12x+27}$$

by direct substitution.

A) $\frac{1}{160}$
B) $-\frac{1}{160}$
C) 0
D) ∞
E) limit does not exist

5. Using the summation formulas and properties, evaluate the following expression.

$$\sum_{i=1}^{60} 4$$

A) 7320
B) 1830
C) 236
D) 240
E) 64

6. Use the graph to find $\lim_{x \to 3} \frac{4x^2 - 36}{x - 3}$.

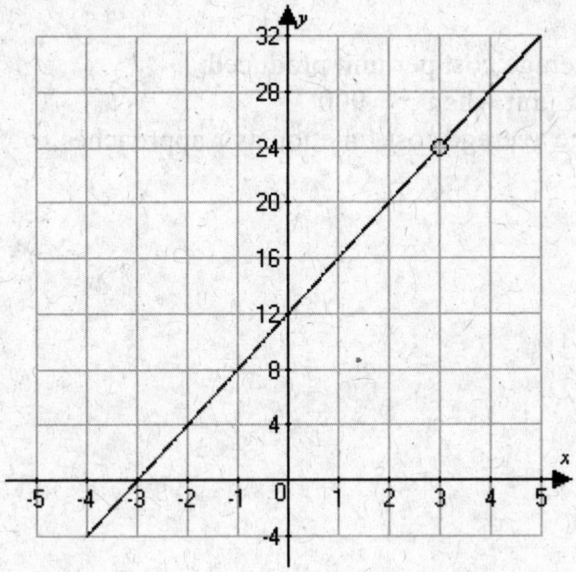

A) 24
B) ∞
C) 12
D) 0
E) limit does not exist

7. Find $\lim_{x \to \infty} \frac{1 - 3x}{1 + 7x}$ (if it exists).

A) 0
B) $-\frac{3}{7}$
C) $\frac{3}{7}$
D) $-\infty$
E) limit does not exist

8. The cost function for a certain graphing calculator is given by $C = 11.35x + 53{,}000$ where C is in dollars and x is the number of calculators produced.

   a. Write a model for the average cost per unit produced.
   b. Find the average cost per unit when $x = 900$.
   c. Determine the limit of the average cost function as x approaches ∞.

9. Find

$$\lim_{x \to 6} \frac{\sqrt{x+7}}{x-2}$$

by direct substitution.

A) $\frac{\sqrt{6}}{4}$
B) $\frac{\sqrt{13}}{2}$
C) $\frac{\sqrt{13}}{4}$
D) $\frac{1}{4}$
E) $\sqrt{13}$

10. If $f(x) = -3x^2 - 4x$, find the following limit, if it exists.

$$\lim_{h \to 0} \frac{f(x+h) - f(x)}{h}$$

A) $-6x - 4$
B) $-3x - 4$
C) $2x - 4$
D) $-6x$
E) The limit does not exist.

11. Find $\lim_{x\to 0^-} \dfrac{x}{\sqrt{x+15}-\sqrt{15}}$
    A) 30
    B) $\dfrac{15}{2}$
    C) $\dfrac{\sqrt{15}}{2}$
    D) $2\sqrt{15}$
    E) limit does not exist

12. Use the limit process to find the slope of the graph of $\sqrt{x+12}$ at $(4,4)$.
    A) $\dfrac{1}{4}$
    B) 4
    C) ∞
    D) $\dfrac{1}{8}$
    E) the slope is undefined at this point

13. Consider the following graph of the function and approximate $\lim_{x\to 0} \dfrac{e^{-4x}-1}{x}$, if it exists.

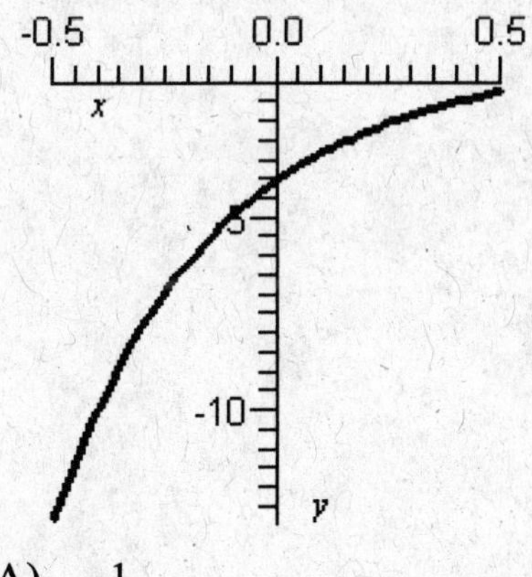

    A) –1
    B) –4
    C) 0
    D) 1
    E) The limit does not exist.

14. Approximate the area of the indicated region under the given curve using five rectangles.

$f(x) = 5 - x^2$

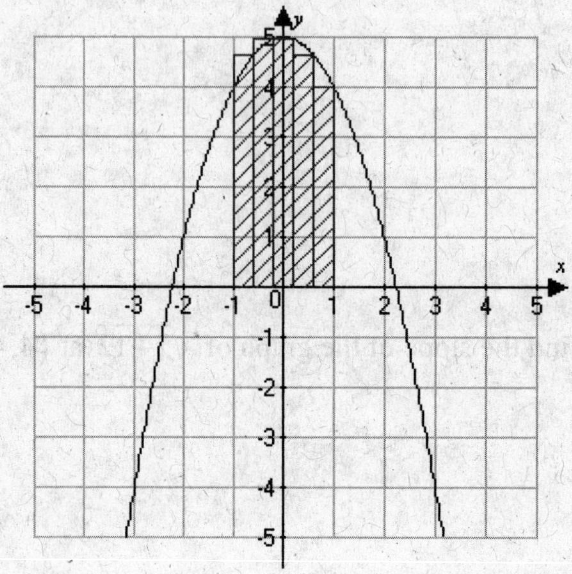

A) 9.28
B) 4.64
C) 5.48
D) 4.83
E) 9.25

15. Estimate the following limit numerically, if it exists.

$$\lim_{x \to 1} \frac{x-1}{x^2+6x-7}$$

A) $\frac{1}{8}$
B) 0
C) $-\frac{1}{6}$
D) $\frac{1}{6}$
E) The limit does not exist.

16. Find the following limit, if it exists.

$$\lim_{x\to\infty}\frac{5}{x^8}$$

A) $\frac{5}{8}$
B) –5
C) –1
D) 0
E) The limit does not exist.

17. Find

$$\lim_{x\to 3}\left[g(x)-f(x)\right]$$

for $f(x)=3x^3$ and $g(x)=\dfrac{\sqrt{x^2+2}}{6x^2}$.

A) $\dfrac{\sqrt{11}}{54}-81$
B) $\dfrac{\sqrt{3}}{6}-81$
C) $\dfrac{3\sqrt{11}}{2}$
D) $\dfrac{\sqrt{3}}{6}+81$
E) limit does not exist

18. Find $\displaystyle\lim_{h\to 0}\frac{f(x+h)-f(x)}{h}$ for $f(x)=\sqrt{x-6}$.

A) $\dfrac{1}{6\sqrt{x-6}}$
B) $\dfrac{1}{2\sqrt{x-6}}$
C) $\dfrac{\sqrt{x-6}}{2}$
D) $\sqrt{x-6}$
E) limit does not exist

19. Find $\lim_{x\to\infty} \frac{6}{5x}$ (if it exists).
   A) 0
   B) $\frac{6}{5}$
   C) ∞
   D) $\frac{1}{30}$
   E) limit does not exist

20. Graph $f(x)=\begin{cases} 3x+2 & x<-3 \\ x-4 & x\geq -3 \end{cases}$ and find the limit of $f(x)$ as x approaches –3.

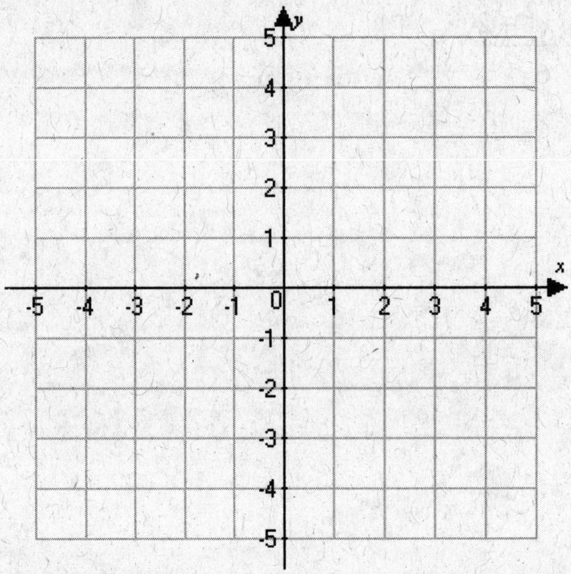

   A) –35
   B) –7
   C) –28
   D) 7
   E) limit does not exist

Answer Key

1. C
2. B
3. D
4. B
5. D
6. A
7. B
8.
   a. $\dfrac{11.35x + 53{,}000}{x}$

   b. \$70.24

   c. \$11.35
9. C
10. A
11. D
12. D
13. B
14. A
15. A
16. D
17. A
18. B
19. A
20. B

Name: ____________________________ Date: _____________

1. Find

$$\lim_{x \to -3} \left(\tfrac{1}{3} x^3 - 2x \right)$$

by direct substitution.

A) –15
B) 33
C) –3
D) –29
E) limit does not exist

2. Find the following limit, if it exists.

$$\lim_{x \to -3} x^8$$

A) –3
B) –24
C) 3
D) 6561
E) The limit does not exist.

3. Find $\lim_{x \to -\infty} \left[\frac{x}{(7+x)^2} - 6 \right]$ (if it exists).

A) $-\frac{5}{49}$
B) 6
C) –6
D) ∞
E) limit does not exist

4. Use the limit process to find the area of the region between $f(x)=\frac{1}{8}\left(x^2+8x\right)$ and the x-axis on the interval $[1,8]$.

A) $\frac{1087}{2}$
B) $\frac{2023}{24}$
C) $\frac{191}{2}$
D) 574
E) $\frac{1267}{24}$

5. Use the graph below to find $\lim_{x\to -2}\frac{|x+2|}{x+2}$, if it exists.

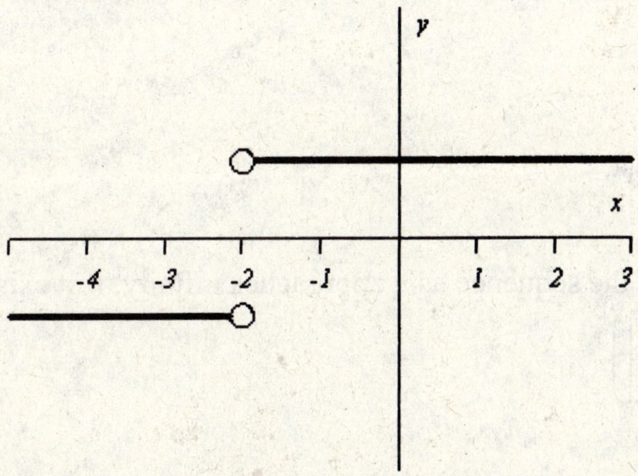

A) –1
B) –2
C) 2
D) 1
E) The limit does not exist.

6. Determine any points on the graph of the following function at which the tangent line is horizontal.

   $f(x) = x^2 + 6x - 8$

   A) $(3,19)$
   B) $(3,0)$
   C) $(-3,0)$
   D) $(1,-1)$
   E) $(-3,-17)$

7. Estimate the following limit numerically, if it exists.

   $\lim_{x \to 1} \frac{x-1}{x^2+4x-5}$

   A) $\frac{1}{6}$
   B) 0
   C) $-\frac{1}{4}$
   D) $\frac{1}{4}$
   E) The limit does not exist.

8. Find the following limit of the sequence as n approaches infinity, if it exists.

   $a_n = \frac{1}{n}\left(n + \frac{4}{n}\left[\frac{n(n+1)}{6}\right]\right)$

   A) $\frac{2}{3}$
   B) $\frac{8}{3}$
   C) $\frac{5}{3}$
   D) $\frac{11}{6}$
   E) The limit does not exist.

9. Find $\lim_{h \to 0} \dfrac{f(x+h)-f(x)}{h}$ for $f(x)=8-x$.
   A) 1
   B) 9
   C) –1
   D) 8
   E) limit does not exist

10. Find $\lim_{x \to 15^-} \dfrac{x-15}{x^2-225}$.
   A) $\dfrac{1}{30}$
   B) $\dfrac{1}{15}$
   C) $\dfrac{15}{2}$
   D) $\dfrac{2}{15}$
   E) limit does not exist

11. Complete the table and use the result to estimate
$\lim_{x \to -8} \dfrac{x+8}{x^2+4x-32}$
numerically.

x	–8.1	–8.01	–8.001	–8	–7.999	–7.99	–7.9
$f(x)$				?			

   A) $-\dfrac{1}{12}$
   B) 12
   C) ∞
   D) –12
   E) limit does not exist

12. Use the figure below to approximate the slope of the curve at the point (x, y).

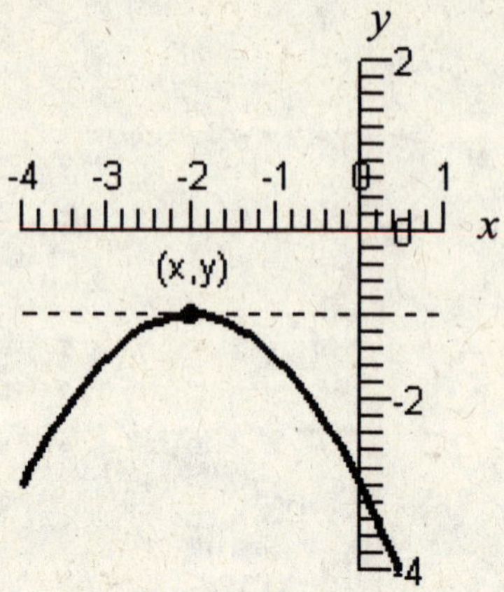

A) $-\frac{1}{2}$
B) The slope is undefined.
C) 1
D) $\frac{1}{2}$
E) 0

13. Complete the table and use the result to estimate $\lim_{x \to 5} (2x+4)$ numerically.

x	4.9	4.99	4.999	5	5.001	5.01	5.1
$f(x)$				?			

A) 9
B) 10
C) 14
D) 6
E) limit does not exist

14. Find the following limit, if it exists.

$$\lim_{x \to -3} \frac{9 - 6x - 3x^2}{3 + x}$$

A) –6
B) 12
C) 6
D) –12
E) The limit does not exist.

15. Find

$$\lim_{x \to 7} \frac{\sqrt{x+3}}{x-3}$$

by direct substitution.

A) $\frac{\sqrt{7}}{4}$

B) $\frac{\sqrt{10}}{3}$

C) $\frac{\sqrt{10}}{4}$

D) $\frac{1}{4}$

E) $\sqrt{10}$

16. Use the figure below to approximate the slope of the curve at the point (x, y).

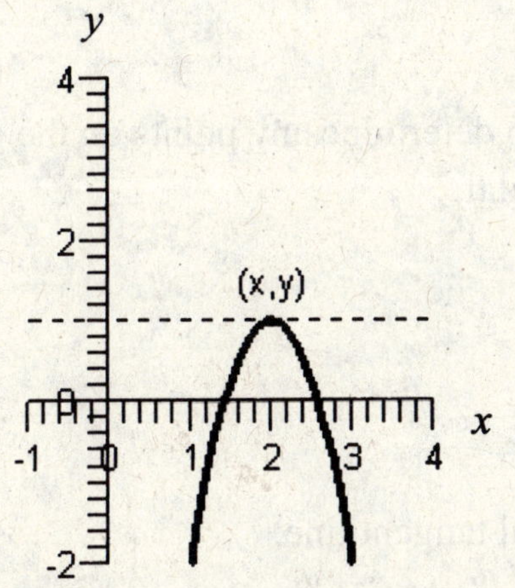

A) 1

B) $\frac{1}{2}$

C) -1

D) 0

E) The slope is undefined.

17. Consider the following graph of the function and approximate $\lim_{x\to 0}\dfrac{e^{-4x}-1}{x}$, if it exists.

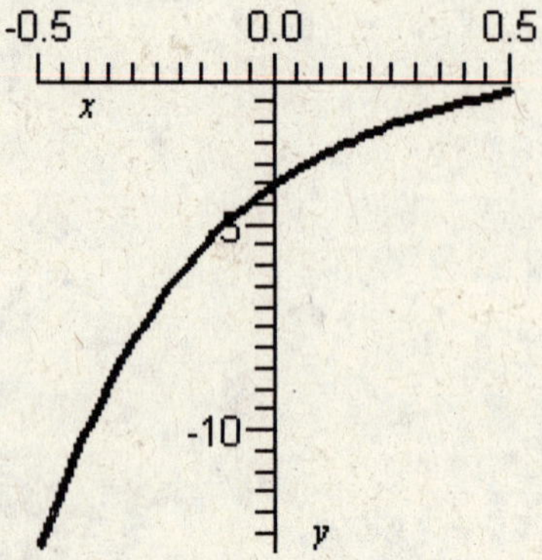

A) –1
B) –4
C) 0
D) 1
E) The limit does not exist.

18. Use the derivative of $f(x)=5x^3+15x$ to determine any points on the graph of $f(x)$ at which the tangent line is horizontal.

A) $(1,20)$
B) $(1,20)$ and $(-1,-20)$
C) $(5,700)$ and $(-5,700)$
D) $(0,0)$
E) $f(x)$ has no points with a horizontal tangent line.

19. Consider the graph of the function and approximate $\lim_{x \to 0} \frac{\sin(8x)}{x}$, if it exists.

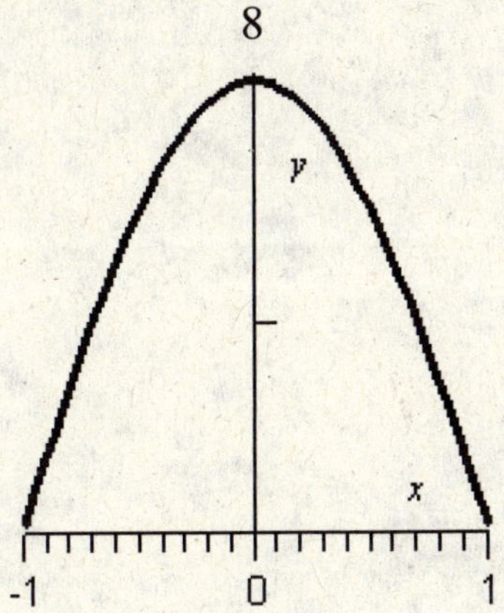

A) 1
B) 4
C) 8
D) −1
E) The limit does not exist.

20. Use the first six terms to predict the limit of the sequence $a_n = \frac{5n^3}{n^3 + 3}$ (assume n begins with 1).
A) 4
B) 5
C) $\frac{23}{6}$
D) $\frac{32}{5}$
E) the sequence diverges

Answer Key

1. C
2. D
3. C
4. E
5. E
6. E
7. A
8. C
9. C
10. A
11. A
12. E
13. C
14. B
15. C
16. D
17. B
18. E
19. C
20. B

Name: ______________________________ Date: _____________

1. Find an equation of the tangent line to the graph of the following function at the point $(-1,-5)$.

   $-3x^2-2$

   A) $y=6x-11$
   B) $y=6x+11$
   C) $y=6x+1$
   D) $y=30x-25$
   E) $y=30x+25$

2. Graph $f(x)=\begin{cases}4x+3 & x<0\\ x+3 & x\geq 0\end{cases}$ and find the limit of $f(x)$ as x approaches 0.

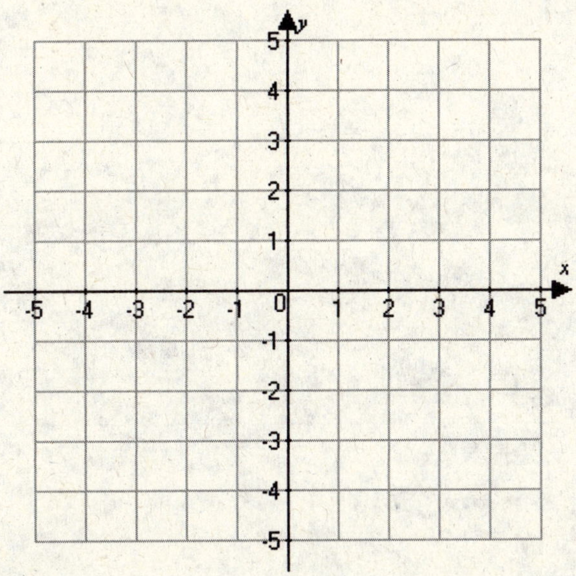

   A) 15
   B) 3
   C) 12
   D) –3
   E) limit does not exist

3. Use the graph to find $\lim_{x \to 2} \frac{3x^2 - 12}{x - 2}$.

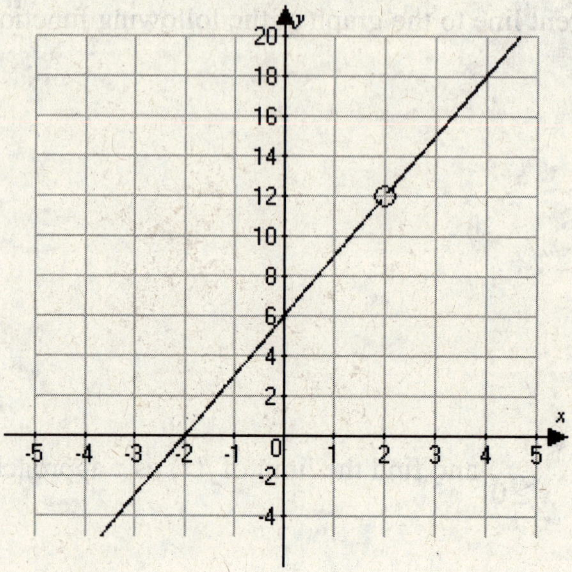

A) 12
B) ∞
C) 6
D) 0
E) limit does not exist

4. Use asymptotes to match $f(x) = \dfrac{4x^2}{x^2 + 1}$ with its graph.

A)

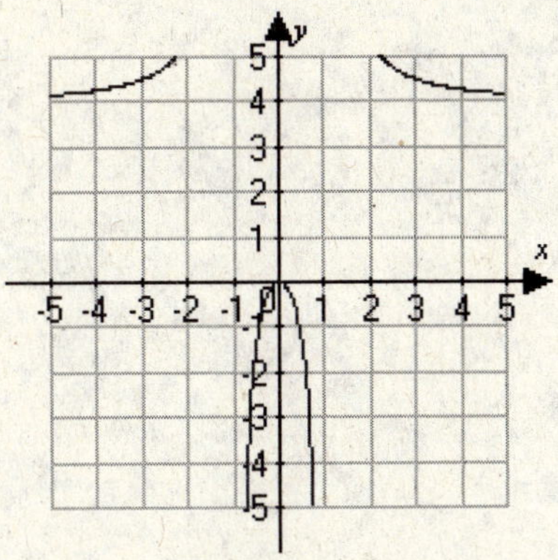

B)

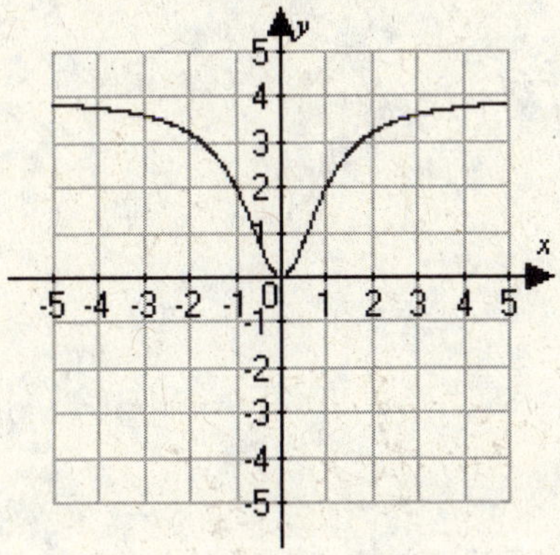

C)

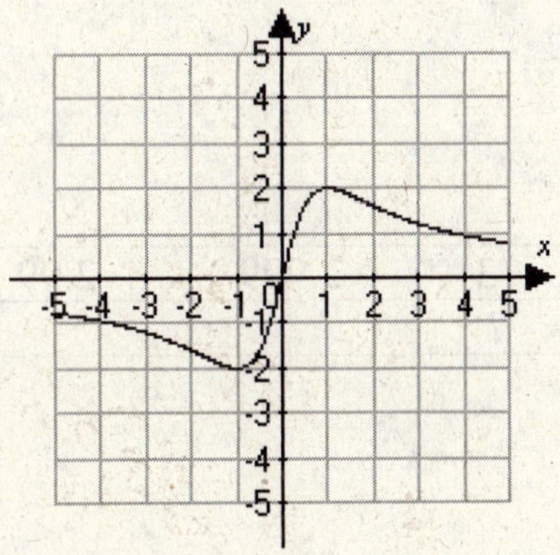

D)

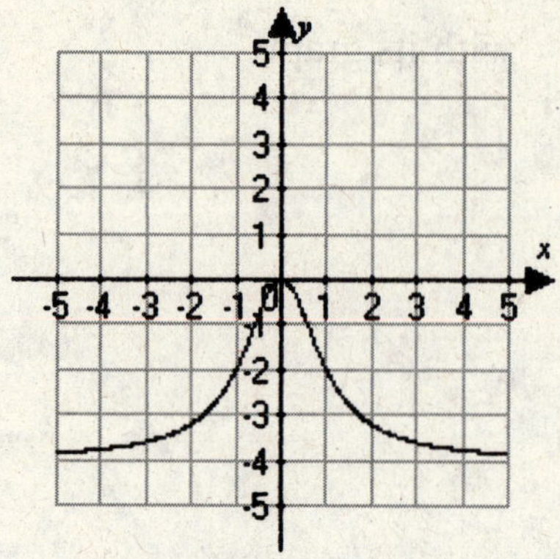

E)

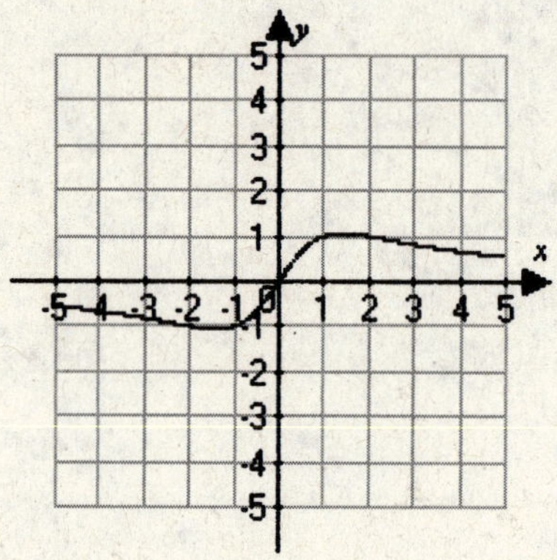

5. Complete the table and use the result to estimate

$$\lim_{x \to -3} \frac{x+3}{x^2 - x - 12}$$

numerically.

x	–3.1	–3.01	–3.001	–3	–2.999	–2.99	–2.9
$f(x)$				?			

A) $-\frac{1}{7}$
B) 7
C) ∞
D) –7
E) limit does not exist

6. Find $\lim_{h \to 0} \dfrac{f(x+h)-f(x)}{h}$ for $f(x)=4-6x$.
   A) 6
   B) 10
   C) –6
   D) 4
   E) limit does not exist

7. Use the graph below to find $\lim_{x \to -4} \dfrac{|x+4|}{x+4}$, if it exists.

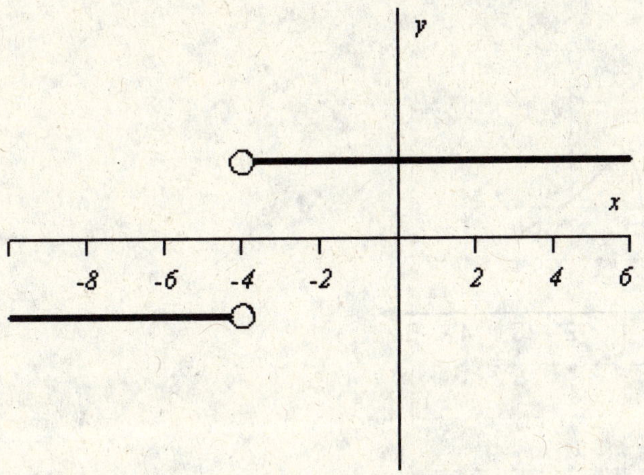

   A) 1
   B) –4
   C) 4
   D) –1
   E) The limit does not exist.

8. Use the limit process to find the slope of the graph of $4x-8x^2$ at $(3,-60)$.
   A) –12
   B) 52
   C) 28
   D) 48
   E) –44

9. If $f(x) = -2x^2 + x$, find the following limit, if it exists.

$$\lim_{h \to 0} \frac{f(x+h) - f(x)}{h}$$

A) $-4x+1$
B) $-2x+1$
C) $2x+1$
D) $-4x$
E) The limit does not exist.

10. Use the graph to find $\lim_{x \to 0} 3\sin\left(\frac{\pi}{3x}\right)$.

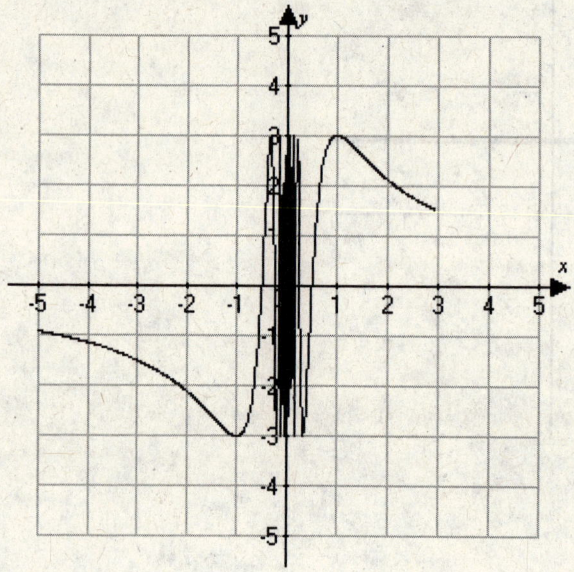

A) $\frac{3\sqrt{3}}{2}$
B) ∞
C) 0
D) 3
E) limit does not exist

11. Use the figure below to approximate the slope of the curve at the point (x, y).

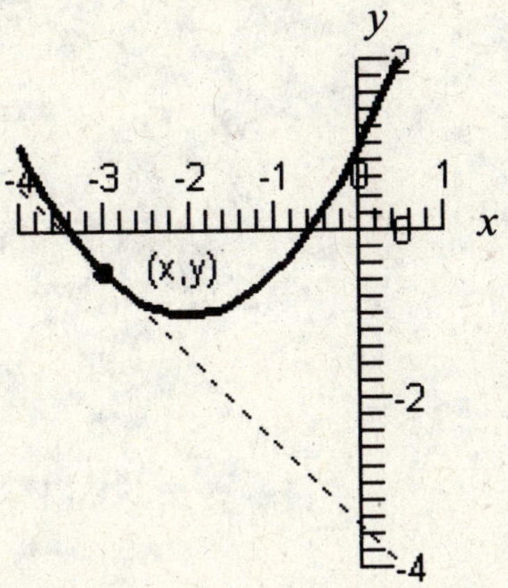

A) $-\frac{3}{2}$
B) 0
C) 1
D) $\frac{3}{2}$
E) -1

12. Rewrite $\sum_{i=1}^{n}\left[\frac{12}{n}+\left(\frac{11i}{n^2}\right)\right]\left(\frac{8i}{n}\right)$ as a rational function $S(n)$ and find $\lim_{n\to\infty} S(n)$.

A) $S(n)=\frac{4\left[36n+11(n+1)\right]}{3n^2}$, $\lim_{n\to\infty} S(n)=\frac{232}{3}$
B) $S(n)=\frac{4\left[36n(n+1)+11(n+1)(2n+1)\right]}{3n^2}$, $\lim_{n\to\infty} S(n)=\frac{232}{3}$
C) $S(n)=\frac{4\left[36n(n+1)+11(n+1)(2n+1)\right]}{3n^2}$, $\lim_{n\to\infty} S(n)=\frac{5}{6}$
D) $S(n)=\frac{4\left[36+11n\right]}{3n}$, $\lim_{n\to\infty} S(n)=\frac{232}{3}$
E) $S(n)=\frac{4\left[36n(n+1)+11(n+1)(2n+1)\right]}{3n^2}$, limit does not exist

13. Find $\lim_{x\to\infty}\frac{1-7x}{1+9x}$ (if it exists).
A) 0
B) $-\frac{7}{9}$
C) $\frac{7}{9}$
D) $-\infty$
E) limit does not exist

14. Find the following limit, if it exists.
$\lim_{x\to -2} x^8$
A) –2
B) –16
C) 2
D) 256
E) The limit does not exist.

15. Approximate the area of the region under the function below on the interval $[0,2]$ using 8 rectangles. Round your answer to two decimals.
$f(x)=\frac{1}{3}x^4$
A) 5.71 square units
B) 2.86 square units
C) 3.69 square units
D) 8.57 square units
E) 17.13 square units

16. Evaluate $\sum_{j=1}^{10}\left(j^3-4j^2\right)$ using the summation formulas and properties.
A) 3,025
B) 2,625
C) 3,021
D) 2,985
E) 1,485

17. Use the function below and its derivative to determine any points on the graph of f at which the tangent line is horizontal.

    $f(x)=3x^4-6x^2,\ f'(x)=12x^3-12x$

    A) $(-1,-3),\ (0,0),\ (1,-3)$
    B) $(-1,3),\ (1,12)$
    C) $(-1,-3),\ (1,-3)$
    D) $(0,0),\ (1,0)$
    E) $(0,0),\ (1,-3)$

18. Consider the following graph of the function and approximate $\lim_{x\to 0}\frac{e^{-2x}-1}{x}$, if it exists.

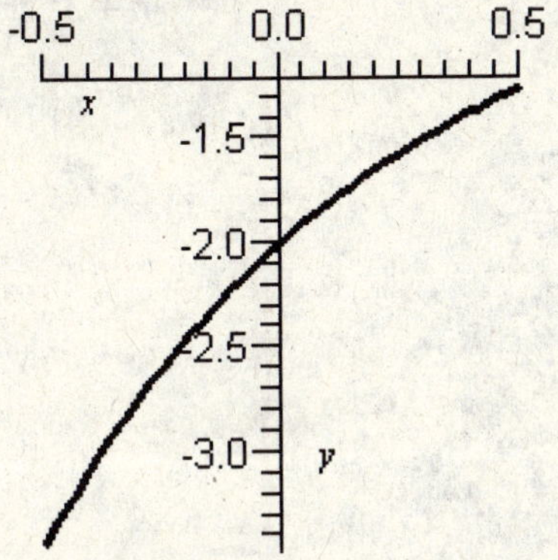

A) –1
B) –2
C) 0
D) 1
E) The limit does not exist.

19. Find $\lim\limits_{x \to 13^-} \dfrac{x-13}{x^2-169}$.

A) $\dfrac{1}{26}$

B) $\dfrac{1}{13}$

C) $\dfrac{13}{2}$

D) $\dfrac{2}{13}$

E) limit does not exist

20. Use the first six terms to predict the limit of the sequence $a_n = \dfrac{8n^3+6}{n+3}$ (assume n begins with 1).

A) $\dfrac{8}{3}$

B) 183

C) 126

D) 0

E) the sequence diverges

Answer Key

1. C
2. B
3. A
4. B
5. A
6. C
7. E
8. E
9. A
10. E
11. E
12. B
13. B
14. D
15. B
16. E
17. A
18. B
19. A
20. E

Name: ___________________________ Date: _____________

1. Use the limit process to find the slope of the graph of $\sqrt{x+21}$ at $(4,5)$.

A) $\dfrac{1}{5}$
B) 5
C) ∞
D) $\dfrac{1}{10}$
E) the slope is undefined at this point

2. Use the limit process to find the slope of the graph of the following function at the point $(3,-19)$.

$g(x) = -10 - 3x$

A) –13
B) –19
C) –7
D) –3
E) 3

3. Find $\lim\limits_{y\to 0} \dfrac{\sqrt{15+y}-\sqrt{15}}{y}$.

A) $\dfrac{\sqrt{15}}{30}$
B) $\dfrac{\sqrt{15}}{15}$
C) 0
D) $\dfrac{\sqrt{15}}{2}$
E) limit does not exist

4. Use the limit process to find the area of the region between $f(x) = x^2 + 5$ and the x-axis on the interval $[0, 6]$.
   A) 246
   B) 72
   C) 221
   D) 102
   E) 77

5. Find the derivative of $f(x) = 7x^2 - 9x + 4$.
   A) $14x - 9$
   B) $7x - 9$
   C) $14x - 5$
   D) $7x - 5$
   E) $14x$

6. Find $\lim_{t \to 3} \frac{t^3 - 27}{t - 3}$.
   A) 9
   B) 18
   C) 27
   D) 3
   E) limit does not exist

7. Evaluate $\sum_{j=1}^{15} \left(j^3 - 5j^2 \right)$ using the summation formulas and properties.
   A) 14,400
   B) 13,275
   C) 14,395
   D) 14,325
   E) 8,200

8. Find the following limit of the sequence as n approaches infinity, if it exists.

$$a_n = \frac{1}{n}\left(n + \frac{3}{n}\left[\frac{n(n+1)}{5}\right]\right)$$

A) $\frac{3}{5}$
B) $\frac{13}{5}$
C) $\frac{8}{5}$
D) $\frac{9}{5}$
E) The limit does not exist.

9. Determine any points on the graph of the following function at which the tangent line is horizontal.

$$f(x) = x^2 + 8x - 8$$

A) $(4,40)$
B) $(4,0)$
C) $(-4,0)$
D) $(0,-8)$
E) $(-4,-24)$

10. Rewrite $\sum_{i=1}^{n} \frac{9i^3}{n^5}$ as a rational function $S(n)$ and find $\lim_{n\to\infty} S(n)$.

A) $S(n) = \frac{9(n+1)^2}{4n^2}$, $\lim_{n\to\infty} S(n) = \frac{9}{4}$

B) $S(n) = \frac{9n^2(n+1)^2}{4}$, $\lim_{n\to\infty} S(n) = 9$

C) $S(n) = \frac{n^2(n+1)^2}{36}$, $\lim_{n\to\infty} S(n) = 0$

D) $S(n) = \frac{9(n+1)^2}{4n^3}$, $\lim_{n\to\infty} S(n) = 0$

E) $S(n) = \frac{9(n+1)^2}{4n^3}$, the limit does not exist

11. Use the graph to determine $\lim_{x\to 0} \frac{x^2-3x}{x}$ (if it exists).

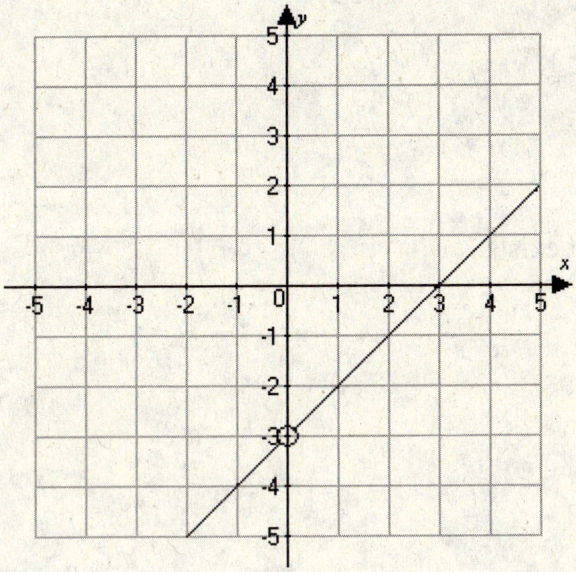

A) 0
B) 3
C) −3
D) 6
E) limit does not exist

12. Find $\lim_{h\to 0} \frac{f(x+h)-f(x)}{h}$ for $f(x)=\sqrt{x-7}$.

A) $\frac{1}{7\sqrt{x-7}}$
B) $\frac{1}{2\sqrt{x-7}}$
C) $\frac{\sqrt{x-7}}{2}$
D) $\sqrt{x-7}$
E) limit does not exist

13. Find $\lim_{h\to 0} \dfrac{f(x+h)-f(x)}{h}$ for $f(x) = -4 + 7x$.
   A) –7
   B) –11
   C) 7
   D) –4
   E) limit does not exist

14. Find the following limit, if it exists.
   $$\lim_{x\to\infty} \frac{6x^2-4}{-3-x^2}$$
   A) $-\dfrac{1}{2}$
   B) –6
   C) 6
   D) –4
   E) The limit does not exist.

15. Find $\lim_{x\to 0^-} \dfrac{x}{\sqrt{x+3}-\sqrt{3}}$
   A) 6
   B) $\dfrac{3}{2}$
   C) $\dfrac{\sqrt{3}}{2}$
   D) $2\sqrt{3}$
   E) limit does not exist

16. Find

$$\lim_{x\to 3}\left[g(x)-f(x)\right]$$

for $f(x)=3x^3$ and $g(x)=\dfrac{\sqrt{x^2+2}}{5x^2}$.

A) $\dfrac{\sqrt{11}}{45}-81$
B) $\dfrac{\sqrt{3}}{5}-81$
C) $\dfrac{9\sqrt{11}}{5}$
D) $\dfrac{\sqrt{3}}{5}+81$
E) limit does not exist

17. Complete the table and use the result to estimate

$$\lim_{x\to -2}\frac{x+2}{x^2-x-6}$$

numerically.

x	–2.1	–2.01	–2.001	–2	–1.999	–1.99	–1.9
$f(x)$				?			

A) $-\dfrac{1}{5}$
B) 5
C) ∞
D) –5
E) limit does not exist

18. Use the function below and its derivative to determine any points on the graph of f at which the tangent line is horizontal.

$$f(x)=3x^4-6x^2,\quad f'(x)=12x^3-12x$$

A) $(-1,-3)$, $(0,0)$, $(1,-3)$
B) $(-1,3)$, $(1,12)$
C) $(-1,-3)$, $(1,-3)$
D) $(0,0)$, $(1,0)$
E) $(0,0)$, $(1,-3)$

19. Use the figure below to approximate the slope of the curve at the point (x, y).

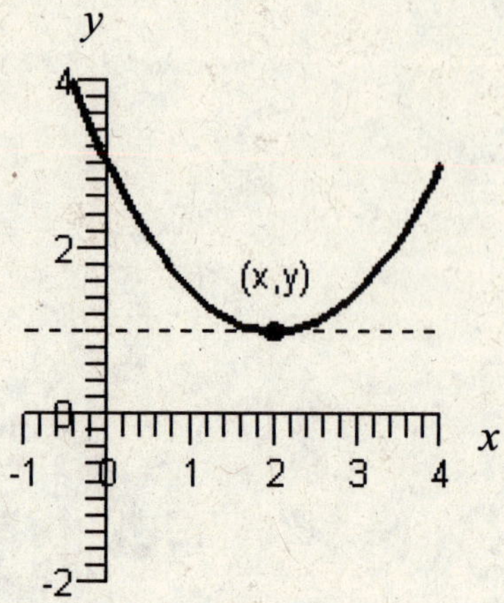

A) 3
B) $\frac{1}{2}$
C) –3
D) 0
E) The slope is undefined.

20. Use the limit process to find the slope of the graph of $6x-4x^2$ at $(7,-154)$.
A) –2
B) 62
C) 34
D) 56
E) –50

Answer Key

1. D
2. D
3. A
4. D
5. A
6. C
7. E
8. C
9. E
10. D
11. C
12. B
13. C
14. B
15. D
16. A
17. A
18. A
19. D
20. E